APOLLO OVER THE MOON
A View From Orbit

APOLLO
OVER THE MOON
A View From Orbit

Editors

HAROLD MASURSKY, G. W. COLTON, and FAROUK EL-BAZ

With contributions by

Frederick J. Doyle, Richard E. Eggleton, Maurice J. Grolier, James W. Head III,
Carroll Ann Hodges, Keith A. Howard, Leon J. Kosofsky, Baerbel K. Lucchitta,
Michael C. McEwen, Henry J. Moore, Gerald G. Schaber, David H. Scott,
Laurence A. Soderblom, Mareta West, and Don E. Wilhelms

Scientific and Technical Information Office   1978
NATIONAL AERONAUTICS AND SPACE ADMINISTRATION
Washington, D.C.

Library of Congress Cataloging in Publication Data

Main entry under title:

Apollo over the moon.

   (NASA SP ; 362)
   Bibliography: p. 251.
   1. Moon—Photographs from space. 2. Project Apollo. I. Masursky, Harold, 1922-   II.
Colton, George Willis, 1920-   III. El-Baz, Farouk. IV. Series: United States. National
Aeronautics and Space Administration.
NASA SP ; 362.
QB595.A66   559.9'1'0222   77-25922

☆ U.S. GOVERNMENT PRINTING OFFICE : 1978 O—253-286

For sale by the Superintendent of Documents
U.S. Government Printing Office, Washington, D.C. 20402
Stock No. 033-000-00708-6

Foreword

Man has always wondered and dreamed about the landscape of the distant and intriguing Moon. The first step in deciphering surface details of Earth's only satellite was initiated when Galileo Galilei trained his crude telescope toward our closest neighbor in the sky. A greater step came with the advent of the space age when automated spacecraft telemetered their intelligence to Earth. Yet a longer step was taken by Apollo, when photographic equipment captured the Moon's surface in intimate detail and greater accuracy than ever before.

When we decided to add the scientific instrument module (SIM) bay to Apollo missions 15 through 17, photography from orbit was high on our list of scientific objectives. The Apollo metric camera system was flown to acquire photographic data with high accuracy to aid the effort of Moon mapping, both for operational reasons and for future study and research. To complement this photography, we selected the panoramic camera to provide high-resolution (nearly 1 m) photography of lunar surface features for detailed analysis and photointerpretation.

This book presents only a fraction of the great volume of the acquired photographs. It is our hope that this selection from some 18 000 metric, panoramic, and other camera views of the Moon will inspire further interest in Apollo photographs. Through a better understanding of our neighbors in the solar system, we aim to achieve a better understanding of our own planet, its history, and evolution. With the Apollo lunar missions we have barely opened the door to the study of our solar system. Beyond that opened door lie many puzzles to be deciphered, mysteries to be unraveled, and secrets to be challenged.

Dr. Rocco A. Petrone
Apollo Program Director

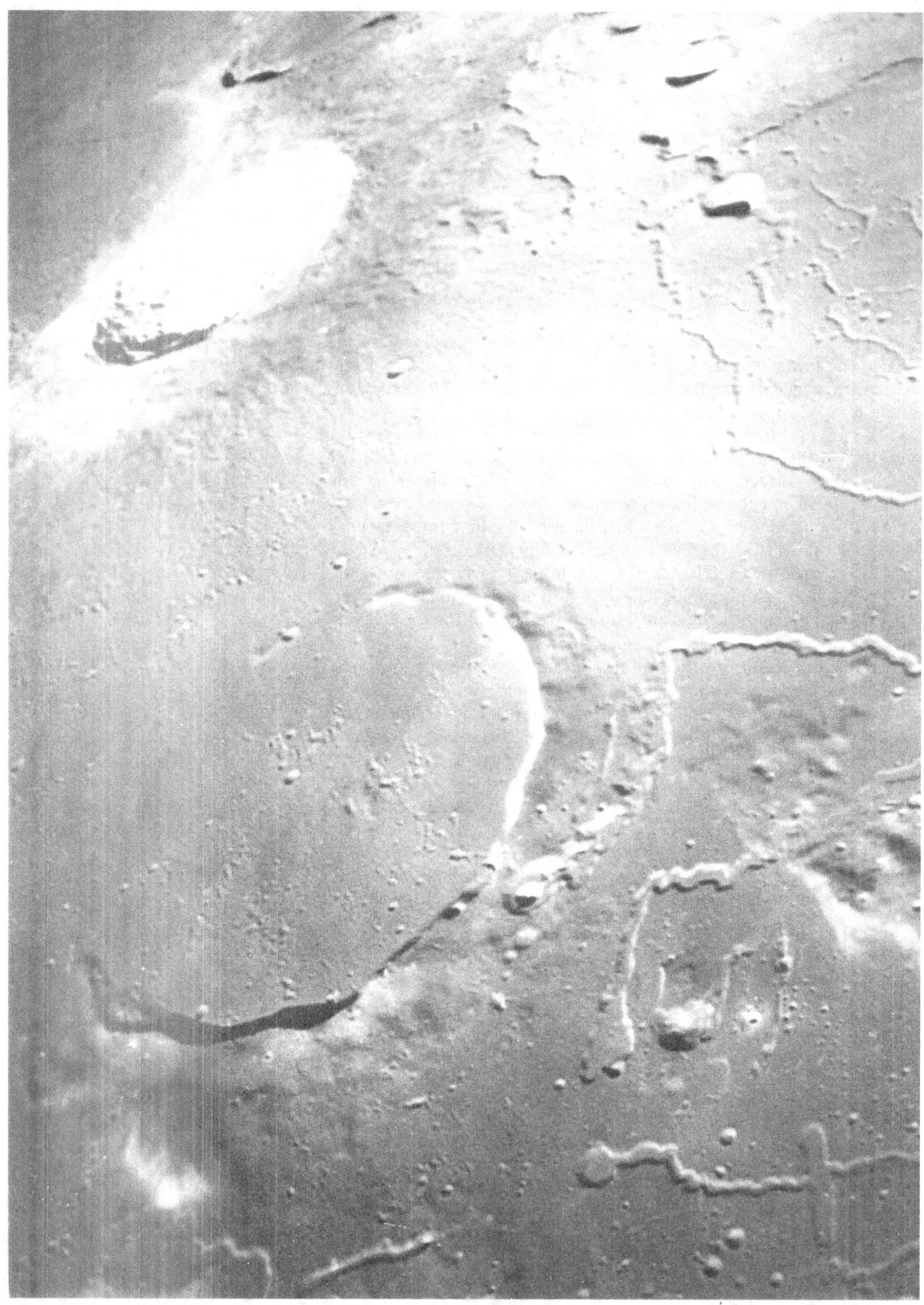

Contents

Chapter   *Page*

1 Introduction   1
  Objectives   1
  Orbital Science Photographic Team   2
  Orbital Camera Systems   5
  Detailed Camera Description and Stereograms   9
  Photographic Mission Plans and Accomplishments   16
  Cartographic Mapping Products   17
  Selected Orbital Experiments   20
  Scientific Results   21
  Acknowledgments   27

2 Regional Views   31

3 The Terrae   45

4 The Maria   67

5 Craters   105

6 Rimae   179
  Sinuous Rimae   179
  Straight Rimae   199

7 Unusual Features   211

Appendix

A Photographic Data   233

B Lunar Probes, Attempted and Successful 239

C Basic Data Analysis Scheme for Preparation of Lunar Maps . . 245

Glossary   247

References and Bibliography   251

Epilog   253

1

Introduction

Harold Masursky,
Farouk El-Baz,
Frederick J. Doyle,
and
Leon J. Kosofsky

Objectives

Photography of the lunar surface was considered an important goal of the Apollo program by the National Aeronautics and Space Administration. The important objectives of Apollo photography were (1) to gather data pertaining to the topography and specific landmarks along the approach paths to the early Apollo landing sites; (2) to obtain high-resolution photographs of the landing sites and surrounding areas to plan lunar surface exploration, and to provide a basis for extrapolating the concentrated observations at the landing sites to nearby areas; and (3) to obtain photographs suitable for regional studies of the lunar geologic environment and the processes that act upon it. Through study of the photographs and all other arrays of information gathered by the Apollo and earlier lunar programs, we may develop an understanding of the evolution of the lunar crust.

In this introductory chapter we describe how the Apollo photographic systems were selected and used; how the photographic mission plans were formulated and conducted; how part of the great mass of data is being analyzed and published; and, finally, we describe some of the scientific results.

Historically most lunar atlases have used photointerpretive techniques to discuss the possible origins of the Moon's crust and its surface features. The ideas presented in this volume also rely on photointerpretation. However, many ideas are substantiated or expanded by information obtained from the huge arrays of supporting data gathered by Earth-based and orbital sensors, from experiments deployed on the lunar surface, and from studies made of the returned samples. These ideas are still evolving. The reader will notice that in some cases the authors of captions for the photographs in this volume interpret similar features differently, or place different emphasis on the relative importance of the various processes involved in the formation of such features. These differences in interpretation reflect in large part the evolving state of lunar data analysis and demonstrate that much work remains to be done before our goal of understanding the history of the Moon is reached. One of our goals with this volume is to convey an impression of the many exciting scientific results still emerging from the study of the photographs and other data already gathered.

At the termination of the Apollo program in December 1972, nearly 20 percent of the surface of the Moon had been photographed in detail

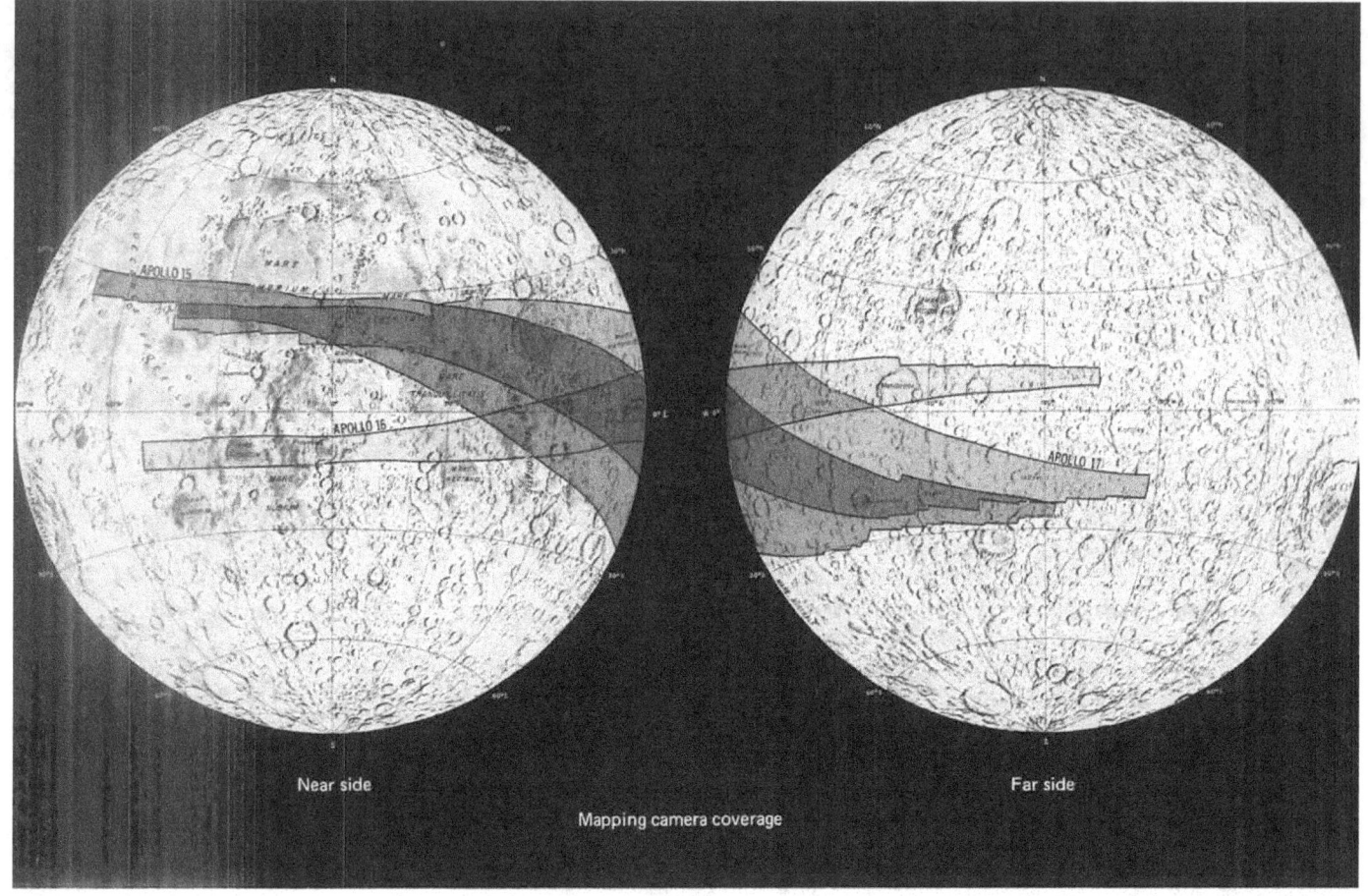

Near side

Far side

Mapping camera coverage

FIGURE 1.–*Above:* Mapping camera coverage obtained by the Apollo 15, 16, and 17 missions. Only areas photographed in sunlight in the vertical mode are shown. Excluded are areas photographed obliquely, those in darkness beyond the terminator, and regional scenes taken after leaving lunar orbit. *Facing page:* Panoramic camera coverage obtained under the same conditions. [Base map courtesy of the National Geographic Society]

from orbit by a variety of camera systems (fig. 1). A selected sample of the nearly 18 000 orbital photographs so acquired is shown in this volume. (See app. A for an index of photographs.) The tremendous success of the Apollo photographic mission must be attributed ultimately to the great dedication, enthusiasm, and ability of the members of the Apollo program. Their efforts have resulted in a great harvest of new scientific information and a consequent increase in knowledge for mankind.

Orbital Science Photographic Team

Early in the Apollo program, then directed by Lt. Gen. Sam C. Phillips, it was realized that adequate planning for the acquisition of photographs would require the participation of many individuals acting in an advisory capacity. On September 6, 1968, prior to the flight of Apollo 8, an Ad Hoc Lunar Science Working Group was convened at the Manned Spacecraft Center (now Johnson Space Center), Houston,

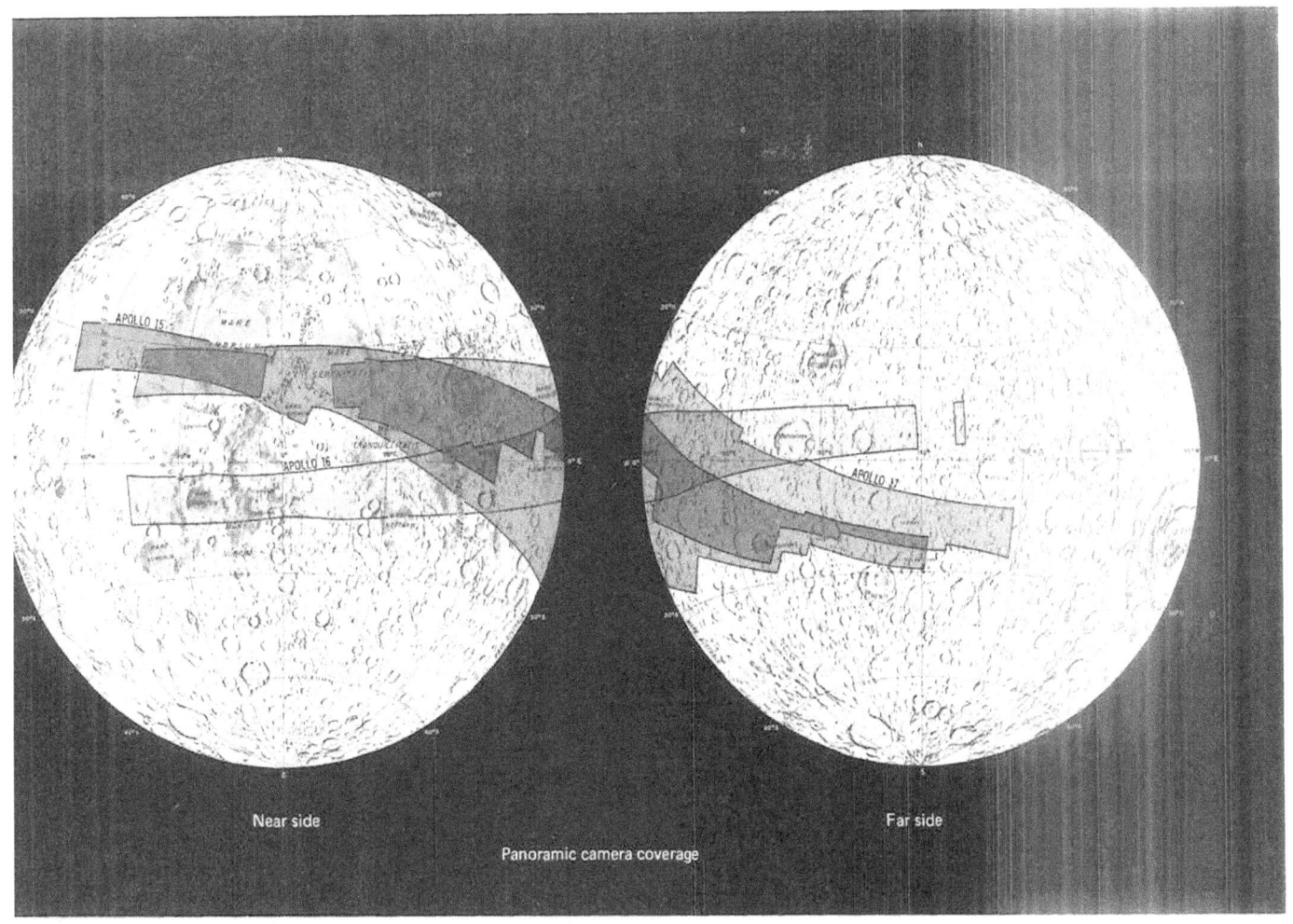

Near side   Far side

Panoramic camera coverage

Tex. The Working Group's function was to recommend target areas of scientific interest for lunar photography. This group planned and supported the photography done on the first five lunar orbital missions (Apollos 8 and 10 through 13; Apollo 9 was an Earth orbital mission) working with the Mapping Science Branch of the Science and Applications Directorate of the Manned Spacecraft Center. Hasselblad cameras with 80- and 250-mm lenses constituted the major photographic system employed on these early flights (table 1). (App. B contains a list of lunar probes.)

On December 14, 1969, a year before Apollo 14 was launched, Dr. Rocco Petrone, Apollo Project Director, established the Apollo Orbital

TABLE 1.—*Camera Systems Carried on Board the Command and Service Module (CSM) on Apollo Lunar Missions 8 Through 17*

Camera	Mission								
	8	10	11	12	13	14	15	16	17
Data acquisition (Maurer)	X	X	X	X	X	X	X	X	X
70-mm EL (Hasselblad)	X	X	X	X	X	X	X	X	X
70-mm DC (Hasselblad)	X	X	X	X	X	X	X	X	X
Lunar topographic (Hycon)					X	X			
35-mm (Nikon)							X	X	X
Mapping and stellar (Fairchild)							X	X	X
Panoramic (Itek)							X	X	X

3

Science Photographic Team because Apollo 14 was the first mission to include an orbital mapping camera. The purpose of the team was to provide "scientific guidance in the design, operation, and data utilization of photographic systems for Apollo program lunar orbital science." The team included a chairman and 12 members who were specialists in the fields of geology, geodesy, photogrammetry, astronomy, and space photographic instrumentation:

> Frederick J. Doyle, Chairman (photogrammetry), U.S. Geological Survey
>
> Lawrence Dunkelman (astronomy), NASA Goddard Space Flight Center
>
> Farouk El-Baz (geology), Bellcomm, Inc.; later, the National Air and Space Museum, Smithsonian Institution
>
> William Kaula (geodesy), Institute of Physics and Planetary Physics, University of California at Los Angeles
>
> Leon J. Kosofsky (space photography), NASA Headquarters
>
> Donald Light (photogrammetry), Defense Mapping Agency Topographic Center
>
> Douglas D. Lloyd (space photography), Bellcomm, Inc.
>
> Harold Masursky (geology), U.S. Geological Survey
>
> Robert D. Mercer (astronomy), Dudley Observatory, Albany, N.Y.
>
> Lawrence Schimerman (photogrammetry), Defense Mapping Agency, Aerospace Center
>
> Helmut H. Schmid (geodesy), U.S. Environmental Science Services Administration
>
> Ewen A. Whitaker (astronomy), Lunar and Planetary Laboratory, University of Arizona

As specified in the charter of the Apollo Orbital Science Photographic Team, its responsibilities included providing NASA with recommendations in the following areas:

(1) Equipment functional specifications:
   (a) The team shall recommend functional requirements for orbital photographic systems.
   (b) The team shall provide technical advice during the procurement phase for photographic systems.
   (c) The team shall be represented at photographic equipment reviews.
(2) Equipment use:
   (a) The team shall participate in preparation of plans for scientific use of the photographic systems and support mission operations planning.
   (b) The team shall participate in planning for coordinated use of photographic systems to support other experiments.
   (c) The team shall participate in planning for other experiments that will support photography.
   (d) The team shall participate in operations planning photographic requirements to support Apollo lunar landing site selection.

(3) Crew training: The team shall provide technical advice for and will support, as requested by the Manned Spacecraft Center, astronaut training related to photographic tasks.

(4) Real-time operations: The team shall support operations as requested by providing real-time scientific and technical advice to astronauts for photographic and related tasks.

(5) Data processing:

(a) The team shall provide technical advice in selection of film and film processing requirements to optimize post-mission scientific analysis by photographic users.

(b) The team shall recommend major data reduction equipment and analysis procedures to assure that optimum scientific use is made of the photographic data obtained.

Analysis of the photographic data was carried out by a broad spectrum of scientists representing the following institutions: the U.S. Geological Survey, Bellcomm, Inc., the University of Arizona, Ames Research Center, and the Smithsonian Institution. H. Masursky directed and collated for publication the results of the studies relating to the scientific interpretation of the photographs. He and his colleagues, T. A. Mutch of Brown University and G. W. Colton, K. A. Howard, and C. A. Hodges of the U.S. Geological Survey, performed the same function for the last three Apollo missions. Many of the studies were performed as part of the later NASA-funded experiment S-222, which analyzed data after the completion of the flights.

Orbital Camera Systems

In preparation for the Apollo program of landing men on the surface of the Moon, the Lunar Orbiter project inserted five spacecraft into lunar orbit. Their cameras photographed almost all of the lunar surface (Kosofsky and El-Baz, 1970; NASA Langley Research Center, 1971). However, the nature of the camera's electronic readout system was such that it was difficult to measure positions on the lunar surface accurately from the Lunar Orbiter pictures. Another disadvantage was that little stereoscopic coverage was obtained, making three-dimensional measurements impossible. Furthermore, most Lunar Orbiter pictures were taken under low Sun angles so that few high Sun pictures showing differences in albedo were available. The Apollo photographic systems succeeded in correcting these conditions.

Two groups of cameras were used for Apollo orbital photographs: those in the command module (CM) which were handheld or mounted on brackets and operated by the astronauts, and those that were stowed in the scientific instrument module (SIM) bay in the service module (SM) and remotely operated by the astronauts from the CM. Table 1 lists the photographic systems carried on these missions, and table 2 lists their characteristics.

Hasselblad cameras, using both black-and-white film and color film, were carried on all lunar orbital flights of the Apollo program. On Apollo missions 8 through 14 the bulk of the photogeologically useful pictures acquired was taken with the Hasselblad systems. Hasselblad

TABLE 2.—*Characteristics of Camera Systems Carried*

Camera	Features
SIM bay:	
Mapping (Fairchild)	Electric; 76-mm focal length lens; reseau crosses, fiducial marks, time, altitude, shutter speed, and forward motion control setting recorded on each frame.
Stellar (Fairchild)	Part of mapping camera system; 76-mm focal length lens; axis oriented at 96° to mapping camera axis; exposure synchronous with mapping camera exposures; reseau crosses, fiducial marks, time, and altitude recorded on each frame.
Panoramic (Itek)	Electric; 610-mm focal length lens; optical bar design for high-resolution and image-motion compensation; frame number, fiducial marks, time, mission data, velocity/height ratio, and camera-pointing attitude recorded on each frame.
CM:	
Hasselblad EL	Electric; interchangeable 80-, 105-, 250-, and 500-mm focal length lenses; 105-mm lens transmits ultraviolet (UV) wavelengths.
Hasselblad DC	Electric; 80-mm focal length lens; reseau plate.
Lunar topographic (Hycon, later Actron)	Vacuum platen and image-motion compensation; 457-mm focal-length, $f/4$ lens; fiducial marks, time, and shutter speed recorded on each frame.
Data acquisition (Maurer)	Electric; interchangeable 5-, 10-, 18-, 75-, and 200-mm focal length lenses.
35-mm (Nikon)	Mechanical; through-the-lens viewing and metering; 55-mm focal length lens.

[a]Different combinations of the films listed for the Hasselblad and data acquisition cameras were used on different missions.

[b]Specific uses and emphasis upon various uses changed somewhat from mission to mission.

Film size and type	Uses
127 mm; type 3400 Panatomic-X aerial.	To provide mapping quality photography (25- to 30-m resolution)—when used stereoscopically and in conjunction with auxiliary data, the geometry of the lunar surface can be reconstructed with a higher degree of precision than possible with earlier systems.
35 mm; type 3401 Plus-X aerial.	To record star field at a fixed point in space relative to mapping camera axis so that orientation of the latter can be accurately determined for each mapping camera frame.
127 mm; type EK 3414.	To provide strips of high-resolution (about 2-m) stereoscopic coverage for relatively large-scale topographic mapping and for detailed photogeologic analysis.
70 mm; color reversal films SO-121 and SO-368 Ektachrome MS; black and white films 3400 Panatomic X, 2485, SO-349 aerial, and 3414 aerial; spectroscopic film (UV-sensitive) IIa-0.[a]	To document operations and maneuvers; to obtain convergent stereoscopic coverage of candidate landing sites, particularly with the 500-mm lens; to photograph preselected orbital science targets, different terrain types near the terminator, astronomical phenomena, views of the Moon after transearth injection, views of Earth, and to acquire special UV spectral photographs of the Moon and Earth.[b]
70 mm; black and white films 2485, 3400 Panatomic X, and SO-349 aerial.[a]	To obtain strips of stereoscopic coverage of the approach paths to candidate landing sites, as well as of orbital science targets.[b]
127 mm; black and white films 3400 Panatomic X and SO-349 aerial.[a]	Primarily to obtain high-resolution coverage of the Apollo 16 candidate landing site area near Descartes on the Apollo 14 mission. Camera malfunction prevented achieving this and most other goals.[b]
16 mm; color reversal films SO-368 Ektachrome MS, SO-168 Ektachrome EF, and SO-368 Ektachrome EF; black and white films 2845, SO-164 Panatomic X, and AEI 16.[a]	To document engineering and operational data, experiment records within the CSM, maneuvers with the lunar module (LM), CM entry, and landmark tracking; to photograph general targets inside and outside the CSM; to obtain continuous sequence terrain photography.[b]
35 mm; black and white film type 2485; SO-168 Ektachrome EF.[a]	To photograph, under dim light conditions, astronomical phenomena and lunar targets illuminated by Earth shine.[b]

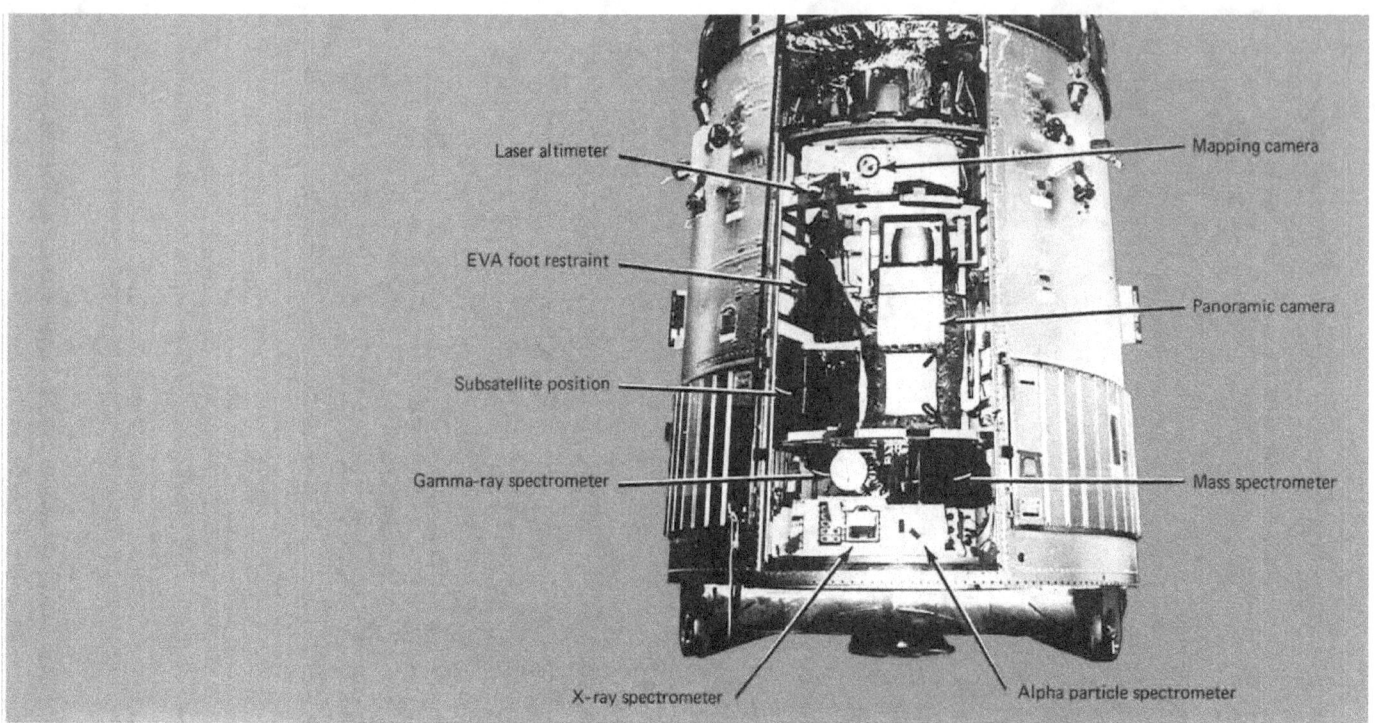

Laser altimeter

EVA foot restraint

Subsatellite position

Gamma-ray spectrometer

X-ray spectrometer

Mapping camera

Panoramic camera

Mass spectrometer

Alpha particle spectrometer

FIGURE 2.–Arrangement of the equipment used for the orbital experiments conducted from the SIM bay during Apollos 15 and 16 is shown. A largely different assortment of instruments was carried on Apollo 17. The extravehicular activity (EVA) foot restraint is used by the astronaut who retrieves the exposed film from the cameras during the return of the spacecraft to Earth.

photographs were used to study future landing sites, to perform geologic mapping, to conduct geodetic studies, to study the regional and local geology of the Moon, and to train astronauts for later lunar missions.

On Apollo missions 13 and 14, a lunar topographic camera (the Hycon camera) was carried in the CM. A very limited amount of photography was obtained with it, however, because the Apollo 13 mission was aborted before the photographic phase was to have begun and on Apollo 14 the camera malfunctioned early in lunar orbit, just as the spacecraft approached the Apollo 16 candidate landing site in the Descartes region. The necessary photographs of the Descartes area were obtained with the Hasselblad (500-mm lens) camera by pitching the spacecraft to compensate for its forward motion in orbit so that no image smear appeared on the photographs.

On Apollos 15 to 17 the much-more-sophisticated mapping and panoramic camera systems were used. These, and other remote sensing instruments, were mounted in the SIM bay that was added to the CSM for the last three missions (fig. 2). Higher resolution and much more systematic coverage of the ground track areas traversed by the spacecraft were obtained with these cameras. Personnel from the Flight Operations Directorate, Johnson Space Center, monitored the performance of the instruments and were essential to the successful acquisition of the data.

The mapping camera system consisted of a terrain camera coupled to a stellar camera and a laser altimeter. Each exposure of the terrain camera was accompanied by a stellar camera exposure of the star field to provide a means of determining the orientation of the spacecraft in space. Simultaneously the laser altimeter recorded the height of the spacecraft above the Moon's surface. At nominal orbital altitude the terrain camera was capable of resolving objects on the surface as small as 25 to 30 m on a side. The geometry of the optical system of the terrain camera permitted the metric pictures to be used for precise triangulation and cartography. Thus they can be used for three purposes: (1) to construct a geodetic network of control points for topographic mapping with both terrain and panoramic camera photographs, (2) to compile medium-scale topographic maps, and (3) to perform photogeologic interpretation.

The panoramic camera, capable of attaining 1-m resolution on the surface from orbital altitude, provided high-resolution stereoscopic photographs of the surface during periods of varying Sun angle conditions. High Sun angle pictures show differences in albedo to advantage, and low Sun angle pictures delineate small and low relief features more clearly.

Detailed Camera Description and Stereograms

The Metric Camera System

The coordinated operation of the three components of the mapping camera system is illustrated schematically in figure 3. Except when in use, the mapping camera system is in a retracted position in the SIM bay. During photographic sequences the system is extended on rails to a point where the terrain and stellar cameras have clear fields of view. The components of the system are illustrated in different degrees of detail in figures 4 through 6. The terrain camera component has a 76-mm, $f/4.5$ lens to cover a square frame 114 mm on a side. The shutter consists of a pair of continuously rotating disks and a capping blade. An automatic exposure device selects the correct disk speed to provide a range of exposures from 1/15 to 1/240 s in duration. The film is held precisely in place by pressure against a movable glass stage plate that contains reseau marks. Fiducial marks are flashed on the film marking the optical axis at midexposure time. Simultaneously, the time of exposure, to the nearest millisecond, according to the spacecraft clock is recorded on the data block on each frame. The film magazine holds 460 m of 127-mm film—enough for 3600 frames. The terrain camera was designed to compensate for the image blur that would otherwise be present in a picture taken from a rapidly moving camera; the stage plate that controls the exposure is driven forward in the direction of flight at a velocity proportional to the ground speed of the spacecraft. The velocity over height value (v/h) is set by the astronaut from the CM. Successive frames overlap by 78 percent (fig. 3) and alternate frames by 57 percent; thus each point on the lunar surface appears on at least four successive frames. In most cases successive orbital passes provided for 60 percent sidelap.

The stellar camera exposes 35-mm film with a 76-mm $f/2.8$ lens. Each exposure, of 1.5-s duration, is made simultaneously with an exposure of the terrain camera. In the stellar camera the film is pressed against a glass stage plate that bears an array of 25 reseau crosses. The

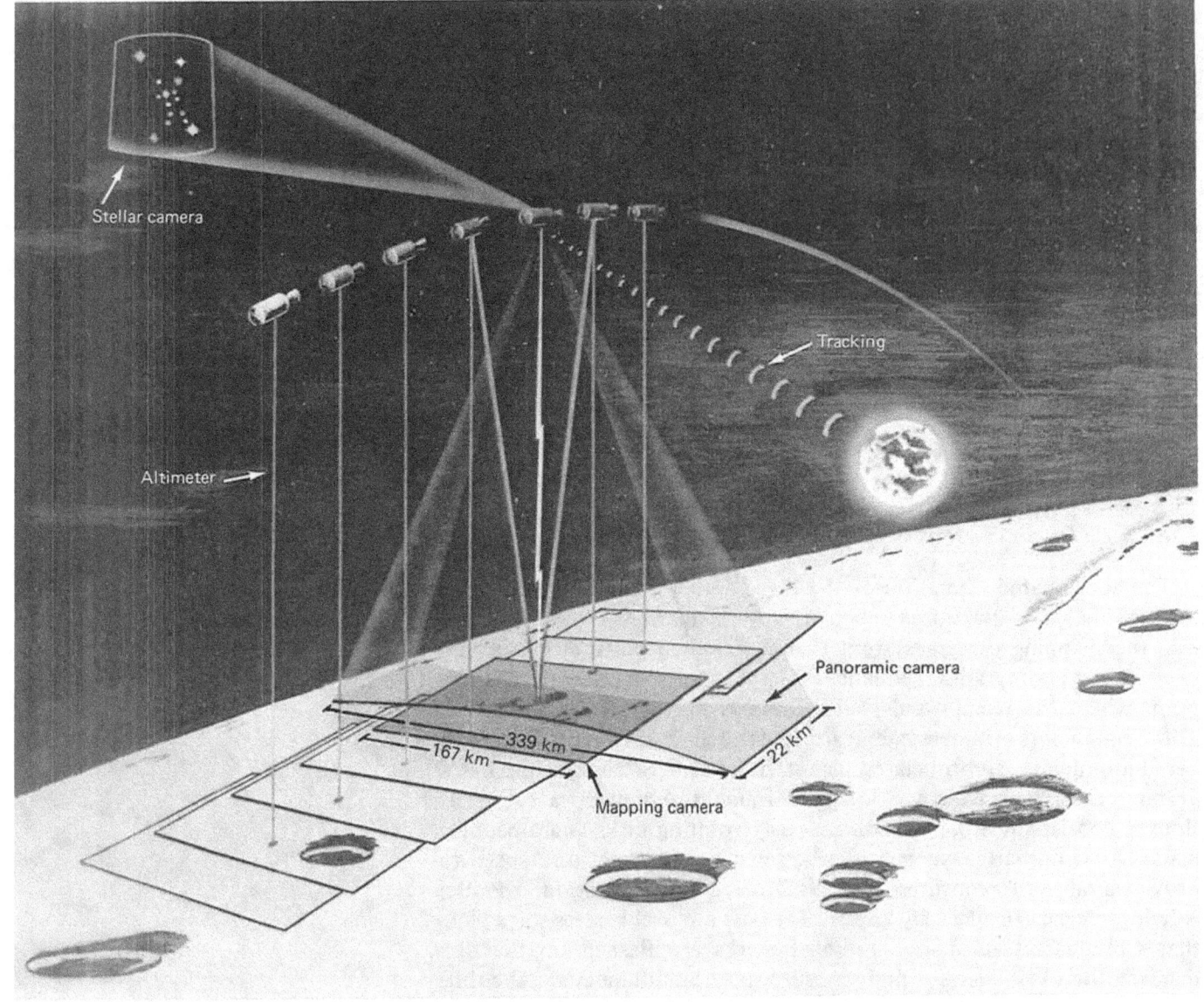

Stellar camera

Tracking

Altimeter

Panoramic camera

339 km

167 km

22 km

Mapping camera

FIGURE 3.—The mapping camera system in operation. During normal photographic operations the terrain camera automatically exposes a series of pictures of the Moon's surface. When the camera axis is perpendicular to the surface (red lines), each exposure outlines a square area (the blue parallelogram in this perspective view). The areas covered by each exposure overlap to form a continuous strip across the surface. The position of the spacecraft in orbit is recorded continuously by radiotracking stations on Earth; however, for precision photography the orientation as well as the position of the spacecraft must be known. This is accomplished with the stellar camera component, which takes a picture of part of the star field each time a terrain camera picture is taken. The stellar camera thus provides the data needed to orient the spacecraft and the terrain camera. The distance to the lunar surface is measured with the laser altimeter. At each exposure of the terrain camera, a beam of light (red lines) from the altimeter is pulsed to the surface (to the center of the area being photographed), and its time of return is recorded. Because the velocity of light is known, the distance between the spacecraft and the Moon is easily calculated. This diagram also shows the relationship in size and shape of areas covered by the terrain (blue) and panoramic (tan) cameras.—G.W.C.

FIGURE 4.–The Apollo mapping camera system, which consists of a terrain or mapping camera, a stellar camera, and a laser altimeter. The three components can be located by referring to figure 5. [Photograph courtesy of the Fairchild Camera and Instrument Corporation]

FIGURE 5.–A schematic drawing showing the major components and selected parts of the mapping camera system. The angular fields of view (FOV) are given in degrees for the stellar and mapping (or terrain) camera components.

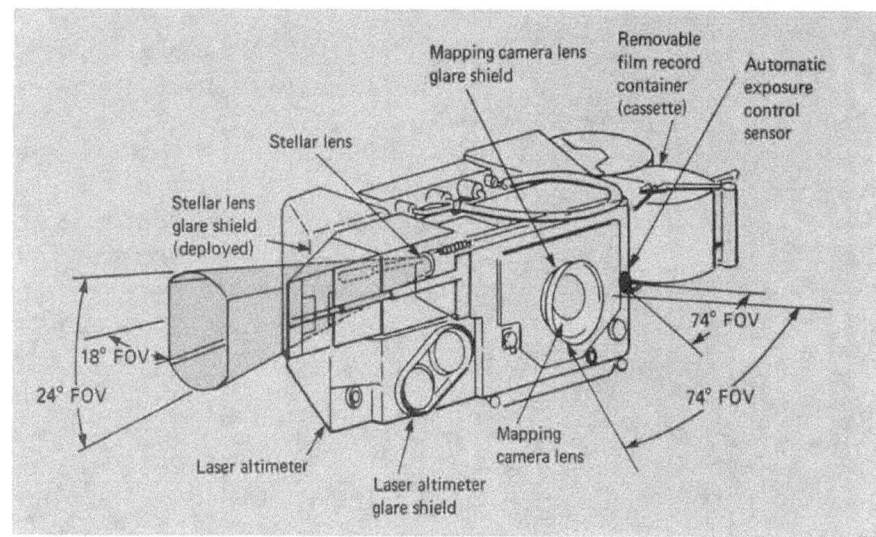

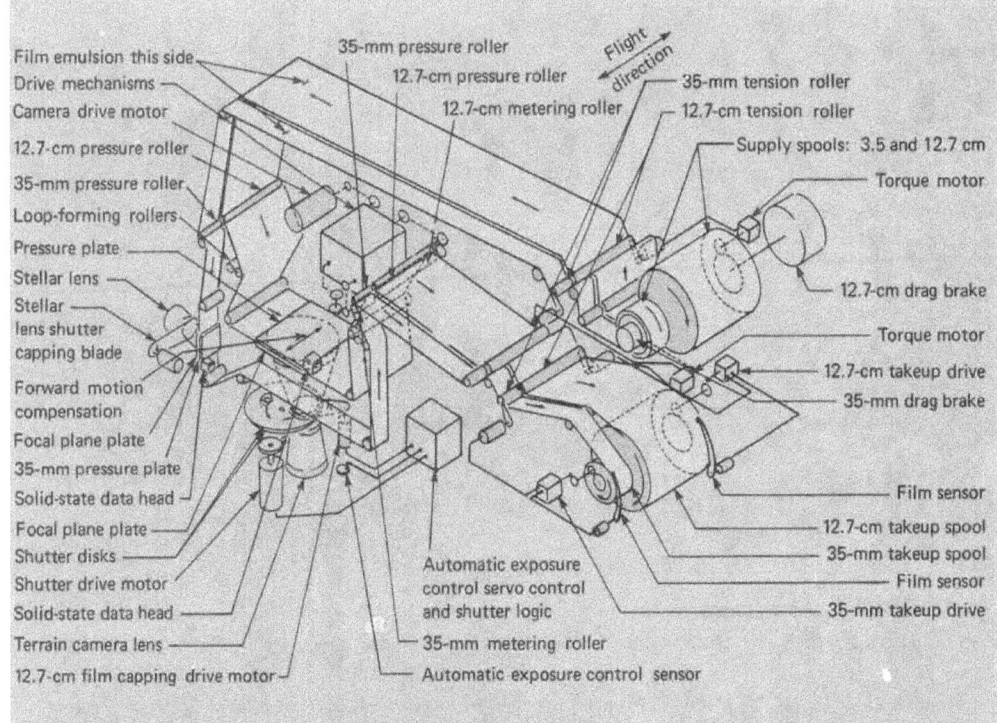

FIGURE 6.–The terrain camera mechanism. (For a more complete description, see Kosofsky (1973).)

film is 155.4 m long, allowing 3600 exposures to be made. The stellar camera is rigidly mounted with its optical axis at an angle of 96° to that of the terrain camera. Thus the orientation of the latter (as well as of the spacecraft) can be determined from the known position of the stars recorded by the stellar camera.

The laser altimeter is alined parallel to the optical axis of the terrain camera. A ruby laser emits a very short pulse of red light at the time of each mapping camera exposure and a photomultiplier tube detects the portion of the pulse that is reflected from the Moon. Measurement of the round trip traveltime of the pulse multiplied by one-half the speed of light gives the precise altitude of the spacecraft above the Moon's surface. With the orientation and altitude of each metric photograph thus determined, the metric photographs can be tied together to create a geodetic network of control points.

The Panoramic Camera

High-resolution stereoscopic coverage of large areas of the lunar surface was provided by the panoramic camera. The camera used on the Apollo missions, a modified version of the U.S. Air Force's KA-80A "optical bar" camera, was manufactured by Itek Corp.

The panoramic camera mechanism allowed an exceptionally wide area to be covered with a narrow-angle lens. This is accomplished by

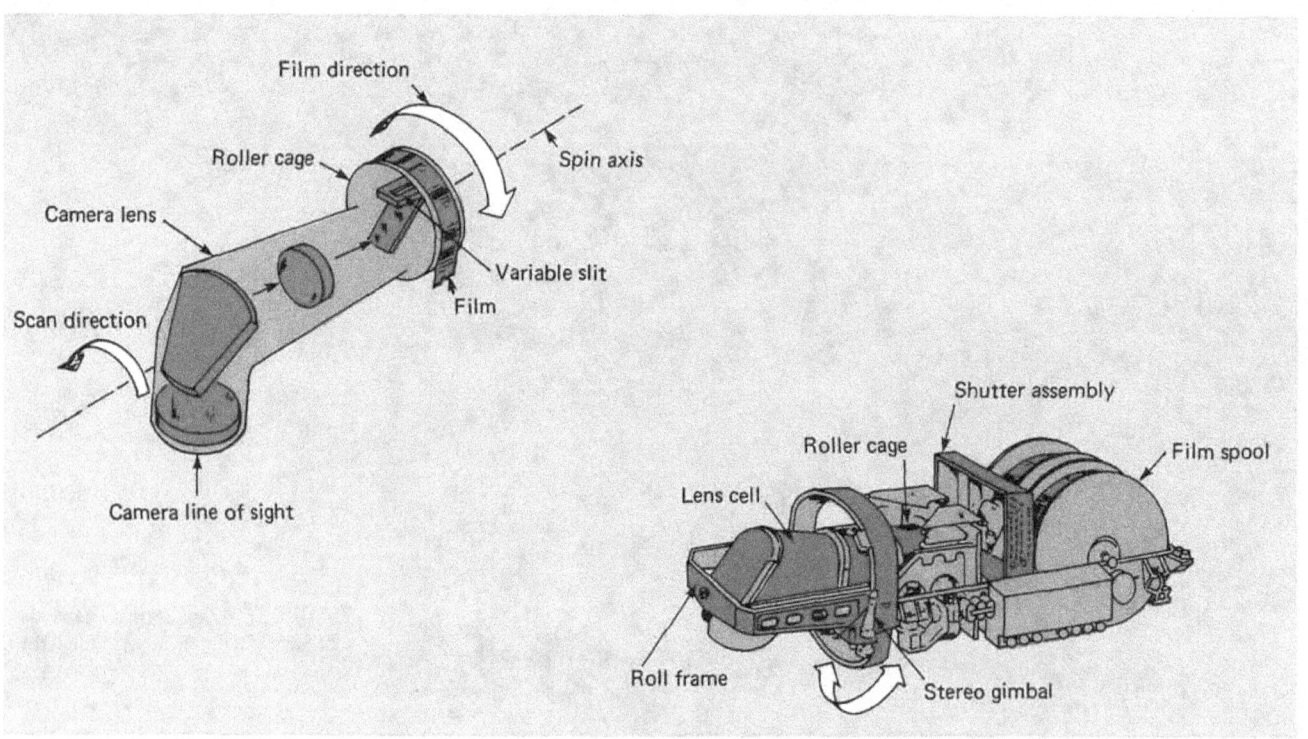

FIGURE 7.—The panoramic camera shown diagrammatically. *Left:* The optical bar concept. *Right:* The camera mechanism with covers removed.

rotating the lens during the exposure. The 610-mm $f/3.5$ lens has eight lens elements and two folding mirrors. The optical bar, consisting of this optical system, an exposing slit, and a roller cage that supports the film, rotates continuously during camera operation. Film exposure starts at 54° on one side of the flight line and extends to 54° on the other side for a total scan of 108° perpendicular to the flight path. In the direction of flight, the field of view is 10.6°. The photographic exposure of the film is determined by the rate of rotation of the optical bar and the width of the slit. To prevent image smear due to the rotation of the lens, the film is pulled across the slit in the opposite direction.

The optical bar and the motor that spins it are mounted in a roll frame that is connected to the camera's main frame by a gimbal structure (fig. 7). By rocking the roll frame about this gimbal, the camera provides both stereoscopic overlap and forward motion compensation. The exposures are made with the roll frame rocked alternately 12.5° forward and 12.5° aft (fig. 8). The camera cycle rate is controlled so that the ground covered by a forward-looking photograph is covered again five frames later by an aft-looking photograph, thus providing a stereo pair. During the time (about 2 s) that each of these exposures is being made, the same gimbal mechanism "freezes" the ground image by matching the rocking motion to the angular rate at which the ground passes beneath the spacecraft. The camera's v/h sensor, which measures this rate continuously, is the pacemaker for the entire operation.

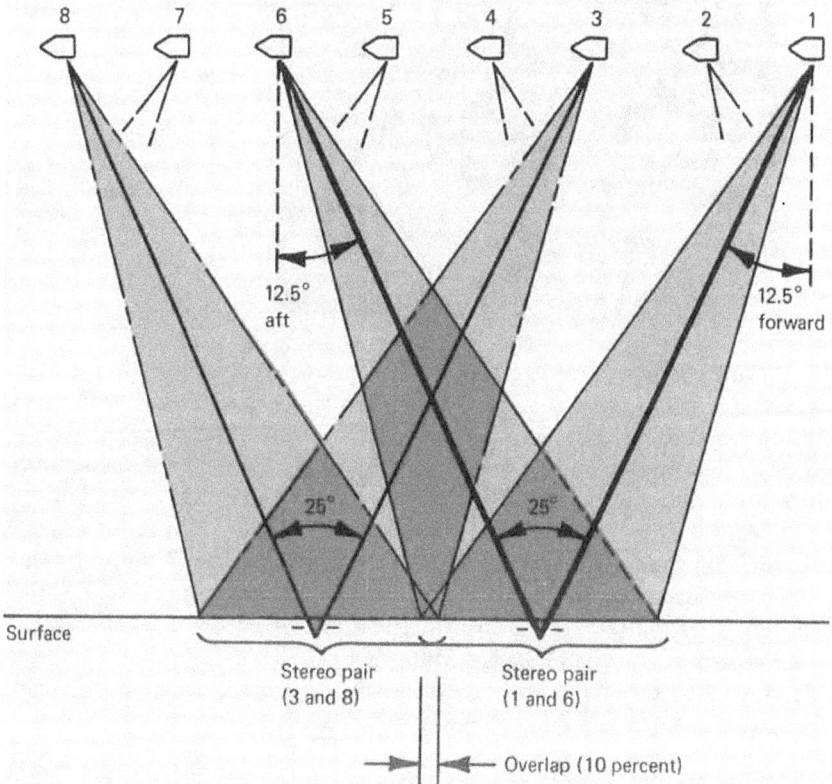

FIGURE 8.—Diagram showing how alternate exposures of the panoramic camera are taken with the roll frame rocked forward and aft by 12.5° in each direction. The two heavy solid lines converging on a common point at an angle of 25° indicate how stereoscopic coverage is obtained with every fifth exposure. In this case it is with frames 1 and 6.

13

The main frame of the camera holds all film handling mechanisms. The panoramic camera must move considerable lengths of film rapidly. During the 2 s it takes for one exposure, 1.2 m of film must be pulled smoothly over the roller cage and across the exposure slit. This action is

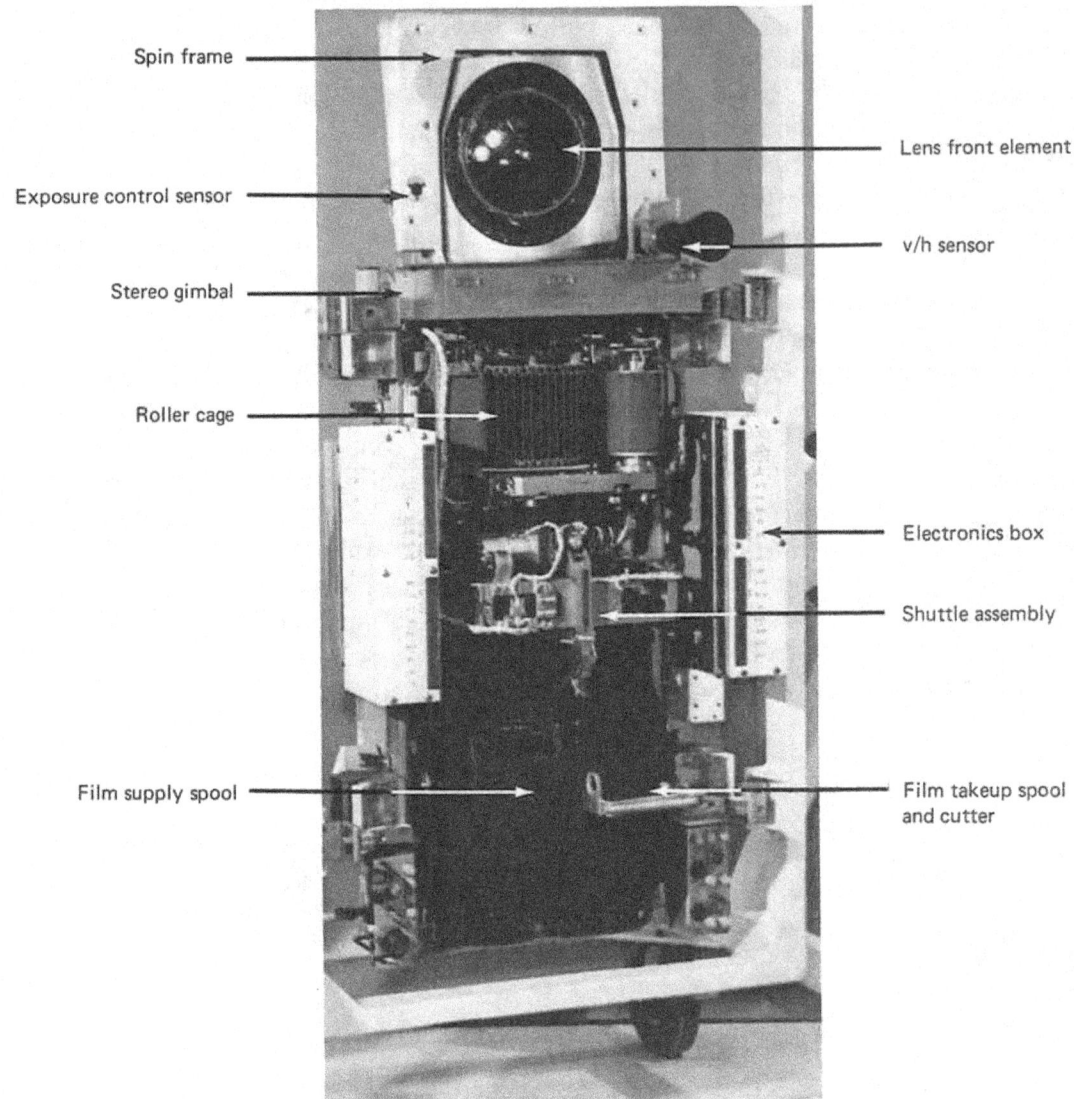

FIGURE 9.—This photograph of the Apollo panoramic camera, supplied by the Itek Corp., was taken at the camera plant. The side facing the reader would face the Moon in actual operation. The covers that normally protect the film from light have been removed to show the internal mechanism. The parts that are not coated with white paint in the photograph would normally be under the covers. The half of the optical bar that includes the lens front elements is necessarily outside the covers, while the half with the roller cage and the rear elements is inside. The roller cage, which is an open drum composed of 60 small rollers, appears dark toned in the photograph. It hides the lens' rear elements. The film shuttle assembly is the silver-colored frame located between the two white-painted electronics boxes. The very dark objects at the bottom of the photograph are the large film spools. A bright metal film cutter can be seen on the takeup side.—L.J.K.

14

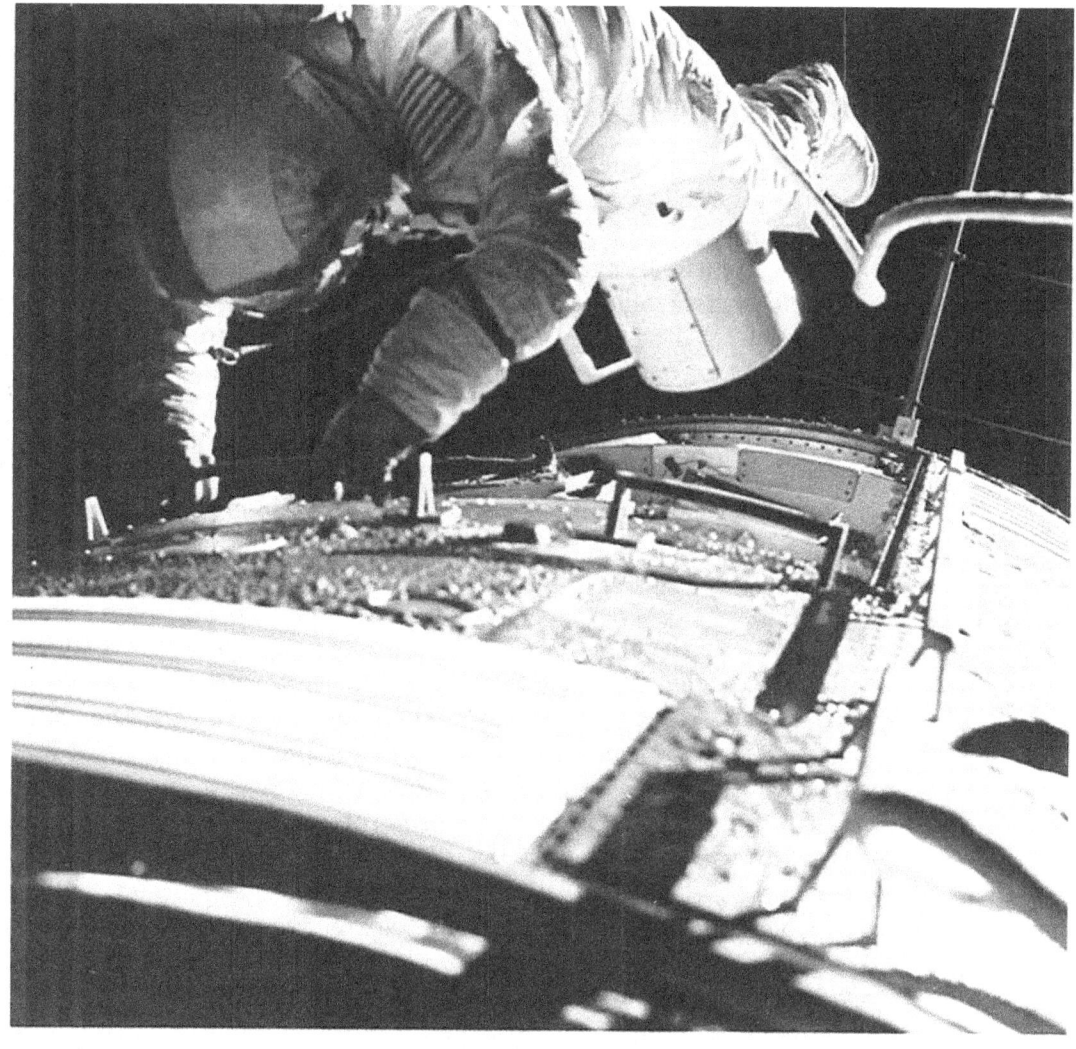

FIGURE 10.—Astronaut Ron Evans retrieving film from the SIM bay cameras while the Apollo 17 spacecraft was returning to Earth.

repeated every 6 s. A load of film sufficient for 1600 exposures is 2 km long and weighs 25 kg. Because the forces that would be required to start and stop such a mass of film intermittently are prohibitively large, the supply and takeup spools actually rotate continuously during camera operation. An ingenious "shuttle assembly" functions as a buffer between the continuous and the intermittent film movements. In the interval between exposures, the supply side of the shuttle accumulates enough film for the next frame while the takeup spool empties the takeup side of the shuttle.

The arrangement of the camera parts can best be studied in the accompanying photograph (fig. 9), which shows the camera as it would be seen from outside the SIM bay if the light-excluding covers were removed. A more detailed description of the camera systems is given in Kosofsky (1973).

The film cassettes from both the metric and panoramic cameras were recovered by the command module pilot (CMP) on the way back to Earth (fig. 10) and stored in the CM for later processing at the Photographic Technology Laboratory, Johnson Space Center.

Photographic Mission Plans and Accomplishments

Personnel from the Science and Applications Directorate and the Apollo Spacecraft Programs Office were responsible for designing the mission plans. The final plans incorporated many of the recommendations made by the Apollo Orbital Science Photographic Team.

Metric camera photography for Apollo 15 had several limiting constraints set by the instrumentation, the requirements of other SIM bay experiments, and the orbital plane of the CM. The high inclination of the Apollo 15 orbital plane (approximately 25° from the equator; see fig. 1) resulted in a relatively wide separation of the ground tracks of succeeding revolutions. Because 60 percent side overlap was required for pictures taken in adjacent orbits, careful planning and budgeting of the film was required.

The performance of other scientific experiments using instruments in the SIM bay prohibited concurrent use of the metric camera during some parts of the mission. Some of the experiments were conducted with the SIM bay pointed away from the Moon. During most of the time devoted to the geochemical experiments, such as the gamma ray spectrometer experiment, the SIM bay was pointed toward the lunar surface but the metric camera was retracted and the lens was covered to prevent radioactive thorium contained in the camera and laser altimeter lenses from interfering with the spectrometric measurements. During one orbit, as a test, the two instruments were run concurrently. Radiation from the lenses was less than anticipated; consequently, during the Apollo 16 mission, the two instruments were run concurrently, with only slight degradation of the spectrometric measurements.

Despite constraints such as these, all objectives of the Apollo 15 photographic plan were met, and an excellent set of vertical metric photographs was acquired. In addition, many oblique photographs looking 40° fore and aft and to the north and south of the ground tracks were obtained; although they are of limited use in making topographic maps, they are extremely useful in photogeologic studies.

The panoramic camera on Apollo 15 obtained excellent, very useful photographs, even though its operation was subject to constraints similar to those placed on the mapping camera system. Furthermore, the panoramic camera could only be run for about one-half hour (half the sunlit portion) of any given orbit because of thermal limitations on the instruments. The area of coverage was limited to the same area covered by the mapping camera system because only the latter could provide the geodetic control necessary for constructing mosaics and orthophoto maps from the panoramic camera pictures.

On the Apollo 16 mission, film budgeting was not difficult because the low angle of the orbital plane (about 9° from the equatorial plane) restricted the spacecraft to a narrow orbit belt and hence limited the width of the area that could be photographed from the vertical mode (fig. 1). The planned coverage for both systems was obtained long before the film loads were consumed. The original mission plans were to take the metric and panoramic photographs very early and very late in the orbital phase when the near-side terminator had progressed far enough west to permit photographing the western limb. However, early in the orbital phase problems developed with the spacecraft orientation mechanism that forced a decision to shorten the mission time by 1 day. Although the time available for orbital photography was less than originally planned, about 90 percent of the planned photographic coverage was obtained. Only the westernmost part of the orbital tracks was not photographed.

The orbital attitude of the Apollo 17 spacecraft was similar (about 20° from the equatorial plane) to that of Apollo 15; because much of the ground track area had previously been photographed by the earlier mission (fig. 1), the 60 percent overlap requirement was waived. In some areas previously photographed during the Apollo 15 mission, photographs were obtained when the Sun angle was different from that of the earlier photographs; additional information was supplied in this way.

Apollo 17 carried several new orbital experiments including the infrared scanning radiometer, the UV spectrometer, and the lunar sounder experiments. Some of these required special spacecraft attitudes and freedom from interference, thereby reducing the amount of time available for using the SIM bay camera systems. However, modifications made in the photographic plan to accommodate these experiments on this and earlier missions proved worthwhile because the results of the experiments have proved useful in supporting photogeologic interpretations.

The panoramic camera was used to extend the Apollo 15 photographic coverage and to fill gaps in both the Apollo 15 and 16 belts.

Cartographic Mapping Products

When the Apollo program ended with the successful completion of mission 17, nearly 10 000 metric and 4800 panoramic photographs had been acquired. Controlled orthophotomaps at a scale of 1:250 000 (fig. 11) are being made from the metric photos; the panoramic camera photographs are being used to make large-scale maps at 1:50 000 and 1:10 000 scale for particular areas of scientific interest (fig. 12). The basic data analysis scheme for converting the raw photographs into finished map products is described in Appendix C (Doyle, 1972). The production of these maps is a massive undertaking and will take many years to complete.

Copies of individual Apollo photographs may be ordered from—

National Space Science Data Center
Goddard Space Flight Center
Greenbelt, Md. 20771

Orthophotomaps and other cartographic products may be obtained from—

Lunar and Planetary Programs Office
National Aeronautics and Space Administration
Washington, D.C. 20546

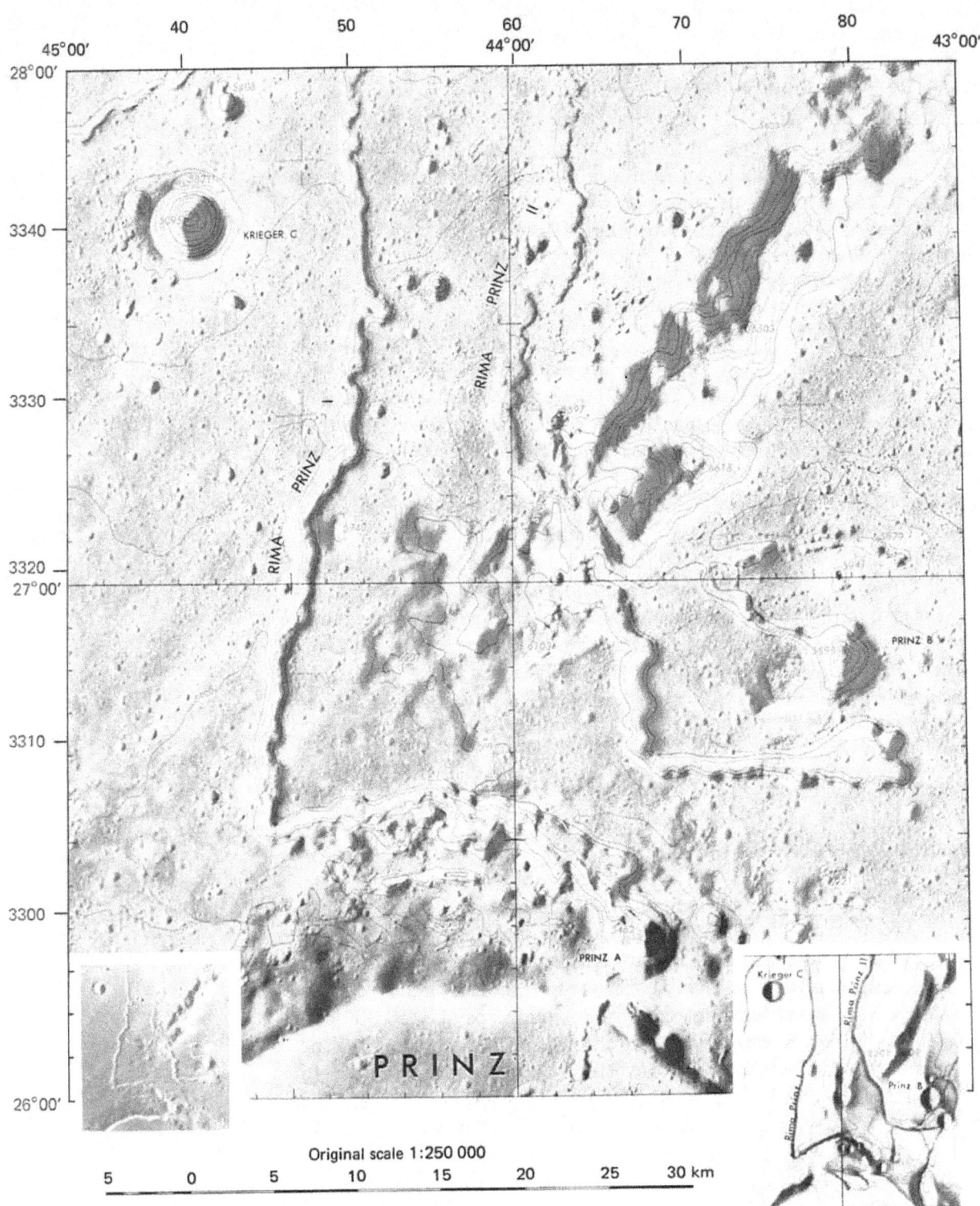

Original scale 1:250 000

5   0   5   10   15   20   25   30 km

Contour interval: 100 m
Supplementary contours at 50-m intervals

Transverse Mercator Projection

Map and contour elevations are derived from radius
vectors from the mass center of the Moon as referred to
an arbitrary zero vertical datum of 1 730 000 m; for
example, the map elevation of a point with a radius
vector length of 1 735 200 m is derived by subtracting
1 730 000 m to obtain 5200 m.

The Apollo 15 datum was established by
photogrammetric triangulation based on the Apollo 15
mission ephemeris dated December 1971.

FIGURE 11.–An example of the detailed topographic portrayal made possible by mapping camera photographs. The sample area shown is part (about one-fifth) of Lunar Orthophoto Map sheet LTO39A3(250), prepared by the Defense Mapping Agency Topographic Center, Washington, D.C., and published in 1973. The rilles are Rima Prinz I and II in northern Oceanus Procellarum. Topographic contour lines (lines of equal elevation) in red are superposed on an orthophotograph version of mapping camera frame AS15-2474 enlarged to the scale of the map. For comparison (or contrast) that portion of the frame corresponding in area to the sample map is shown in the lower left corner. At the lower right is the same area as it appeared on Lunar Aeronautical Chart (LAC) 39, which was one sheet of the earliest series of detailed cartographic maps of the Moon showing surface relief. Compiled entirely from photographs taken through Earth-based telescopes and from direct telescopic observations, it was published in 1963 by the U.S. Air Force Aeronautical Chart and Information Center, St. Louis, Mo.–G.W.C.

Original scale 1:50 000

0   1   2   3 km

Contour interval: 10 m

FIGURE 12.–Another example of the versatility of photographs taken of the lunar surface from orbital altitudes. This topographic contour map was compiled by photogrammetric methods from two Apollo 15 panoramic camera frames. It shows a small area around the head of Rima Prinz I, one of the rilles covered in the preceding map (fig. 11). It was prepared in Flagstaff, Ariz., in support of a geologic study of rilles and their origin. It will not be published separately, hence, many of the stylistics of a map prepared for publication are lacking. Nevertheless, it is a fine example of what can be "seen" in panoramic camera pictures taken, in this case, at an altitude of 119 km.–G.W.C.

Selected Orbital Experiments

In this section we discuss only those orbital experiments that collected data relevant to photointerpretation. All of the instruments used for these experiments were transported in the SIM bay.

The gamma ray spectrometer (GRS) experiment aboard Apollos 15 to 16 detected gamma rays produced by the radioactive decay of materials on the lunar surface; areal variations in the intensity of gamma radiation make it possible to map the distribution of rock types along the ground tracks. The detector, which was mounted on a boom, contains a crystal that responds to an incident gamma ray by emitting a pulse of light. The light-pulse is converted by a photomultiplier tube into an electrical signal the strength of which is proportional to the energy of the gamma ray. The electrical signal is then processed and sent to Earth over the radio telemetry channel. Signals from other charged particles are canceled in the electronics processing so that only the gamma rays signals are returned to Earth.

The X-ray fluorescence (XFE) experiment was used on Apollo missions 15 and 16 to measure the chemical composition of the Moon. The sunlit portion of the Moon is constantly bombarded by solar X-rays. It is the secondary fluorescent X-rays emitted from the lunar surface material that are detected and measured by this experiment. In particular, the characteristic wavelengths and energies of magnesium, aluminum, and silicon were detected. Variation of the ratios of Mg/Si and Al/Si are extremely useful for determining the chemical composition and hence the type of rock constituting the surface materials.

The bistatic radar experiment, conducted on Apollo missions 13 through 16, measured the electromagnetic and structural properties of the outer few meters of the Moon's crust. Radio signals emitted from the CSM were reflected from the Moon and recorded by radiotelescopes on Earth. As the CSM orbited, the impinging beam of radiowaves, which covered a 10-km-square area of the surface, scanned the lunar disk. The characteristics of the areas so measured were then interpreted in terms of dielectric properties, size of the fragments composing the surface debris, and the magnitude and frequency of sloping surface. The experiment thus provides a means of measuring the small-scale "roughness" of the lunar surface.

The particles and fields subsatellite magnetometer experiment was conducted during Apollos 15 and 16. Instrumentation consisted primarily of a magnetometer mounted on a subsatellite transported in the CSM SIM bay. The subsatellite was inserted into lunar orbit shortly before the CSM began its return to Earth. The purpose of the magnetometer was to record variations in time and space of the magnetic field at Apollo orbital altitudes. The resulting data are being used to complement data obtained by the magnetometers placed on the surface of the Moon and the magnetometer carried onboard the much higher orbiting Explorer 35.

The S-band transponder (SBT) experiment was conducted on the last three missions, and gravity field observations were made on all missions.

The transponder was used to measure areal variations in the Moon's near-side gravitational field by recording very slight changes in the velocity of the orbiting CSM, subsatellite, and the LM. Changes in the gravitational field are caused by differences in density of the rocks. The experiment thus provides data for mapping the density of the rocks composing the upper part of the lunar crust. Radio signals transmitted to the three orbiting spacecraft components from Earth were multiplied by a constant (for electronic reasons) and then retransmitted to Earth. The difference in frequency between the transmitted and returned signals is a function of the velocity of the spacecraft.

The Apollo lunar sounder (ALSE) experiment was conducted only on Apollo 17. It used radar techniques to "see" into the Moon to depths as great as 1½ km. The sounder was designed for three primary modes of operation: the sounding mode detected and mapped subsurface features, and the profiling and imaging modes provided quantitative metric and topographic data as well as albedo measurements. Three frequencies of radiowaves were transmitted to the Moon from antennas mounted on the SM. Some of the waves were reflected by the lunar surface while others penetrated to various depths depending upon the type of material encountered. Those that penetrated the Moon were reflected by layers of rock within the Moon. The reflected component of the radiowaves was detected by the spacecraft antennas, delivered to the receiver, amplified and converted to light signals, and recorded optically on photographic film. The character of the reflected waves furnishes information about the nature of subsurface layers, and their return times tell the depth of the reflecting layers.

The infrared scanning radiometer (ISR) on board Apollo 17 was used to measure the emission of heat from the Moon's surface. A sensitive thermometer was mounted at the focus of a telescope. Light from a small area (about 2 km^2) of the Moon's surface enters the telescope through a mirror that oscillates back and forth to scan the surface of the ground track. After passing through various components of the instrument to a detector, the radiant energy of the light beam is changed into an electrical signal. This electrical signal is related to the temperature of the spot that is viewed by the telescope at any instant in time. The thermal properties so measured can then be correlated with known geographic and geologic features.

Scientific Results

Interpretation of the photographic products of the Apollo missions, augmented by the array of data obtained from the geochemical and other orbital experiments, has given rise to many new ideas about the configuration and origin of the lunar crust. The evolution of the Moon's crust is diagramed in figure 13 to show the temporal relationship of the major processes that affected the upper part of the crust.

One of the notable returns from the Apollo program is the radiometric dating of returned samples showing their great antiquity. This information is proof that the lunar surface we see is very old—

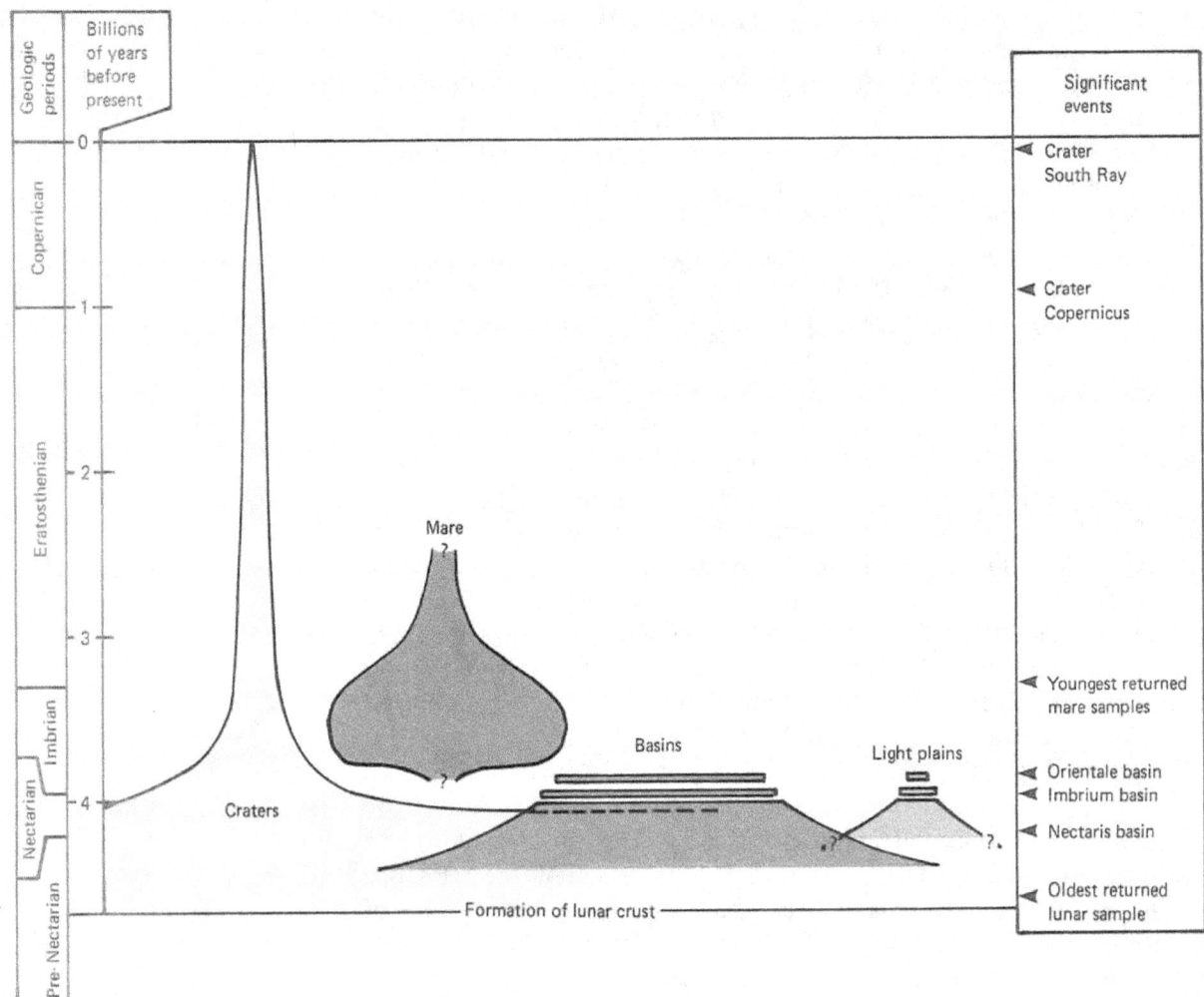

FIGURE 13.—This diagram, conceived by D. E. Wilhelms, graphically summarizes the sequence in which the major rock types of the lunar surface accumulated. Time scales are shown on the left, and the position in time of representative events and of selected returned lunar samples is shown on the right. The horizontal scale has no time connotation. The areas covered on this diagram by each of the four types of deposit are roughly proportional to their present areal extent on the surface of the Moon. Craters probably began to form and crater deposits began to accumulate as soon as the crust had solidified about 4.6 billion years ago. However, the early part of the record is hidden, and the rate at which they formed is unknown. At the end of Nectarian time or the beginning of Imbrian time, about 4 billion years ago, the rate at which craters formed began to decrease abruptly. It has continued to decrease, but much less rapidly, to the present. Mare materials had a very different history. The oldest mare materials to be recognized among the returned samples or on the basis of photogeologic mapping are early Imbrian in age. They apparently postdate, although by little, the formation of the Imbrium basin. The bulk of the mare material accumulated during the Imbrian period, but photogeologic studies (including crater-counting methods) indicate that some is at least as young as Eratosthenian. All basin and light plains deposits are ancient—apparently none being younger than early Imbrian in age.—G.W.C.

much older than the surface of Earth, which has been so degraded and changed by dynamic forces that primary crustal material is now largely unrecognizable. The lunar samples, therefore, provide a clue to the possible composition of original rocks on Earth. Information from the returned samples can also be extrapolated to date rocks underlying the lunar surface far from the landing site areas. In this way a new understanding of the general history of the Moon is emerging.

The mare areas of the near side of the Moon appear to be 2 to 5 km below mean lunar radius; the far side appears to be as much as 5 km above mean radius. Basaltic lava flows (the "maria") fill most of the low-lying basins on the near side (fig. 14); thicker, low density rocks underlie the high-standing ("terra") regions found on both near and far sides. The marked difference between the dark lowland regions (maria) and the bright highland regions (terrae) was first noted by Galileo in 1609. The results from several of the Apollo orbital experiments have confirmed their essential difference (fig. 15). Laser altimetry and gravity tracking results have defined marked differences in elevation. Variations in Al/Si and Mg/Si ratios obtained from the X-ray fluorescence experiment have confirmed the chemical difference between the two terrain types and strongly indicate that there has been little lateral surface transport of material from the highlands to the low-lying mare basins (Adler et al., 1974). Variations in gamma activity determined by gamma ray spectrometry have also documented this essential difference.

Recent estimates of the thickness of the lunar crust based on orbital laser altimetry, S-band tracking data, and surface seismic information indicate the existence of 20 km of basalt overlying 20 km of anorthositic material—crust—under eastern Oceanus Procellarum, 5 km of crust under some of the circular mare basins, 48 km under the near-side highlands, and 74 km of crust under the far-side highlands (Kaula et al., 1974). Variation in thickness of the lunar crust may have been caused by early chemical differentiation of the crust soon after the Moon was locked to Earth by gravitational attraction.

Orbital and surface magnetometer measurements (Coleman et al., 1972a,b,c) correlate closely with the gamma ray highs and lows. The deflection of the solar wind observed over some limb areas by the subsatellite magnetometer are thought to be caused by regions of high magnetization in that part of the Moon.

Tracking of the spacecraft as they respond to gravity has shown that craters up to 100 km in diameter are deficient in mass; that is, they constitute gravity lows. On the near side, mare-filled craters more than 150 km in diameter have positive gravity anomalies (Sjogren, Wimberly, and Wollenhaupt, 1974). The anomalies are probably caused by the dense basaltic lava flows that fill the craters and the underlying denser materials in the lower part of the crust that were thrust upward by the impact.

Basaltic lava flows occupy irregular mare areas, such as the very large one recognized as Oceanus Procellarum, as well as the many circular multiringed structures 150 km or more in diameter that are thought to have originated as impact basins (fig. 14). Among the best examples of

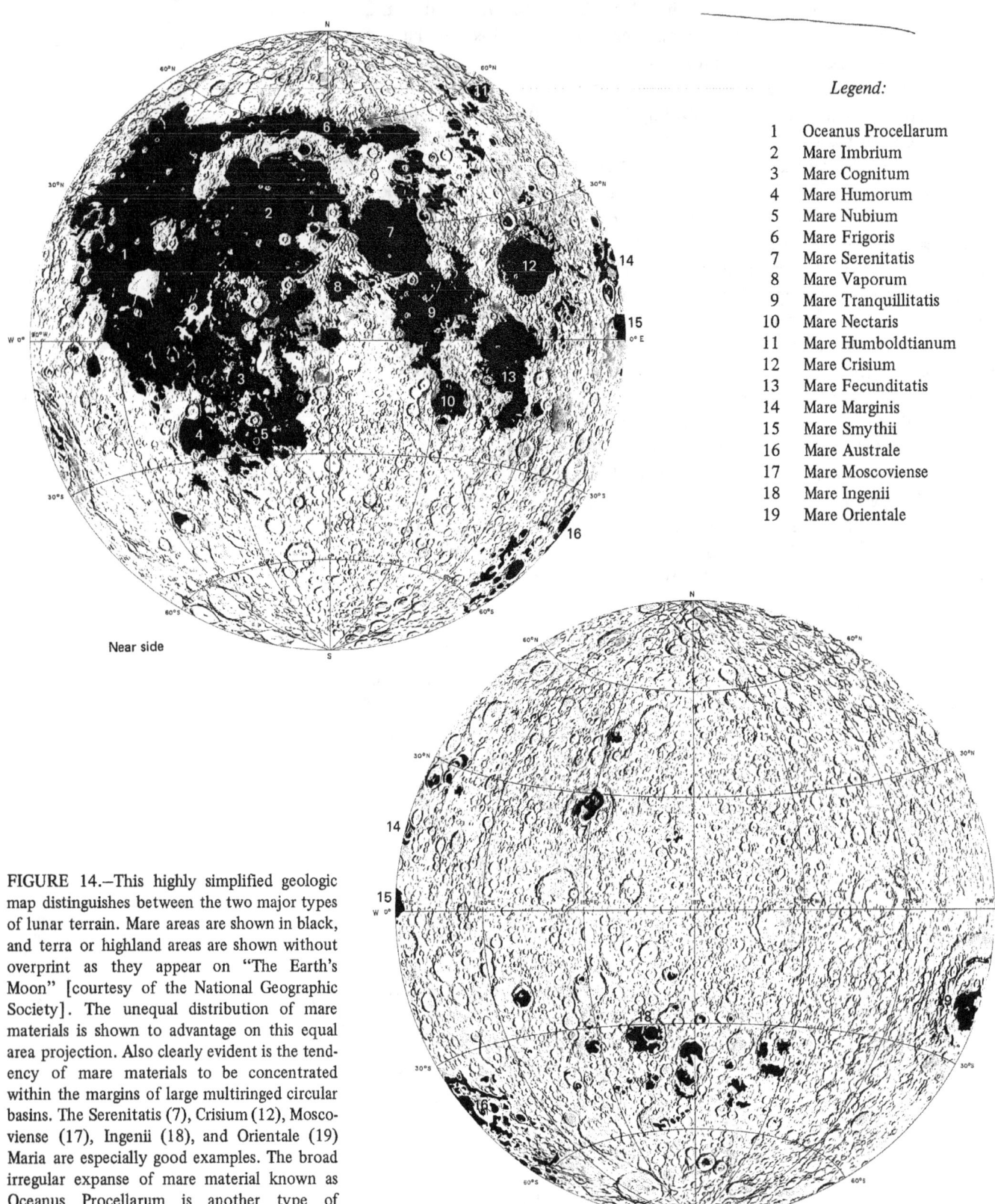

Near side

Far side

FIGURE 14.—This highly simplified geologic map distinguishes between the two major types of lunar terrain. Mare areas are shown in black, and terra or highland areas are shown without overprint as they appear on "The Earth's Moon" [courtesy of the National Geographic Society]. The unequal distribution of mare materials is shown to advantage on this equal area projection. Also clearly evident is the tendency of mare materials to be concentrated within the margins of large multiringed circular basins. The Serenitatis (7), Crisium (12), Moscoviense (17), Ingenii (18), and Orientale (19) Maria are especially good examples. The broad irregular expanse of mare material known as Oceanus Procellarum is another type of occurrence.—G.W.C.

24

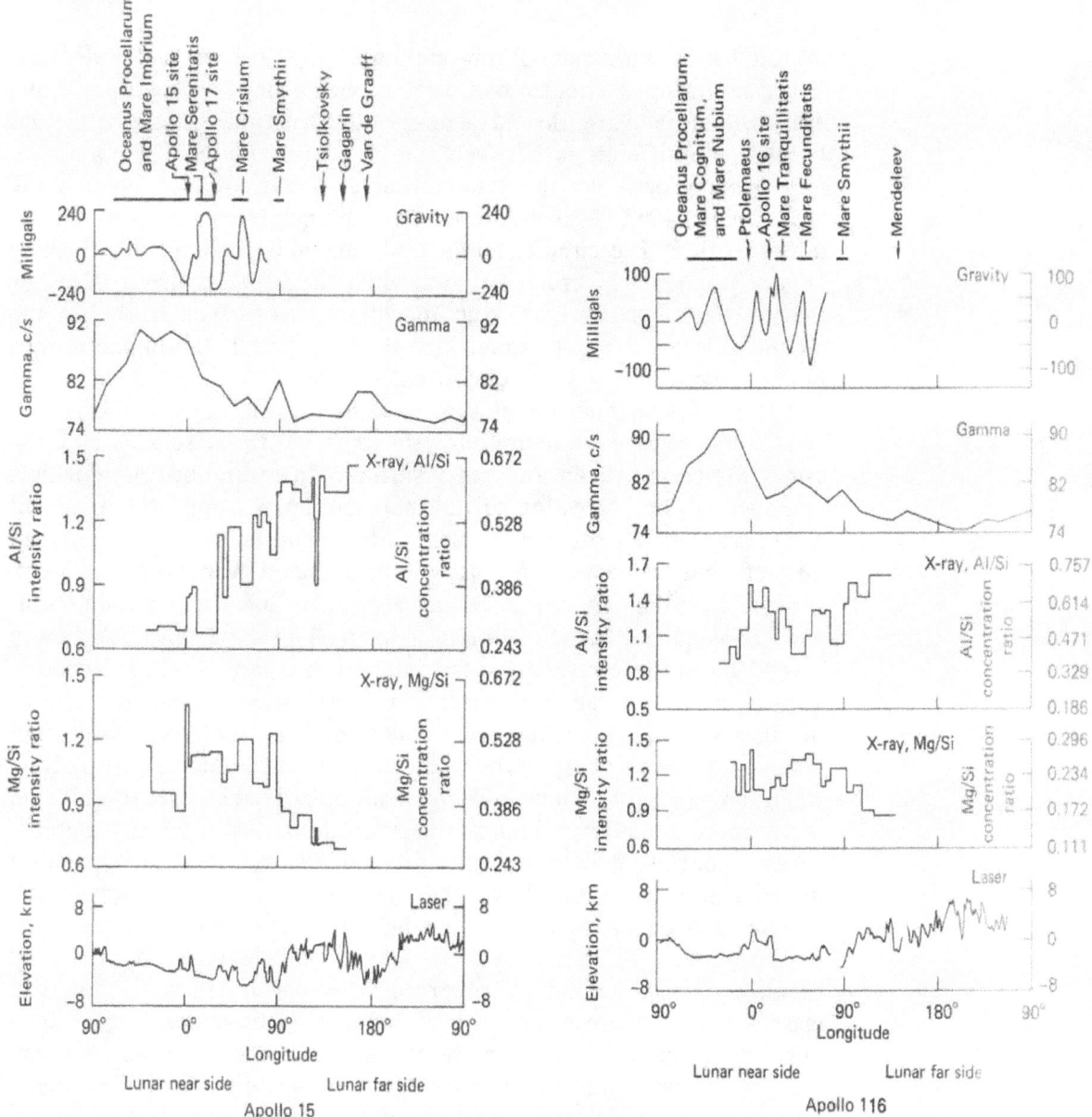

FIGURE 15.—Curves showing correlations of some physical and chemical properties recorded by selected remote sensing instruments carried in the SIM bay with topographic and geologic features. *Left:* Representative sensing curves plotted from data recorded along some of the Apollo 15 ground tracks. *Right:* The same types of data but along some of the Apollo 16 ground tracks. For purposes of location, selected geographic features are shown at the top of the graphs and degrees of longitude along the bottom. These curves show many interesting relations including (1) the positive gravity anomalies (concentrations of mass or "mascons") marking the circular mare basins, (2) the high gamma radiation from the border between Oceanus Procellarum and Mare Imbrium, (3) the inverse relationship between ratios of Al/Si and Mg/Si, (4) the systematic change in these ratios from maria to terrae, and (5) the differences in elevation between mare areas (low) and terra areas (high). Another significant observation is that the east limb of the Moon near Mare Smythii is much lower in elevation than other areas along the Apollo 15 and 16 ground tracks. In fact, the mare areas on the front side decrease in elevation from the west limb at Oceanus Procellarum to the east limb at Mare Smythii.

impact basins with mare filling are Imbrium, Crisium, and Orientale. Other multiringed circular basins of probable impact origin are, however, devoid or nearly devoid of mare fill. Structural, geochemical, and topographic differences between the circular and irregular mare areas have been proven by the laser altimeter, lunar sounder, gamma ray spectrometer, and X-ray fluorescence experiments, as well as by photo-interpretation. The circular maria are bounded by striking cliff-forming arcuate segments of crustal blocks while the irregular maria have low serrate edges. Positive gravity anomalies ("mascons") delineated by the *S*-band transponder experiment are associated with the impact basins but are absent over the irregular areas.

There are many hypotheses to explain the varying distribution of crustal materials documented by the Apollo orbital sensors and the samples returned from the lunar surface. The variation in thickness, composition, and elevation of the mare and terra regions, the increased gamma ray activity observed in the midfront and far sides, the increased magnetic and gamma ray measurements obtained over some limb areas, and the essential differences between irregular and circular mare basins and the highland areas all imply a controlling mechanism. A theory based on mantle convection (the internal circulation of hot material) gives a possible explanation for the observed sensor data and describes a possible controlling mechanism for the chemical, geophysical, and topographic variations. Very early in the history of the Moon's formation, when it was very hot and fluid, the mantle material was separated from the primordial melt by chemical differentiation. Lower density material became concentrated in the upper part of the mantle, whereas denser material settled in the lower part. Convection currents within the mantle then partially stripped the lighter weight material from some areas to cause the marked variation in crustal thickness, density, and chemical composition now observed between the present terra and mare areas. The areas stripped of lighter weight material were then flooded by basalts, which presently lie on a greatly thinned crust. Localized areas of increased concentrations in gamma ray and magnetic activity were caused by internal circulation and concentration of materials with higher magnetic and gamma ray properties. The mantle convection theory, however, is still being debated. True understanding of the development of the lunar crust may require years of additional study.

The mechanics of impact cratering have been studied intensively with the aid of the Apollo data. Craters ranging in size from the giant basins hundreds of kilometers across (like the Imbrium basin), to the smallest craters visible in orbital photographs (1 m in diameter), to microcraters on the surface of minute glass spheres contained in the returned lunar soil samples have been studied.

Crater ejecta material has been studied and classified into two groups: ballistic ejecta that is thrown out to form linear or curved patterns of rays and clusters of secondary craters on the Moon's surface, and fine-grained fluidized ejecta that locally blankets the lunar surface and forms patterned flows extending downrange from the primary

26

impact crater. The continuous ejecta blanket is apparently emplaced by base surge flow on the surface surrounding the crater. A striking example of surface patterns created by ejecta flow and its interaction with the local topography is found near the far-side crater King (fig. 159).

The continued bombardment of the lunar surface by meteoroids and secondary impact material has formed a regolith on the surface composed of breccia fragments and unconsolidated fragmental debris. The thickness and age of the regolith vary systematically. In general, thickness estimates based on crater shape agree well with estimates based on bistatic radar measurements.

A variety of features has been investigated and recognized as being of volcanic origin. Other terrain features have been more equivocally classified as possibly volcanic in origin. A succession of lava flows has been mapped in the Imbrium basin. Dark halo craters have been studied extensively and divided into two classes: The round, smooth-sided craters with no visible blocks in the crater walls are believed to be volcanic, and the dark halo material is thought to be composed of very fine-grained volcanic ejecta. Dark halo craters having irregular outlines may be impact craters from which darker material was exhumed at the time of impact. Lines or chains of craters have also been classified into two groups: volcanic crater chains (Hyginus Rille and Davy crater chain) and secondary impact crater chains extending radially from large craters (such as Copernicus, Kepler, and Aristarchus) and formed by ballistic ejecta from the large craters.

Acknowledgments

This volume was prepared under NASA contract number W-13,130. G. E. Esenwein of NASA Headquarters provided much of the initial encouragement and support needed to begin this effort. L. J. Kosofsky of NASA Headquarters also contributed materially to the completion of the volume. In addition to preparing many of the captions, Don E. Wilhelms and Keith A. Howard of the U.S. Geological Survey and Farouk El-Baz of the Smithsonian Institution participated in the preliminary selection of the photographs. Other U.S. Geological Survey members who contributed significantly to the preparation of this volume are S. S. C. Wu, photogrammetrist; J. S. VanDivier, cartographer-illustrator; and K. A. Zeller, photographer. Captions of the photographs were composed by individual authors who are identified by their initials:

G.W.C.   George W. Colton
         U.S. Geological Survey
         Flagstaff, Ariz.

R.E.E.   Richard E. Eggleton
         U.S. Geological Survey
         Flagstaff, Ariz.

F.E.-B.   Farouk El-Baz
National Air and Space Museum
Smithsonian Institution
Washington, D.C.

M.J.G.   Maurice J. Grolier
U.S. Geological Survey
Reston, Va.

J.W.H.   James W. Head III
Department of Geological Sciences
Brown University
Providence, R.I.

C.A.H.   Carroll Ann Hodges
U.S. Geological Survey
Menlo Park, Calif.

K.A.H.   Keith A. Howard
U.S. Geological Survey
Menlo Park, Calif.

L.J.K.   Leon J. Kosofsky
National Aeronautics and Space Administration
Washington, D.C.

B.K.L.   Baerbel K. Lucchitta
U.S. Geological Survey
Flagstaff, Ariz.

M.C.M.   Michael C. McEwen
National Aeronautics and Space Administration
Johnson Space Center
Houston, Tex.

H.M.   Harold Masursky
U.S. Geological Survey
Flagstaff, Ariz.

H.J.M.   Henry J. Moore
U.S. Geological Survey
Menlo Park, Calif.

G.G.S.   Gerald G. Schaber
U.S. Geological Survey
Flagstaff, Ariz.

D.H.S.   David H. Scott
U.S. Geological Survey
Flagstaff, Ariz.

L.A.S.   Laurence A. Soderblom
U.S. Geological Survey
Flagstaff, Ariz.

M.W.   Mareta West
U.S. Geological Survey
Reston, Va.

D.E.W.   Don E. Wilhelms
             U.S. Geological Survey
             Menlo Park, Calif.

Desiree Stuart-Alexander acted as contract manager for NASA. George W. Colton and Mary E. Strobell of the U.S. Geological Survey compiled the overlays and edited the many preliminary versions of the book. Julienne M. Goodrich and Mary Nelson Rakow of the Smithsonian Institution also assisted in the many editorial tasks.

Our thanks go to all the people associated with the Apollo missions; their hard work and technical expertise resulted in the acquisition of the data that made this volume possible.

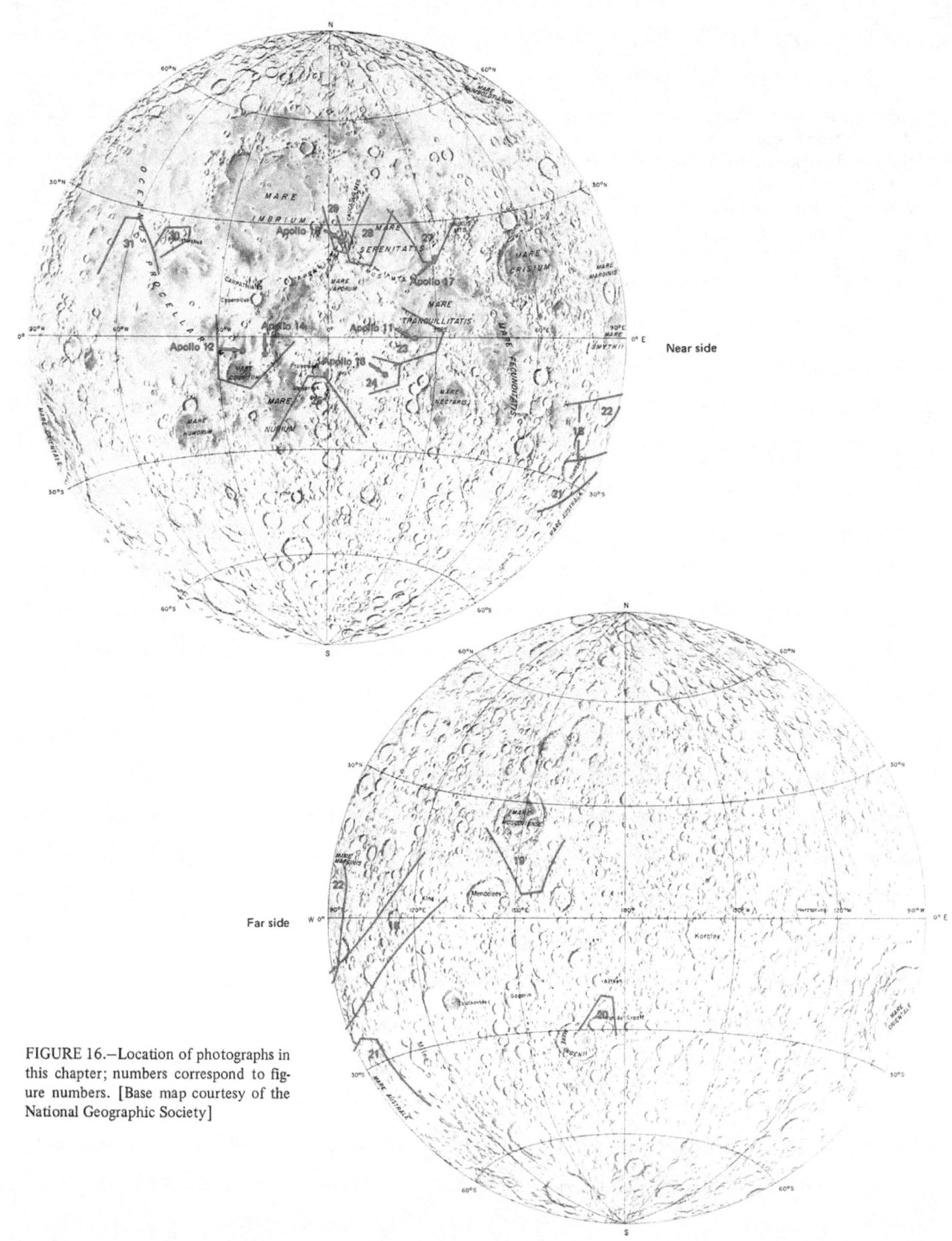

FIGURE 16.—Location of photographs in this chapter; numbers correspond to figure numbers. [Base map courtesy of the National Geographic Society]

Near side

Far side

2

Regional Views

Regional pictures of the sunlit half of the Moon were taken adjacent to the orbital tracks. The pictures in this chapter were selected to show the broader aspects of the lunar surface. Because large features are being shown, many of the pictures in this chapter are oblique views that cover larger areas of the surface. Views taken from the same altitude with the same camera aimed perpendicular to the surface would have been of much smaller areas. Also included are two distant views taken after the Apollo 17 spacecraft had begun its return to Earth.

The terrae dominate the lunar surface, occupying more than about 85 percent of the entire Moon (Stuart-Alexander and Howard, 1970). They are visible from Earth as the brighter parts of the Moon. Viewed through telescopes, they are seen to be more rugged, more densely cratered, and higher in elevation than the mare areas. Terrae have long been proposed to be older than the mare areas on stratigraphic grounds, and now recent studies of samples returned by the astronauts have confirmed this age relationship. Most terra rocks dated by radiometric methods are 3.95 billion years or more in age.

In contrast, the mare areas, including the large irregular area known as Oceanus Procellarum, occupy only about 15 percent of the Moon's surface. They are the dark areas visible from Earth with the unaided eye. In addition to being darker, maria are much less densely cratered and hence smoother than the terrae; most mare areas are relatively level and many occupy depressed areas. These aspects have all suggested that they consist of basaltic lava flows. Such flows occupy extensive areas on Earth, both under the ocean floors and on the continents. Some of the returned lunar samples do consist of basalts that are in many respects similar in chemistry and mineralogy to terrestrial basalts. Radiometric age determinations have shown that most of the basalts sampled to date cooled between 3.2 and 3.8 billion years ago.

Debate about the origin of the mare basalts continues. Perhaps further study of the returned samples, of geophysical data gathered in situ on the lunar surface, and of geophysical and geochemical data obtained from orbit will settle the problem of why the basalts formed and accumulated where they did.—H.M.

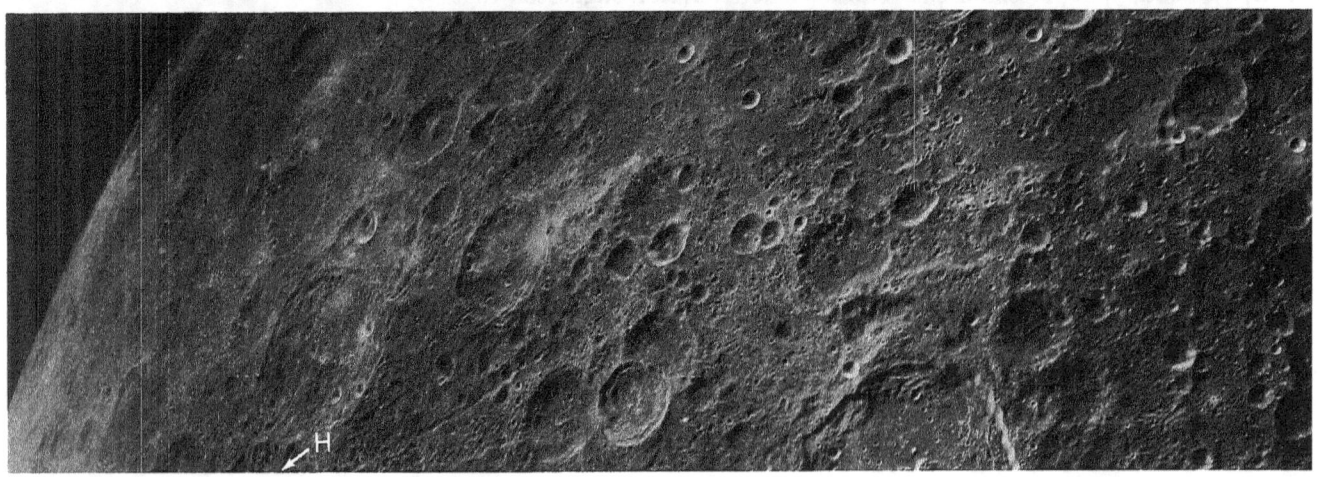

FIGURE 17.—A part of the Moon never seen by man until the space age. The far side was first photographed by the U.S.S.R. spacecraft Luna 3 and later directly viewed by American astronauts from Apollo spacecraft. Taken by the Apollo 16 mapping camera on the flight back to Earth, this view is centered at about the easternmost part of the Moon visible from Earth. The dark smooth area at the left edge of this picture, Mare Crisium, is a conspicuous feature near the right side, or eastern limb, of the Moon, as seen in the evening sky. The other two large maria, Mare Marginis to the right of Crisium and Mare Smythii below it, are seen either not at all or are greatly foreshortened. The right half of this picture, which is nearly devoid of maria, was totally unfamiliar to Earth-bound observers. It is typical of the highly cratered, ancient terrae or uplands that occupy nearly 85 percent of the Moon's surface. Some early astronomers, however, correctly predicted that the Moon's far side would have far fewer maria than the near side and that it would be dominated by craters. The rocks that underlie the maria are now known to consist of basaltic lava, a type of rock distinctly different from the bedrock of the densely cratered highland areas of the Moon. Why there are so few maria on the far side is one of the many puzzles still under discussion.—D.E.W.

AS16-3021 (M)

FIGURE 18.—A panoramic camera frame taken on Apollo 17 after trans-Earth injection; that is, as the astronauts started their journey toward Earth after leaving lunar orbit. As is typical of the terrae or highlands, numerous craters of different sizes and ages are superposed on one another. In general, the larger craters are older than the smaller ones. The dominant feature in the upper right part of the frame is the crater King (K), which displays a lobster-claw-like central peak. This crater will be illustrated in detail later. In the middle part of the frame the dominant feature is the crater Pasteur (P). It is a large, ancient crater, 250 km in diameter, with numerous younger and smaller craters superposed on its rim and flat floor. Around Pasteur, aside from a few small fresh-appearing craters, the terrain appears to be old and subdued as if mantled by a thick layer of debris. In the lower left of the photograph numerous chains of small craters are visible trending in a northeasterly direction. They are part of the ejecta system of the crater Humboldt (H), a part of which appears on the lower left edge of the photograph. Near the lower left edge is the dark mare material of Mare Fecunditatis, which is located near the eastern limb of the Moon when viewed from Earth.—F.E.-B.

AS17-3153 (P)

 Approximate

FIGURE 19.—An oblique view of an area near the center of the far side taken with the Apollo 16 mapping camera from orbit at an altitude of 116 km. Like most of the far side, this lunar highland area is pockmarked with numerous craters tens of kilometers in diameter superposed on older terrain. The dark-floored crater (*K*) in the upper center is Kohlschütter. It is about 60 km in diameter and is filled with basaltlike mare material. This makes it one of the few craters on the lunar far side smaller than 100 km in diameter that contains mare material. The sharp rise near the center of the horizon is the southeastern rim of Mare Moscoviense (*M*). The caduceuslike object protruding into the view from the right edge of the photograph is the boom of the gamma ray spectrometer. A detector at the end of the boom measured the concentration of radioactive materials on the lunar surface along the ground track of the spacecraft. —F.E.-B.

AS16-0729 (M)

FIGURE 21.—This view, one of the most southerly photographed by the Apollo astronauts, straddles the boundary between near and far sides and shows terrain typical of both sides. In the foreground is densely cratered terra or highland terrain typical of the far side. At the top of the picture a large area is covered by mare material, which is easily distinguished from the terrae by its smooth appearance and darker color. This mare area, appropriately named Mare Australe (Southern Sea), occupies a large old circular basin—probably a gigantic impact scar. The rings and arcuate segments of rings projecting through the mare are the rim crests of smaller craters that were created after the basin was formed but before it was filled. On the far side otherwise similar old basins have less mare filling or none, whereas most circular basins on the near side are deeply flooded by mare material.—D.E.W.

AS15-2503 (M)

FIGURE 20.—The very large "twin crater" in this southward-looking oblique view is Van de Graaff, approximately 250 km in length. The dark oval area in the upper right corner near the horizon is the mare-filled crater Thomson, which in turn lies within a much larger mare-filled circular basin that contains Mare Ingenii. The crater Van de Graaff itself is filled with both light plains and mare materials. The darker patch at the near end is mare material, and the hummocky material along its left (east) side is ejecta from the crater Birkeland (B). Light plains materials are also visible in the near field within smaller craters and in areas between them. This region of the lunar far side is of particular interest for two reasons: (1) it has the highest concentration of natural radiation yet recorded on the lunar far side by the gamma ray spectrometer (Metzger et al., 1973), and (2) it is also the site of the most prominent magnetic anomaly associated with the far-side highlands (Coleman et al., 1972a).—F.E.-B.

AS17-150-22959 (H)

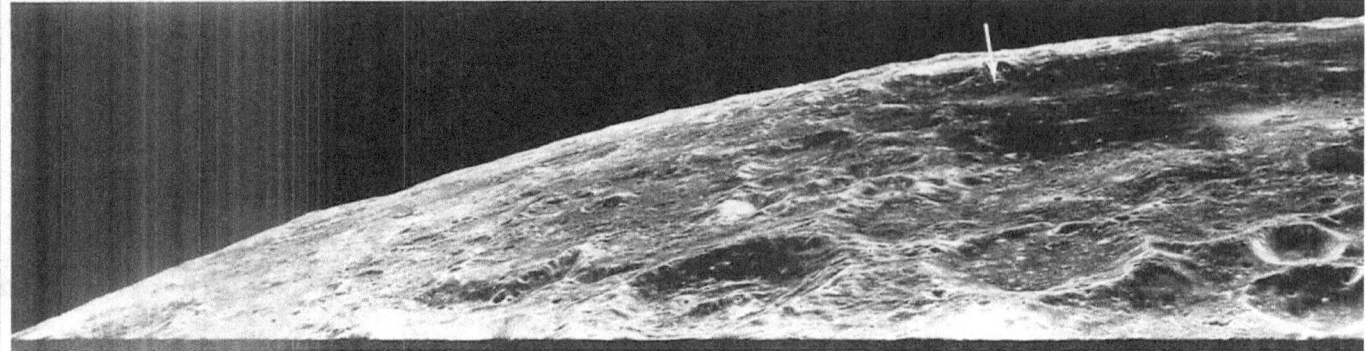

AS17-2871 (P)

FIGURE 22.–The Apollo 17 panoramic camera peers from the far-side hemisphere onto the near side. The large mare extending for a distance of 420 km between the two arrows is Mare Smythii, which is at the transition between hemispheres in several respects. Centered at 95° E longitude, it is at the geographic (or seleno-graphic) boundary of near and far hemispheres. (See fig. 14.) The laser altimeters of Apollos 15 and 16 show that a change in elevation occurs at Mare Smythii from

FIGURE 23.–This view is entirely on the near side where the monotony of the terrae is interrupted by broad expanses of maria. The first two Apollo landing missions were on the relatively smooth maria in areas that were unobstructed by mountains along the approach route. Here, looking westward from the window of the Apollo 11 CM Columbia during its approach, is a view of southwestern Mare Tranquillitatis. Tranquility Base (*TB*), where Neil Armstrong and Edwin ("Buzz") Aldrin aboard Eagle made the first lunar landing, is barely visible along the terminator beneath the arrow. The maria are invariably described as "smooth." In a relative sense this is true, but only when they are compared with the much more rugged terrae, such as those south (to the left) of the landing site. Here, as in most mare areas, the "smoothness" of the mare surface is interrupted by craters, numerous mare ridges (as at *R*), straight rilles or grabens (*G*), sinuous rilles (*S*), and islands of unburied terrae (*T*). Some craters like Maskelyne (*M*) have extensive aprons of ejected debris that further contribute to the roughness of mare surfaces. The dark area in the lower left is not a huge shadow on the lunar surface but is one of the LM thrusters. (It is so close to the camera that it is out of focus.) For an idea of the scale of this picture, the distance from the center of Maskelyne to the landing site is 210 km.–D.E.W.

AS11-37-5437 (H)

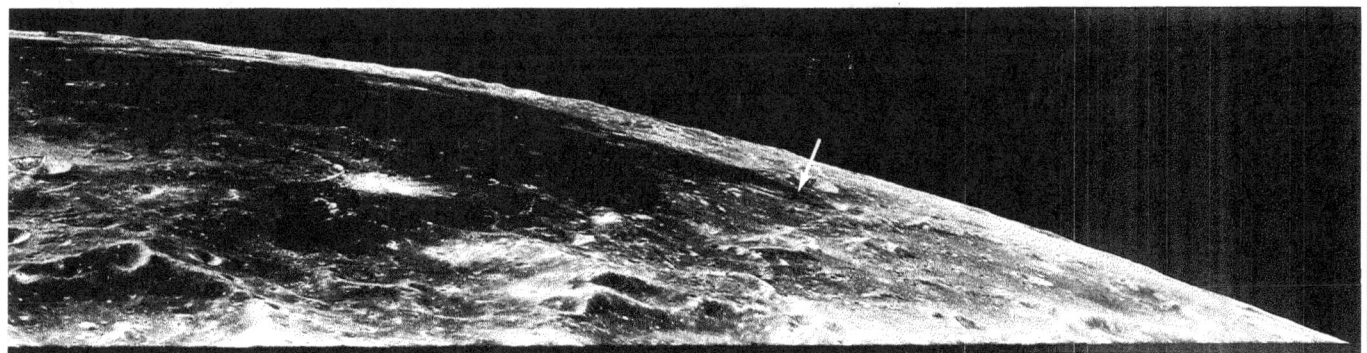

typically high far-side values to lower values typical of the near side. The average decrease in elevation amounts to a substantial 3 km. Smythii is a fairly extensive mare yet it occupies an old degraded basin, the western rim of which is detectable by carefully looking near the horizon and at the left of the mare-filled depression. Such a basin would probably contain little or no mare if located on the far side. Perhaps the higher elevation of the far side indicates a thicker crust that inhibited access of mare magmas to the surface.—D.E.W.

FIGURE 24.—Looking westward from the Apollo 16 CM toward the Descartes landing site southwest of Tranquility Base. To find the site, look straight along the gamma ray spectrometer boom one additional boomlength to the area between two small bright craters. The distance on the surface is 71 km. This part of the Moon's near side, the central highlands, is essentially devoid of maria. Two major goals of this mission were to study the smooth plains immediately west of (above) the site and the peculiar furrowed and hummocky terrain that dominates the center of the frame. Before the mission, both units were thought to have been formed by volcanism, but now, after analysis of the samples, they are interpreted as ejecta deposits resulting from very large impact events. The plains and furrowed materials have partly covered and masked older craters in this region, including Descartes (D) at the left of the picture, so that fewer craters are visible here than in most far-side and many near-side highland areas. The more mountainous terrain along the bottom (east) is part of the Nectaris basin rim. (Dashed lines are used to show the rim crest of crater Descartes, the rim of which has been so severely degraded by erosional and depositional processes that it is otherwise difficult to see. This convention is used frequently throughout this volume.)—D.E.W.

AS16-0566 (M)

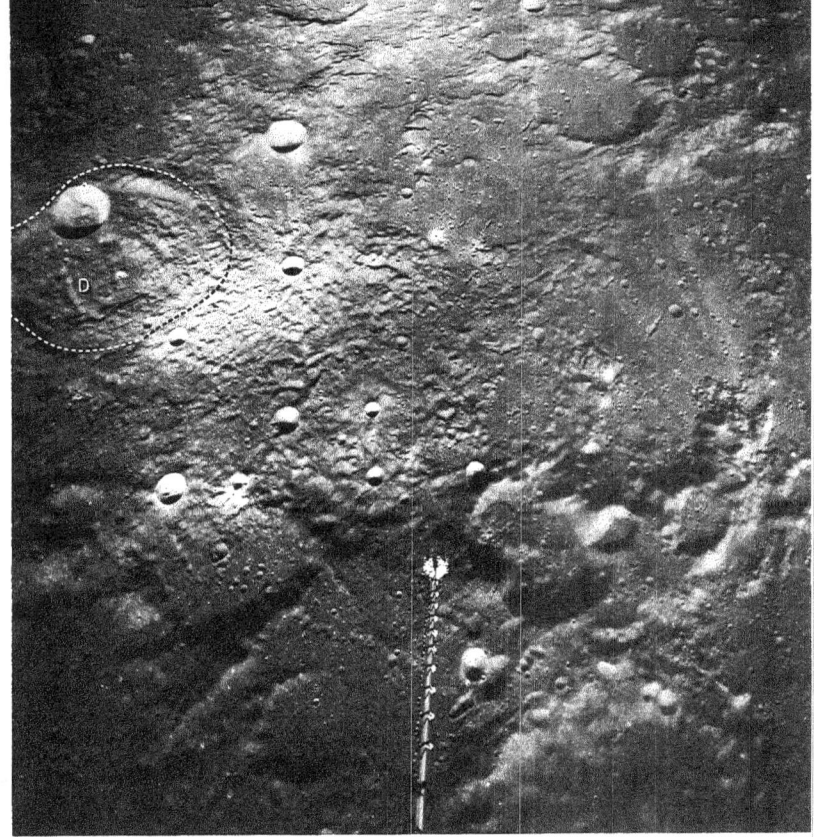

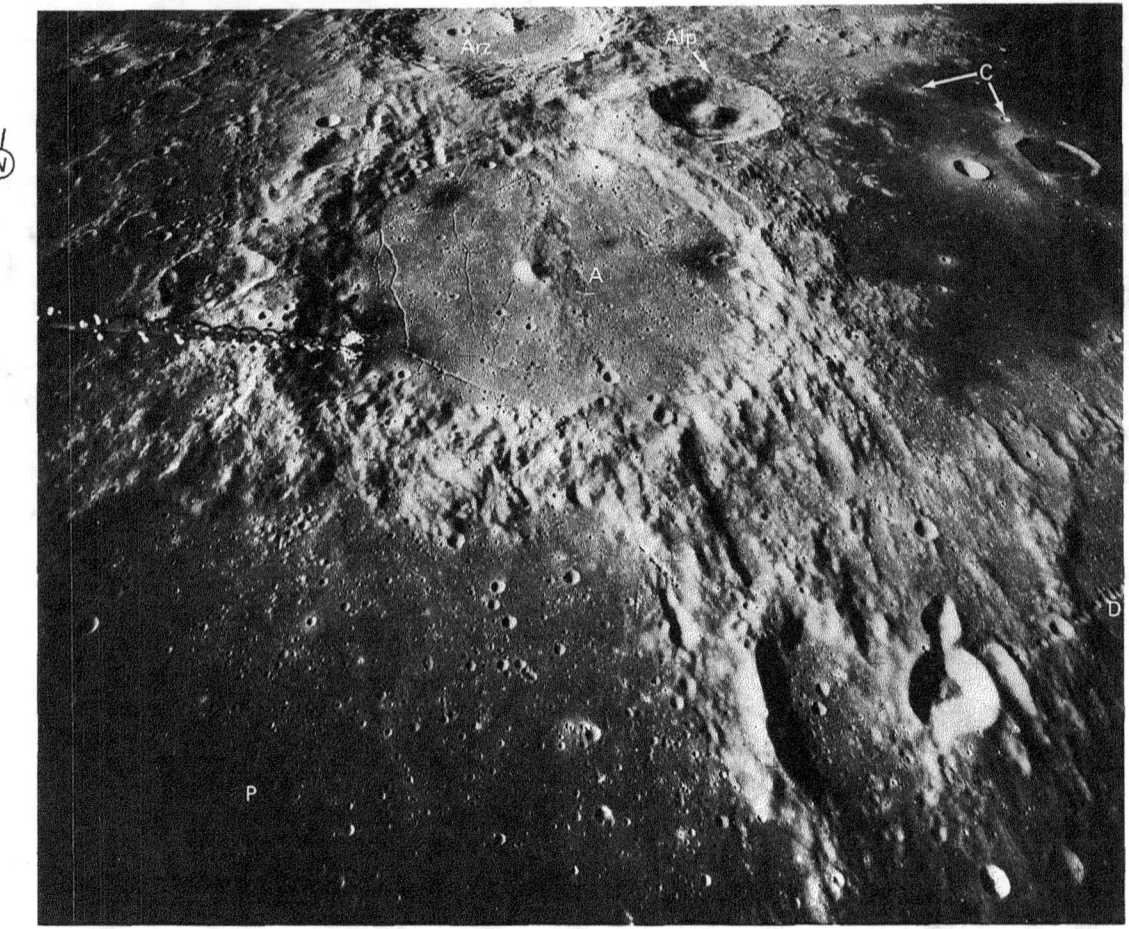

AS16-2478 (M)

FIGURE 25.—This south-looking oblique view shows three of the best-known features of the near-side central highlands: the alined craters Ptolemaeus (*P*), partly in view in the foreground; Alphonsus (*A*), in the center, and Arzachel (*Arz*) beyond. The latter is 100 km in diameter. Alpetragius (*Alp*), with its peculiarly large central peak, is not alined with the others. The three large craters are filled—Ptolemaeus most deeply—by deposits of light plains material. This type of material fills depressions over most of the lunar terrae but is concentrated in this part of the central highlands and in two or three other belts on the near side. In Alphonsus the plains deposits are overlain by dark material in halos surrounding elongated craters. Most of these dark halo craters are situated along rilles and are the sites of reported temporary changes in telescopic appearance. The dark halo craters on the eastern (left) side were the target of the Ranger 9 spacecraft, the last of three successful Ranger missions that impacted the Moon more than 7 years before this photograph was taken.

Some lunar geologists believe that the alinement of the three big craters is controlled by a fundamental structure within the Moon's crust. However, because the three craters are quite different in age, the alinement is more probably coincidental. When Arzachel formed, debris was ejected to form a radial chain of secondary craters on top of the light plains deposits that fill the other two craters. This shows that Arzachel is the youngest of the three craters. Ptolemaeus, considerably more degraded than the other two, is also covered by ejecta from Alphonsus and is the oldest of the three. An extensive additional geologic history can be read from the photograph by extending this type of reasoning. The rims of Ptolemaeus and Alphonsus are cut by a conspicuous system of parallel ridges and grooves called "Imbrium sculpture" because it radiates from the Imbrium basin, behind (north of) the camera. Part of this sculpture is in turn intersected by the Davy crater chain (*D*) in the lower right of the picture. Further, the highland materials are overlain by the obviously younger mare materials in the upper right of the picture. Fresh craters (for example, *C*) and their bright halos overlie the maria. Therefore, from oldest to youngest, the complete historical sequence is Ptolemaeus, Alphonsus, Imbrium sculpture, light plains deposits, Arzachel, Davy chain, mare materials, and rayed craters. The dark halo craters in Alphonsus are as young or younger than the mare materials. Application of similar geologic reasoning based on overlapping and crosscutting relations has unraveled, in terms of relative age, the evolution of most of the lunar surface.—D.E.W.

FIGURE 26.—Apollo 16 crossed the central highlands and then looked northward into southern Oceanus Procellarum. This region contains a mixture of terrae and of younger mare material that has flooded the terrae. The large terra mass in the foreground is Montes Riphaeus, which consist of the unflooded remnants of the rim of a large basin and the rims of two or three inundated craters. Islandlike continuations of the basin rim can be seen arcing eastward (right) from the north end of the mountains. The mare that occupies this basin in the lower right of the picture is called Mare Cognitum, the "Sea That Became Known"—but not until visited by the first successful U.S. lunar mission, Ranger 7, in July 1964. Two years and 9 months later, Surveyor 3 landed on the mare north of Montes Riphaeus along a ray of bright material emanating from the crater Copernicus. Surveyor was joined in November 1969 by Apollo 12. Then, in January 1971, after an earlier aborted try by Apollo 13, Apollo 14 landed on the terra bordering the east side of Mare Cognitum (near the horizon in this picture). The materials at the Apollo 14 landing site had been mapped as the Fra Mauro Formation and interpreted as ejecta from the Imbrium basin long before the landing. Although the site is nearly 1200 km south of the center of the Imbrium basin, the results of the mission substantiate this interpretation and verify the magnitude of the event that formed the basin. Islands and peninsulas of terra in the upper left part of the picture are also probably structures and deposits formed by the Imbrium impact.—D.E.W.

AS16-2518 (M)

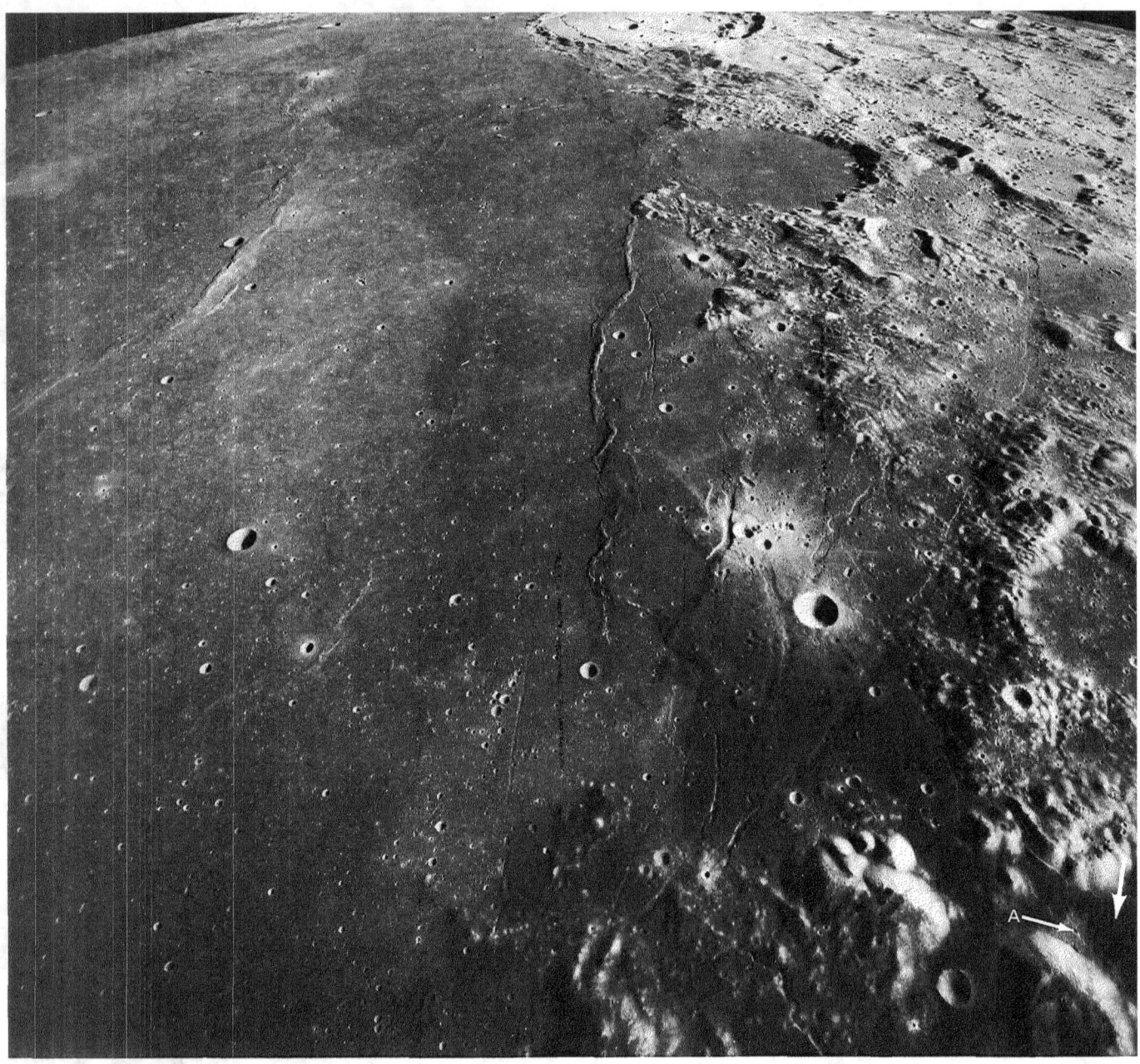

AS17-0940 (M)

FIGURE 27.—On the near side, Apollos 15 and 17 flew over a more northerly strip of the Moon's surface than did Apollo 16 (fig. 1). Here, over the eastern edge of Mare Serenitatis, another mare that fills a circular basin, the Apollo 17 mapping camera recorded the regional setting of the Apollo 17 Taurus-Littrow landing site (arrow, lower right corner). The relatively smooth and unusually dark material around the edge of the basin contrasts sharply with the hummocky bright mountain massifs of the basin rim. Trending along the edge of the basin, an exceptionally fresh system of scarps, mare ridges, and rilles transects both mare and terra. The rilles, mostly straight or gently arcuate, are fault valleys called "grabens." Near the distant horizon is the flat-floored, partly flooded crater Posidonius (*P*), about 100 km in diameter. Just this side of it is a bay formed by the flooding of the older crater Le Monnier, which was visited by the unmanned Soviet roving vehicle Lunokhod 2 in January 1974. Days earlier the Apollo 17 mission had been successfully concluded. During a 3-day stay on the Moon at the Taurus-Littrow site (arrow), the dark material on the valley floor, the avalanche of light-colored debris (*A*), and the mountains surrounding the site were visited and sampled by the astronauts.—D.E.W.

AS17-0953 (M)

FIGURE 28.—A comparable view of the opposite (west) side of Mare Serenitatis taken only a few minutes after figure 27 during the same revolution. Now closer to the terminator, the smaller angle between the Sun's rays and the lunar surface causes an enhancement of topographic relief. Mare ridges appear to be higher and rilles or grabens deeper than on the east side. In both areas the ridges and rilles are grossly parallel to the edge of the mare. Rimae Sulpicius Gallus, the prominent rilles in the lower part of the picture, and the dark plateau into which they are cut are embayed by the younger mare plains. Some of the mare ridges, especially the sharper crested ones, may be younger than the bulk of the mare plains. All of these features are younger than the rugged Apennine Mountains (left) and the Caucasus Mountains (center of the horizon). The two mountain ranges form part of the rim of both the Serenitatis basin and the Imbrium basin (to the left of the ranges).—D.E.W.

41

FIGURE 29.—The Apollo 15 landing site, marked by an arrow, is visible in this north-looking oblique photograph. This is one of the northernmost views photographed during the Apollo missions. The Caucasus Mountains are on the upper right horizon, with Mare Serenitatis off the image on their right and Mare Imbrium on their left. Two large rayed craters, Aristillus (*Ar*) and Autolycus (*Au*), 55 and 40 km in diameter, respectively, penetrate the mare. In September 1959, the first manmade object to strike the Moon, the Soviet Luna 2 spacecraft, impacted just west of Autolycus. The massive mountains in the foreground are the Apennines. At their western slope is the sinuous Hadley Rille, which was examined at close range by the Apollo 15 astronauts who drove to its eastern rim in the first manned roving vehicle used on the Moon.—D.E.W.

AS15-1537 (M)

FIGURE 30.—West of the Apennines (by 1300 km) on the opposite side of Mare Imbrium is the peculiar complex of Montes Harbinger (bottom right) and the Aristarchus plateau (top). These uplifted features probably represent the outermost rim structures of the Imbrium basin. They protrude through a mare that is infested with sinuous rilles and some straight rilles. Features of the rich scene include the head of Vallis Schröteri (*VS*), or Schröter's Valley; the old mare-flooded crater Prinz (*P*); and one of the Moon's youngest and brightest large impact craters, Aristarchus (*A*). Aristarchus, which is 43 km in diameter, has been the site of frequent "transient lunar phenomena" (temporary changes in color or brightness) reported by many telescopic observers throughout the world.—D.E.W.

AS15-93-12602 (H)

FIGURE 31.—During the various Apollo missions, the west limb of the Moon was in shadow behind the terminator and could not be photographed. This south-looking oblique view across western Oceanus Procellarum shows one of the western-most areas visible to the astronauts. As the word "oceanus" (=ocean) implies, Procellarum is larger than a mare. Like the maria however, it is formed by dark mare materials that originated as flows of basaltic lava. The crater Seleucus, 42 km in diameter, is shown in sunrise along the right (west) edge of the frame. The narrowness of its rim and the abrupt contact between its raised rim and the surrounding mare prove that the final mare flooding occurred after the crater was formed. In other words, the crater predates at least the youngest mare basalts in this area. The same age relations hold true at Schiaparelli, the second largest crater in view. Several medium-sized craters and countless small craters are surrounded by finely textured outer ejecta deposits and hence are younger than the mare. The Marius Hills, the most extensive aggregation of volcanic domes and cones on the Moon, are visible on the left horizon. Thus the picture shows the characteristic events of late lunar history—the sporadic formation of impact craters concurrent with volcanic eruptions that form lava plains, hills, and ridges.—D.E.W.

AS15-2618 (M)

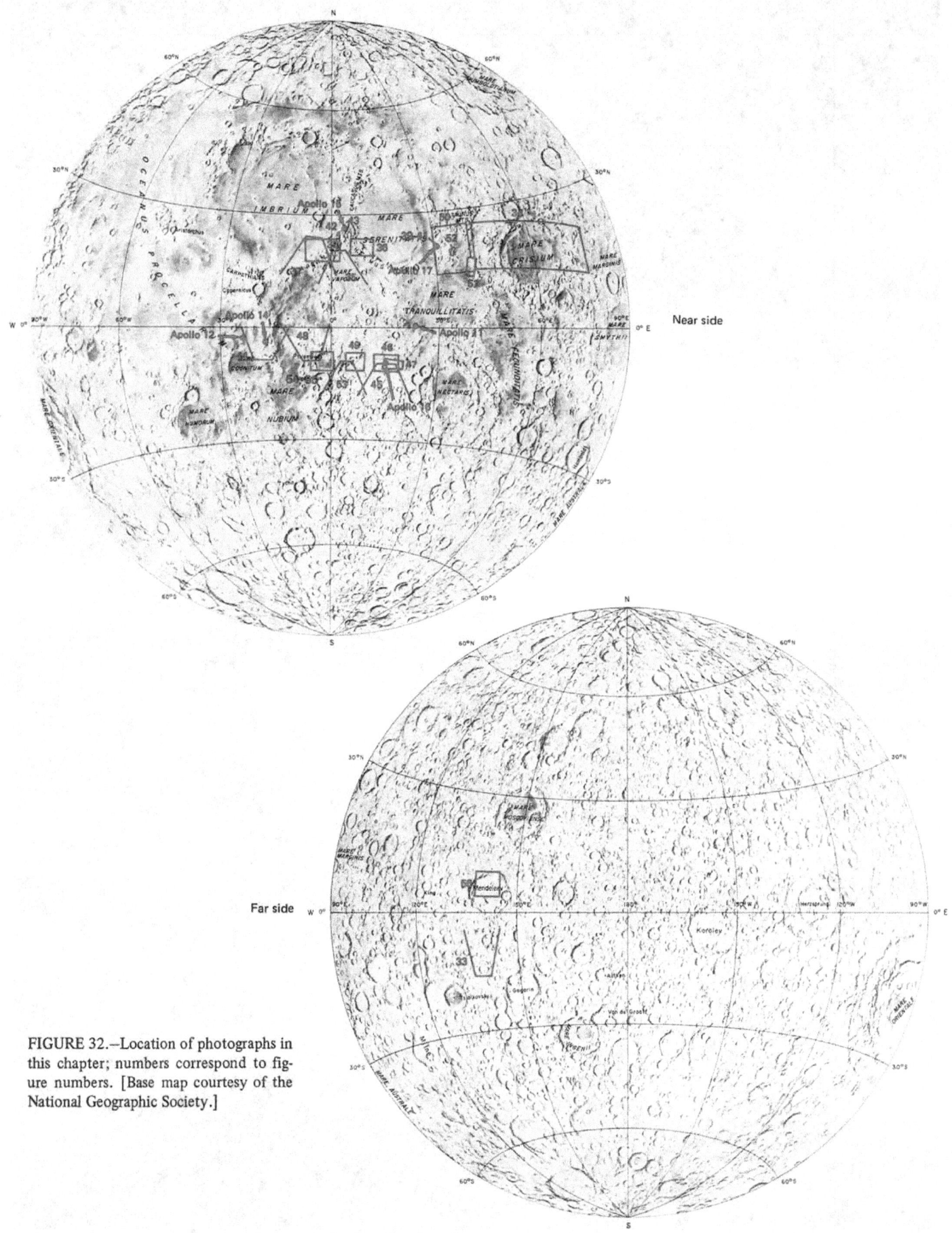

Near side

Far side

FIGURE 32.—Location of photographs in this chapter; numbers correspond to figure numbers. [Base map courtesy of the National Geographic Society.]

3

The
Terrae

In the first chapter the terrae or highlands were described as the brighter, older, and generally higher standing terrain occupying most of the Moon's surface (fig. 14). Because they are older, the terrae are much more densely populated by large craters than are the maria. Even though the terrae occupy two-thirds of the visible or Earth-facing hemisphere (and about 85 percent of the entire surface), less was known about them than about the maria. This is so because of their greater age and apparent complexity and partly because only one of the five successful Surveyor spacecraft landed in the terrae.

Our understanding has, however, increased tremendously as a result of the Apollo missions. The last four missions have been especially rewarding in this respect. Analyses of the returned lunar samples, study of data from instruments emplaced on the lunar surface, and remote sensing instruments in the CSM have filled in many of the information gaps, but have also presented new problems.

Radiometric dates obtained on samples of terrae rocks confirm, as was believed earlier, that the terrae are older than the maria. Although the terrae are highly modified, they are composed of rock material that formed very early in the Moon's history by the process of magmatic differentiation. By this process, minerals formed within an igneous melt become segregated according to differences in their physical properties. Lighter materials rise to the top of the magma body by virtue of their lower specific gravity, and, after solidification, form low density rocks. Among the returned lunar samples thought to represent terra materials not completely altered by subsequent events, varieties of gabbroic anorthosite are the most common. This type of rock is composed largely of plagioclase with varying amounts of olivine and pyroxene. Plagioclase is a common mineral on Earth and one of rather low specific gravity.

The preponderance of anorthositic rocks in the lunar highlands is supported by data from Apollo remote sensing instruments. Some of the chemical differences between anorthositic and basaltic rocks could be determined by the X-ray fluorescence and gamma ray experiments of Apollos 15 and 16. The X-ray fluorescence results show a higher ratio of aluminum to silicon in the terrae than in the maria, corresponding to the known chemical difference between anorthositic and basaltic rocks. Results from the gamma ray spectrometer show that the terrae contain less iron and titanium than do the maria (Metzger et al., 1974). This also is consistent with the chemical compositions of anorthositic versus basaltic rocks.

The lower specific gravity of anorthositic rocks compared to basalts is another characteristic that was measured directly or indirectly by orbital experiments. The *S*-band transponder experiment flown on the last five missions recorded variations in the lunar gravity field along the ground tracks. The results clearly show that the terra materials are less dense than mare materials. Indirect evidence comes from laser altimeters onboard Apollos 15, 16, and 17 that conclusively show that the terra regions are higher in average elevation than the maria. The continuous high-resolution profiles of the Moon's surface provided by the electromagnetic sounder experiment on Apollo 17 substantiated the spot elevations recorded by the laser altimeter. The combined results of these three experiments indicate that most of the Moon's crust—like most of the Earth's crust—is in isostatic equilibrium. In other words, areas of high elevation are underlain by low-density rocks, low-standing areas by high-density rocks, and differences in elevation across broad areas are the result of differences in density, or specific gravity.

The ancient rock materials of the terrae have been drastically modified by various processes since their formation early in lunar history. Repetitive bombardment by impacting bodies has been the most important cause of modification. Countless impact events have resulted in the widespread redistribution of materials over the surface, the brecciation of the rocks so displaced, and the metamorphism by shock of the minerals that make up the rocks. The impact events have been so numerous and their cumulative effect has been so pervasive that few samples recognizable as original crustal material have been returned by the Apollo missions.

Other processes that have modified the terrae are tectonism, volcanism, and mass wasting. Tectonism is visible in numerous linear structures transecting the terrae. Some have been recognized and mapped as normal faults, or as pairs of closely spaced normal faults bordering grabens. Some of the largest linear structures are on the near side, radiating from the edge of the Imbrium basin. They are obviously related to the formation of that basin. However, over the entire Moon, the majority of linear features are oriented in northeast and northwest directions. This arrangement results in a rectilinear gridlike pattern referred to as the "lunar grid" (Strom, 1964). The origin of the lunar grid is unknown. It must have formed at an early stage because parts of it are modified and intersected by patterns of faults and gouges that radiate outward from the circular basins, themselves features of very considerable age.

Volcanism is clearly evident, for example, in the Abulfeda chain of craters extending for more than 200 km southeast from the crater Abulfeda (fig. 45). This chain is closely alined with two crater chains similar in appearance: one is near the crater Ptolemaeus and the other is near Piccolomini. Another area of possible volcanic activity is the Kant plateau, the edge of which was examined by the Apollo 16 astronauts. Both the Ptolemaeus and Kant areas are high and have abnormally high ratios of aluminum to silicon.

Mass wasting has affected the terrae by reducing differences in relief caused by cratering, tectonism, and volcanism. This form of erosion has subdued the inherent ruggedness of the terrae by moving materials from high areas to low areas. The rate of movement ranges from very slow (as by creep) to very rapid (as by avalanching).—H.M.

FIGURE 33.—This oblique view of a small area northeast of Tsiolkovsky on the Moon's far side is occupied entirely by terrae. Like much of the far-side terrae, it is unrelievedly cratered. Thousands of craters are visible, ranging in size from the limits of resolution (a few tens of meters at the lower edge of the picture) to the 75-km crater near the center. This part of the Moon is too far from any young multiringed circular basins to have been mantled by basin ejecta. In areas such as this, if anywhere, are to be found materials of the early lunar crust.—D.E.W.

AS17-155-23702 (H)

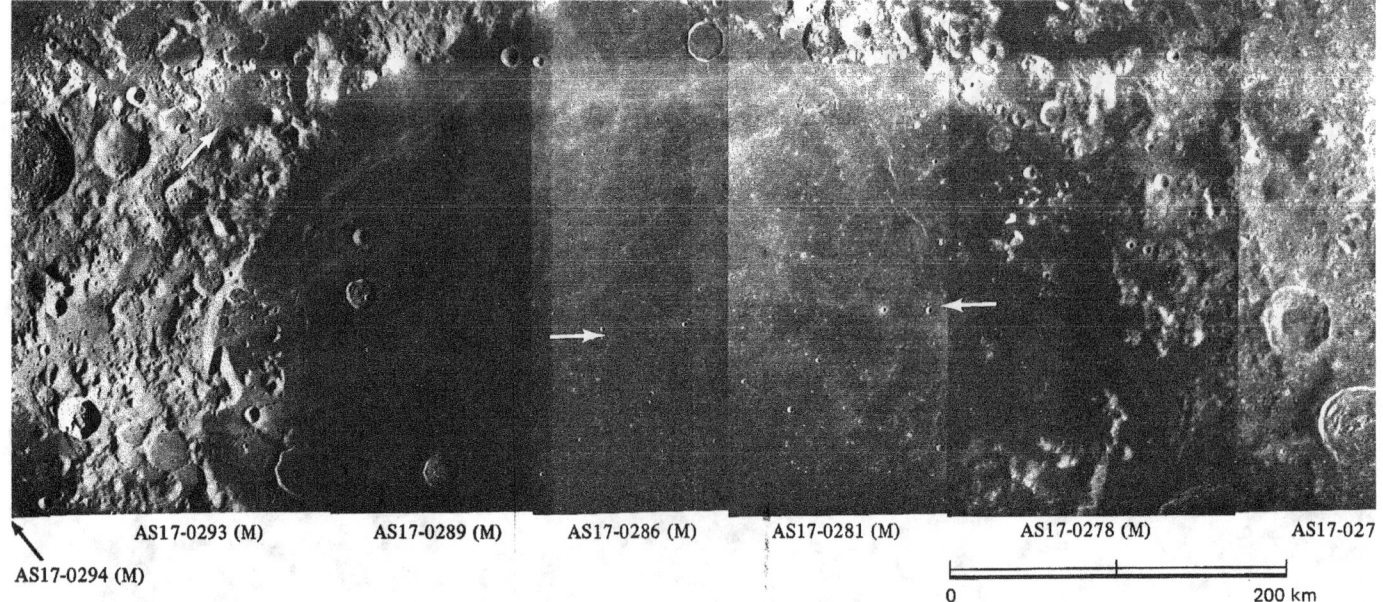

AS17-0294 (M)   AS17-0293 (M)   AS17-0289 (M)   AS17-0286 (M)   AS17-0281 (M)   AS17-0278 (M)   AS17-027

0   200 km

N

FIGURE 34.—A mosaic of seven metric camera frames showing the northern part of the Crisium basin. The inner lava-flooded part of the basin is Mare Crisium. Basins, the principal features of the lunar terrae, are essentially large craters with more than one conspicuous concentric mountain ring. In the west half of this mosaic are two raised basin rings separated by a trough (between arrows) partly filled with mare lavas and light plains deposits. The higher and broader of the two rings forms the "shore" of Mare Crisium. In the east half of the mosaic, the rings are less obvious, partly because the Sun illumination is too high to cast prominent shadows and partly because the rings have been broken up by faulting and flooded by mare material. Craters are less numerous in this picture than in the previous one because many were destroyed or buried during the formation of the basin. Some basins, like Crisium, are deeply flooded by mare material, and others, especially on the far side, are flooded less or not at all. Therefore, the accumulation of mare material is not directly related to the formation of basins, and the terms "mare" and "basin" must always be kept distinct.—D.E.W.

FIGURE 35.—It is common for the circular form of a very old basin to survive even after all basin materials are obliterated. Mare Vaporum, as shown in this south-looking oblique photograph, provides a good example. Its average width is about 200 km. Its circular form marks the outer edge of the ancient, deeply buried Vaporum basin. All the terrae surrounding Mare Vaporum are blanketed by massive ejecta of the Imbrium basin, the center of which lies to the north, behind and to the right of the camera. The ejecta disappears beneath Mare Vaporum. The circularity is enhanced at the left (east) of the picture by a system of mare ridges and scarps that was localized over an old Vaporum basin ring. The cratered, linear Hyginus Rille is near the southern horizon, and the sinuous Conon Rille, to be described later in this book, is in the foreground.—D.E.W.

AS17-1674 (M)

N

AS17-1819 (M)   0   50 km   AS17-1817 (M)

FIGURE 36.—The Haemus Mountains bound the southwestern edge of Mare Serenitatis and form the rim of the Serenitatis basin. They have a strongly lineated pattern that is most apparent in the lower left part of this stereoscopic view. (The width of the stereogram within this mosaic is shown by the bar across the bottom.) The trend of the linear pattern is radial to the Imbrium basin, the margin of which is about 250 km to the northwest of the edge of the picture. Carr (1966) described the mountains as composed mostly of ejecta from the Imbrium basin. The lineation may be due to shattering of the lunar crust by the Imbrium impact event, depositional fluting of the ejecta, gouges made by impacting debris from the Imbrium basin, or a combination of the three.

The prominent rilles in the upper part of the stereogram are grabens or fault troughs transecting both terra and mare surfaces. They are roughly concentric to the edge of the Serenita-tis basin. The rilles become less distinct in the terrae, attesting to the easy destruction of surface features in terra material by mass wasting. Within Mare Serenitatis the rilles are partly flooded by the younger lavas that have filled the basin. A dark mantling material, named Sulpicius Gallus Formation, covers parts of the highland and mare surfaces alike. In the highlands the dark material has been removed from the tops of hills and steep slopes and reveals the underlying bright highland material. The dark mantle is conspicuous in the right center of the stereogram near the small kidney-shaped crater (arrow). In this same crater and in small, young, rayed craters nearby, the Apollo 17 astronauts observed orange material. This suggests that the dark material here is similar to that sampled at the Apollo 17 landing site on the other side of the basin where orange material was found on the rim of a young crater.—B.K.L.

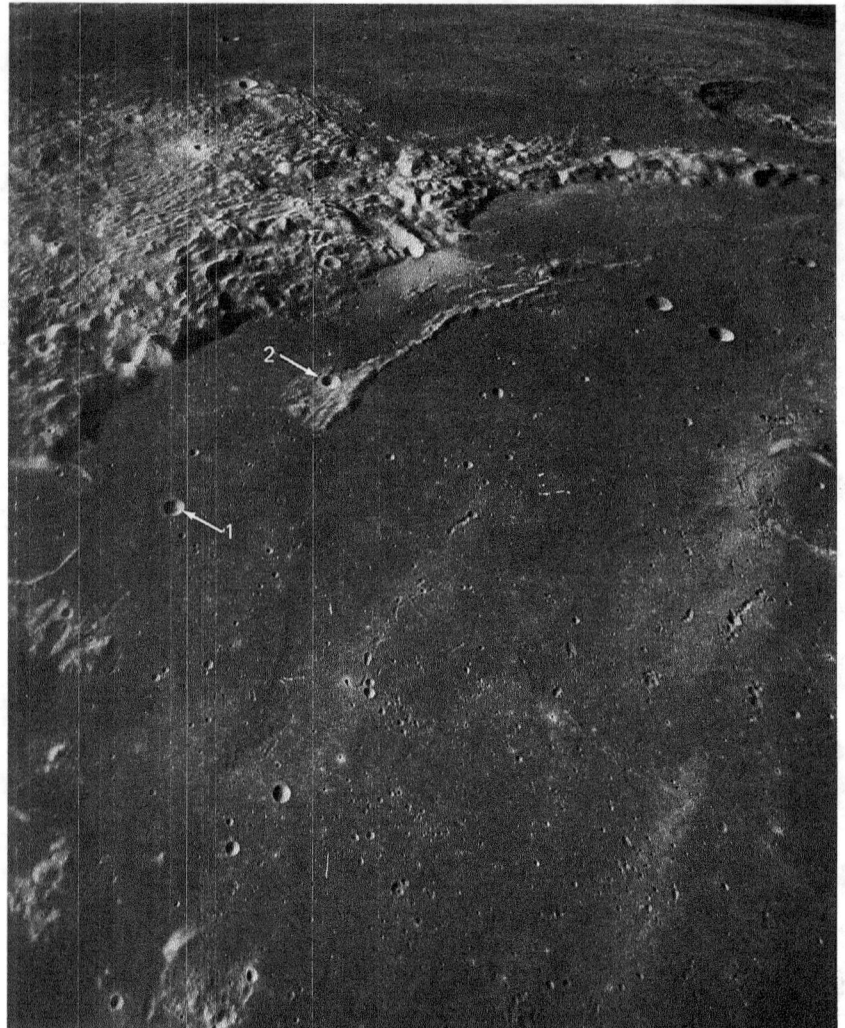

FIGURE 37.—An oblique view of the south-eastern part of the Imbrium basin, one of the largest multiringed, circular basins on the Moon. Most scientists agree that it was formed by the impact of an asteroid, comet, or other planetary body striking the lunar surface at hypersonic velocity. The Imbrium event excavated a depression nearly 1300 km in diameter in the terrae, uplifted and intensely deformed the adjacent terrae, and blanketed much of the lunar surface with debris ejected from the depression. The depressed area was subsequently flooded by lava flows to form the dark relatively smooth surface recognized as Mare Imbrium. The Montes Apenninus form the southeastern rim of the basin. They and other rugged areas of light material visible here are ancient terrae uplifted by the impact event and covered to varying thicknesses by ejecta debris. Material from the Apennine Mountains was collected by the Apollo 15 astronauts who landed near the foot of the mountains not far to the left and below (that is, to the northeast of) the area shown here. The arcuate trends parallel to the margin of the Imbrium basin are mostly faults associated with the formation of the basin. The numbers are explained in the caption for figure 38.—M.W.

AS17-2433 (M)

FIGURE 38.—This mosaic of vertical frames covers part of the same area shown in the preceding oblique view (fig. 37), but shows the Montes Apenninus in much more detail. So that the two pictures can be oriented and compared, the same two craters have been identified in each picture by numbers. The bulk of the mountain chain consists of giant blocks of lunar crust that were lifted and tilted outward by the impact that formed the Imbrium basin. These blocks have been covered by an unknown thickness of debris ejected from the basin. The hummocky deposits (H) probably were formed by the base surge—a turbulent cloud of fluidized debris that moved outward along the surface from the point of impact. The hummocks resemble huge dunes. Their dimensions indicate a velocity of flow in excess of 100 km/hr and a maximum thickness of the deposits of several kilometers.

The Imbrium event is believed to have occurred 3.95 billion years ago. Later the basin was almost completely filled by successive flows of basaltic mare material. The same material also inundated parts of the outer edge of the Apennine Mountain chain, as in the lower right corner of the picture.—H.M.

AS17-1828 (M)   0   50 km   AS17-1825 (M)

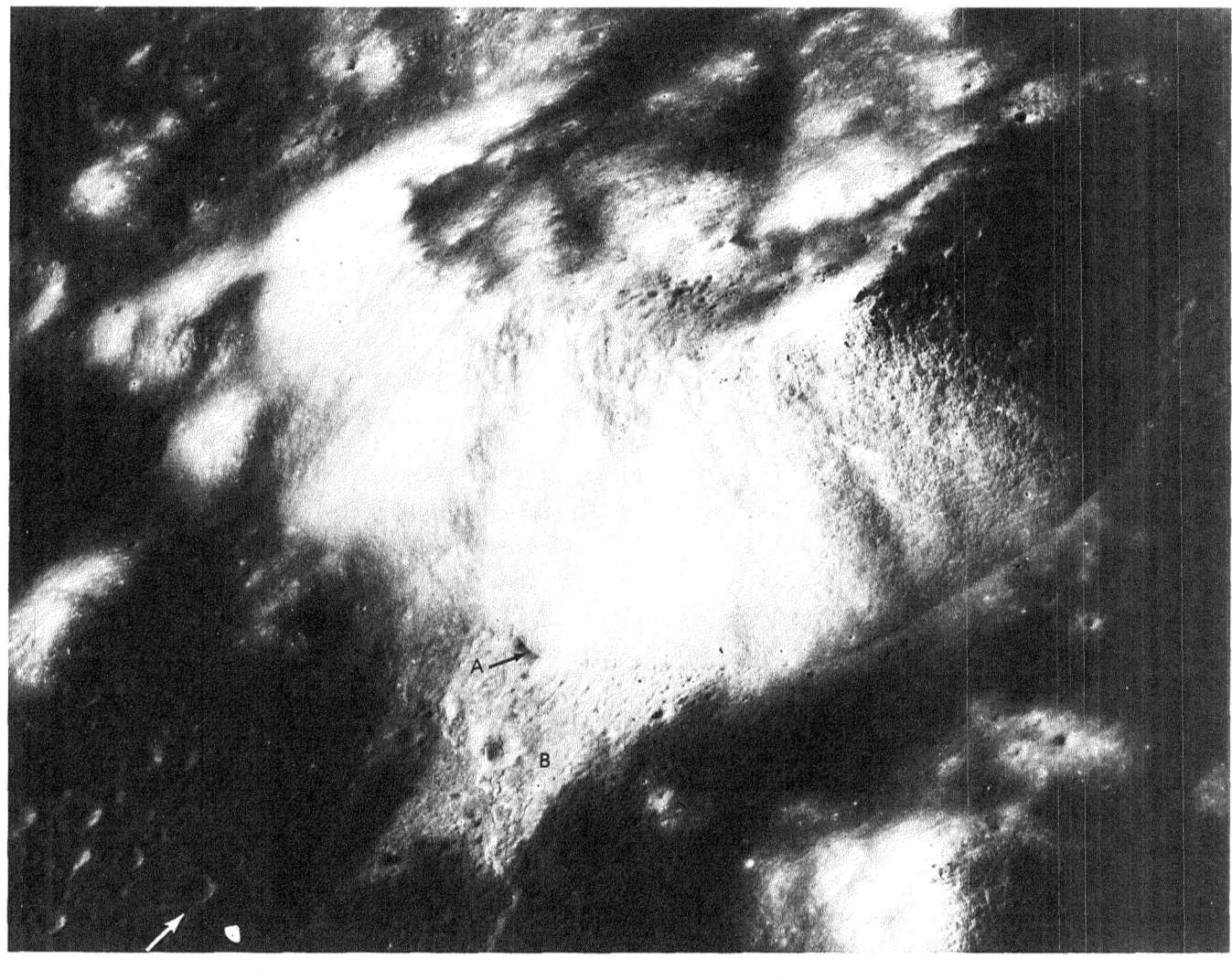

AS15-9297 (P)

0   5 km

FIGURE 39.—This mountain mass, called South Massif, on
the southeastern rim of the Serenitatis basin towers 2000 m
above the Apollo 17 landing site at the bottom left of the
picture (arrow). The mountain is typical of the massifs
forming the main rim of multiringed basins. Most lunar
geologists believe that the massifs are individual fault blocks
uplifted as a result of the impact event that created the
basin. South Massif is composed of highly brecciated rock
that was probably emplaced as ejecta from the Serenitatis
basin, although similar brecciated ejecta from other more
ancient and more distant basins may be present. Rocky
outcrops on the top right of the mountain have shed clearly
visible boulders and blocks, but most of the slopes are
formed of finer mass-wasted debris. Some of this debris has
partially filled a small crater at the base of the mountain
near the center of the picture (A). An avalanche of unconsolidated surficial material propelled by secondary impact
slid off the mountainside and splayed out over the mare
surface leaving a thin blanket of light-colored breccia on the
valley floor (B). The Apollo 17 astronauts traveled across
the slide to the base of the mountain.—B.K.L.

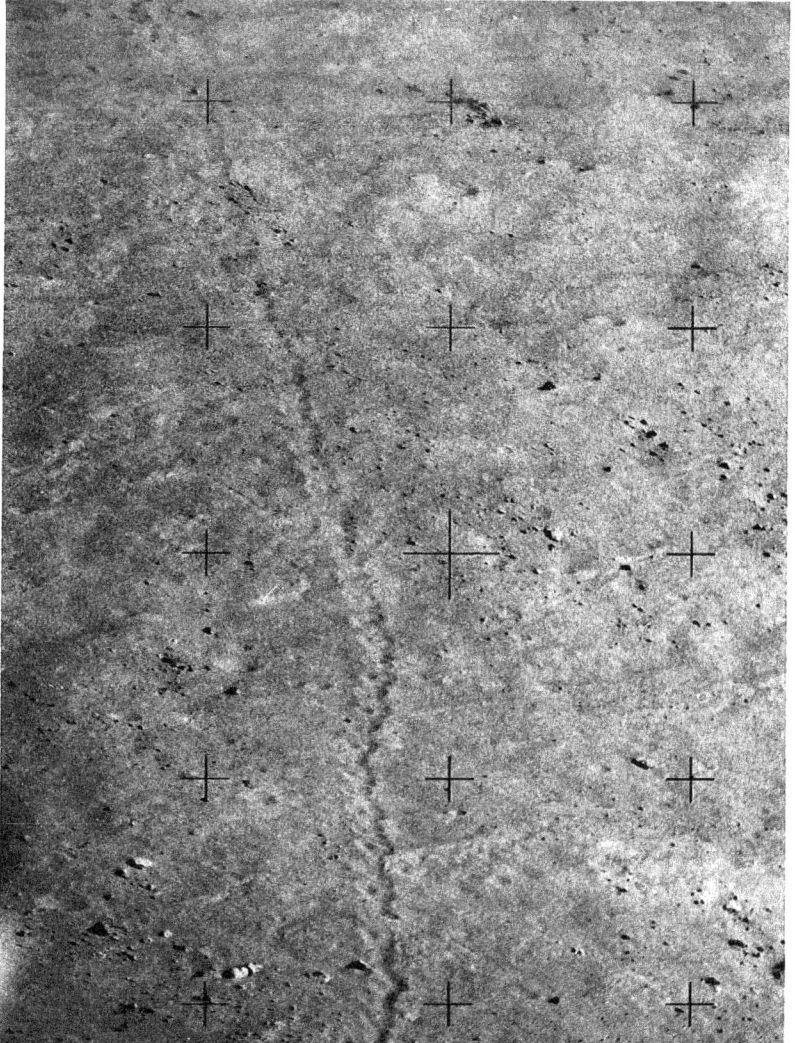

FIGURE 41.—This is a closeup view of part of the boulder track shown in figure 40. The track is about 10 m wide and the boulder that made the track is about 18 m in diameter (Mitchell et al., 1973). The cause of movement, other than the obvious effect of gravity, is uncertain. Various investigators have suggested that movement was initiated by seismic vibrations of internal origin, vibrations caused by repeated impact events, cyclic thermal expansion and contraction, and instability as soil accumulates above the boulder or is removed from below it. It is also possible that some tracks are formed by projectiles, presumably from impact craters, that skid or bounce along the surface before coming to rest. From detailed studies of boulder tracks, some properties related to the strength, density, and thickness of the lunar soil can be measured.—G.W.C.

AS17-144-22129 (H)

FIGURE 40.—Across the valley of Taurus-Littrow from South Massif (from fig. 39) is another big mountain known as North Massif. In this scene of North Massif, an astronaut is kneeling at right. The path traced by a big boulder rolling down North Massif is indicated by arrows. This track is large enough to be visible—but just barely—in some panoramic camera frames taken from orbit at an altitude of slightly more than 100 km. On both North and South Massifs boulder tracks such as this one were used by geologists as markers to find the original positions of boulders that were sampled by the astronauts.—K.A.H.

AS17-136-20694 (H)

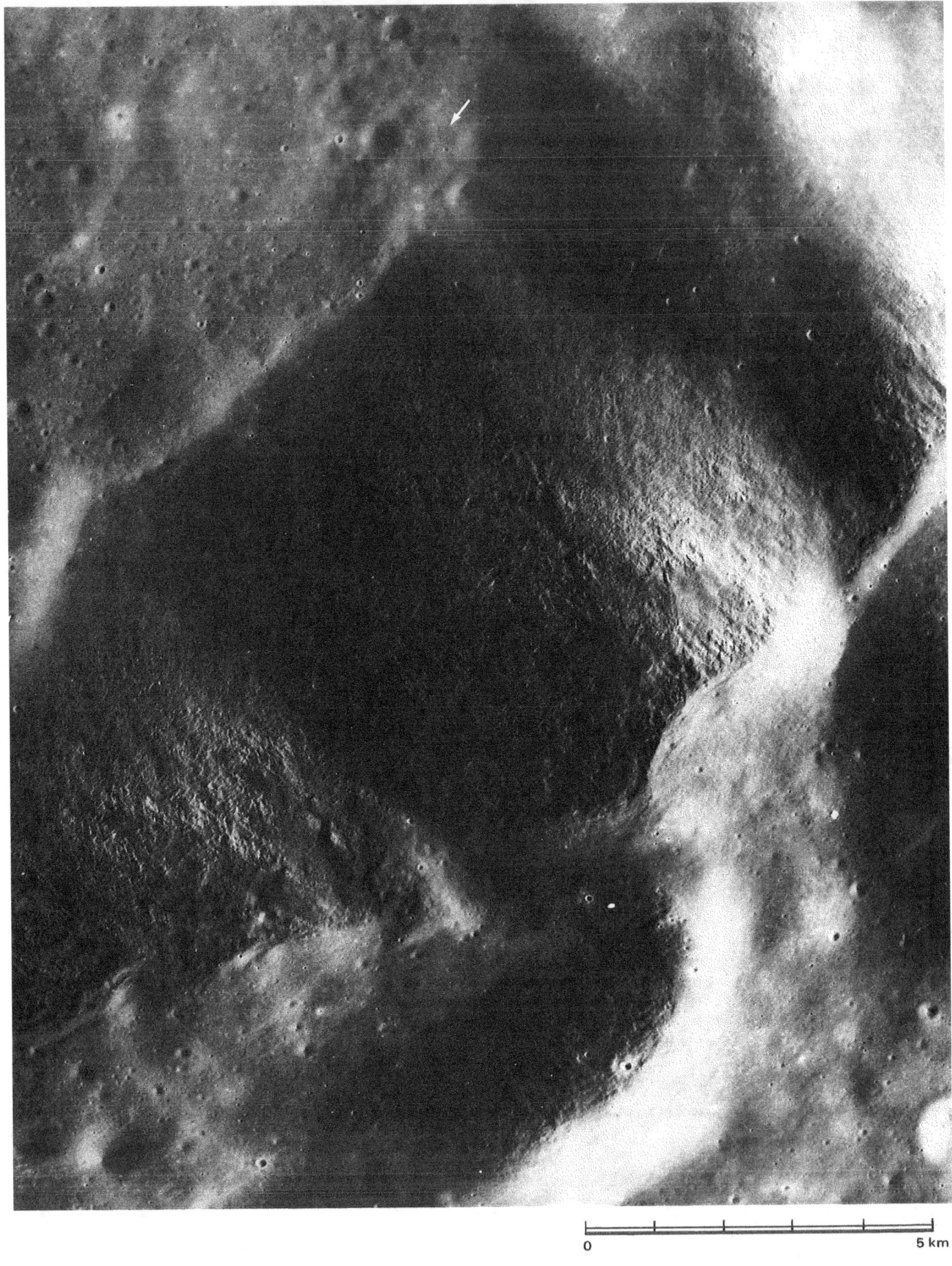

FIGURE 42.—One of the high, steep peaks of the Apennine Mountains, the highest part of the Imbrium basin rim. Lighting is from the east (right). Lunar peaks are normally thickly mantled by their own debris; and most of this peak is mantled, but some outcrops of bedrock are also visible. These include ledges along the ridge top in the center of the picture and, probably, protrusions trending diagonally down the slopes. The fine lineations trending directly downslope are probably tricks of lighting produced by the grazing Sun illumination and not, as was believed during the Apollo 15 mission, edges of strata. Debris from the slopes has accumulated in a smooth convex-upward band all along the base of the mountains but is most noticeable in the area between the two arrows. The photographed area is about 14 km wide.—D.E.W.

AS15-9804 (P)

AS15-90-12187 (H)   AS15-90-12209 (H)

FIGURE 43.—This stereogram showing part of the Apennine Mountain chain is composed of two Hasselblad frames taken from the Moon's surface. To facilitate the postmission geologic study of the landing site, the astronauts took a panoramic series (a clockwise sequence) of horizontal photographs at each major sample collecting station. Both these photographs look northeast from stations about 250 m apart along one of the EVA traverses at the Apollo 15 site. The foregrounds, which do not cover a common area, have been eliminated from the stereogram. The impression of depth in the background is greatly enhanced in comparison with the direct view of the astronauts. The mountain at the left is Mount Hadley at a distance of 20 km. The faint lineations trending downward to the left on Mount Hadley's face are the features that were believed by some investigators, at the time of the mission, to be steeply dipping strata; however, the explanation given in figure 42 now seems more likely.—L.J.K.

AS16-1420 (M)

FIGURE 44.—A north-looking oblique view of the region surrounding the Apollo 14 landing site (arrow), about 600 km south of the Imbrium basin, which is just beyond the horizon. The fine, hummocky material extending through the center of the frame from the lower edge to the horizon has been mapped as ejecta from the Imbrium basin and designated the Fra Mauro Formation. The formation is most easily distinguished in the western half of Fra Mauro, the large (95 km) crater near the center. In the eastern half of the same crater the formation is either absent or is too thin to be visible. The straight rilles trending toward the lower right corner of the frame may be related to the radial stresses generated by the Imbrium event, but detailed mapping has shown that they are much younger in age.—M.W.

AS16-0700 (M)   ←

FIGURE 45.—These terrae of the central highlands are near the Apollo 16 landing site. The landing point is indicated by the arrow just below the lower (northern) edge of the picture. Here the terrae are less rugged and less densely cratered than those on the far side, shown in figure 33. The subdued appearance of these near-side terrae is attributed to accumulations of successive blankets of ejecta from impact basins and possibly to the accumulation of volcanic materials in this, the topographically highest region on the Moon's near side. Descartes, for which the landing site was named, is the ancient, highly subdued crater near the center. It is 47 km in diameter and so indistinct in this view that its rim has been outlined by dashes. The unusual Abulfeda crater chain extends for a distance of 225 km across the upper part of the picture.—G.G.S.

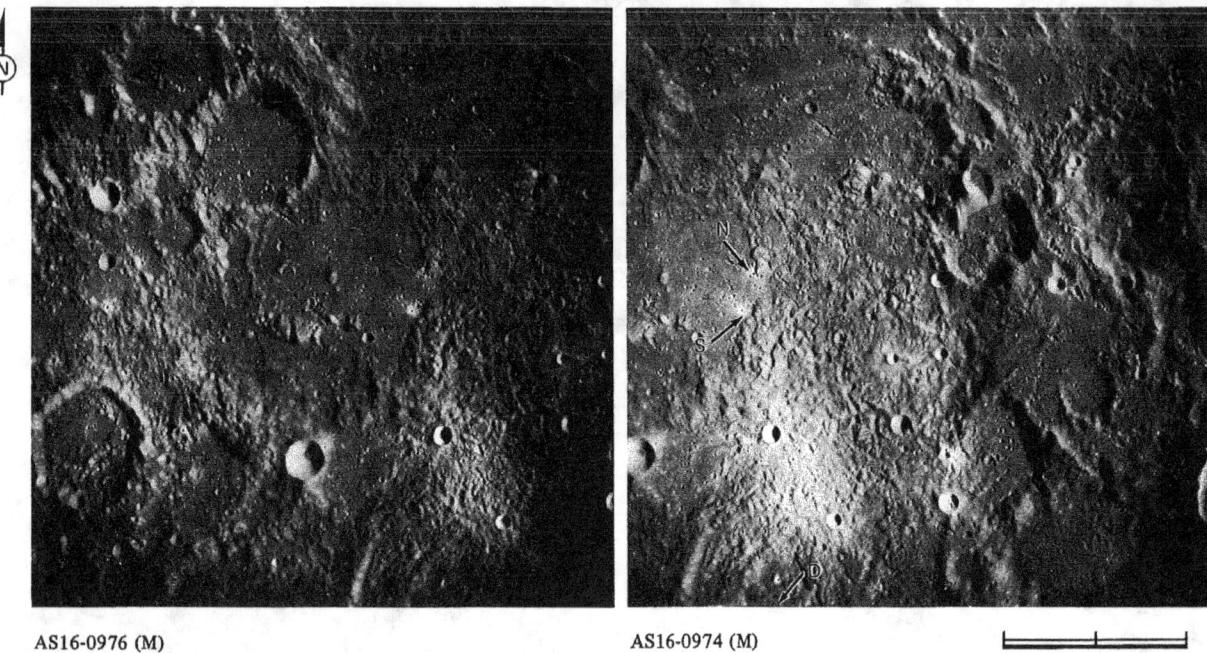

AS16-0976 (M)

AS16-0974 (M)

0   50 km

FIGURE 46.—In contrast to figure 45, which is an oblique view looking southward, this is a vertical view arranged for stereoscopic viewing. North is at the top and the crater Descartes (D) is partly visible at the lower edge. The craters North Ray (N) and South Ray (S) are prominent landmarks used during the planning and conduct of the mission. The LM landed between the two. The age of South Ray is discussed in more detail in figure 105. The smooth materials filling most topographically low areas are mapped as the Cayley Formation. Samples of the Cayley excavated from the craters were collected by the astronauts. Most of the samples consisted of feldspar-rich breccia. This is consistent with an early interpretation (Eggleton and Marshall, 1962) that the Cayley accumulated as ejecta from large multiringed basins, probably Imbrium (1400 km to the northwest) and Orientale (2800 km to the west). The more rugged material immediately east of the landing site is designated as the Descartes Formation. Originally interpreted as volcanic deposits, it is now thought to consist mostly of breccia, although the samples collected may not be representative of the entire formation. From the evidence presently available, its origin as basin ejecta, possibly from the Nectaris basin 450 km to the east, seems to be the most likely explanation.—G.G.S.

FIGURE 47.—The ruggedness of the grooved and furrowed terrain of the central lunar highlands is emphasized here under low Sun illumination of about 3° compared with an illumination angle of about 27° in figure 46. In the north-south direction, this mosaic covers a distance of 210 km. While these photographs were taken, the Apollo 16 crew was completing its first revolution in orbit and described the scene as follows: "In this lighting, you can see the crater Descartes, and it stands out much bigger than you would expect because of the low Sun angle. In fact, I had to look up my map to make sure that that was what I was looking at. . . . It looks very much like a big clinkery cinder field, . . . a big, rounded surface of clinkers. It is fantastic . . . boy, is that rough!" The crater Descartes itself (D) is subdued and partly covered by the rough topography. The Apollo 16 landing site (arrow) is in a gap within the grooved and furrowed unit. Contrary to premission beliefs that this unit was of volcanic origin, returned samples from the site are breccias formed by shock metamorphism of the highland rock.—F.E.-B.

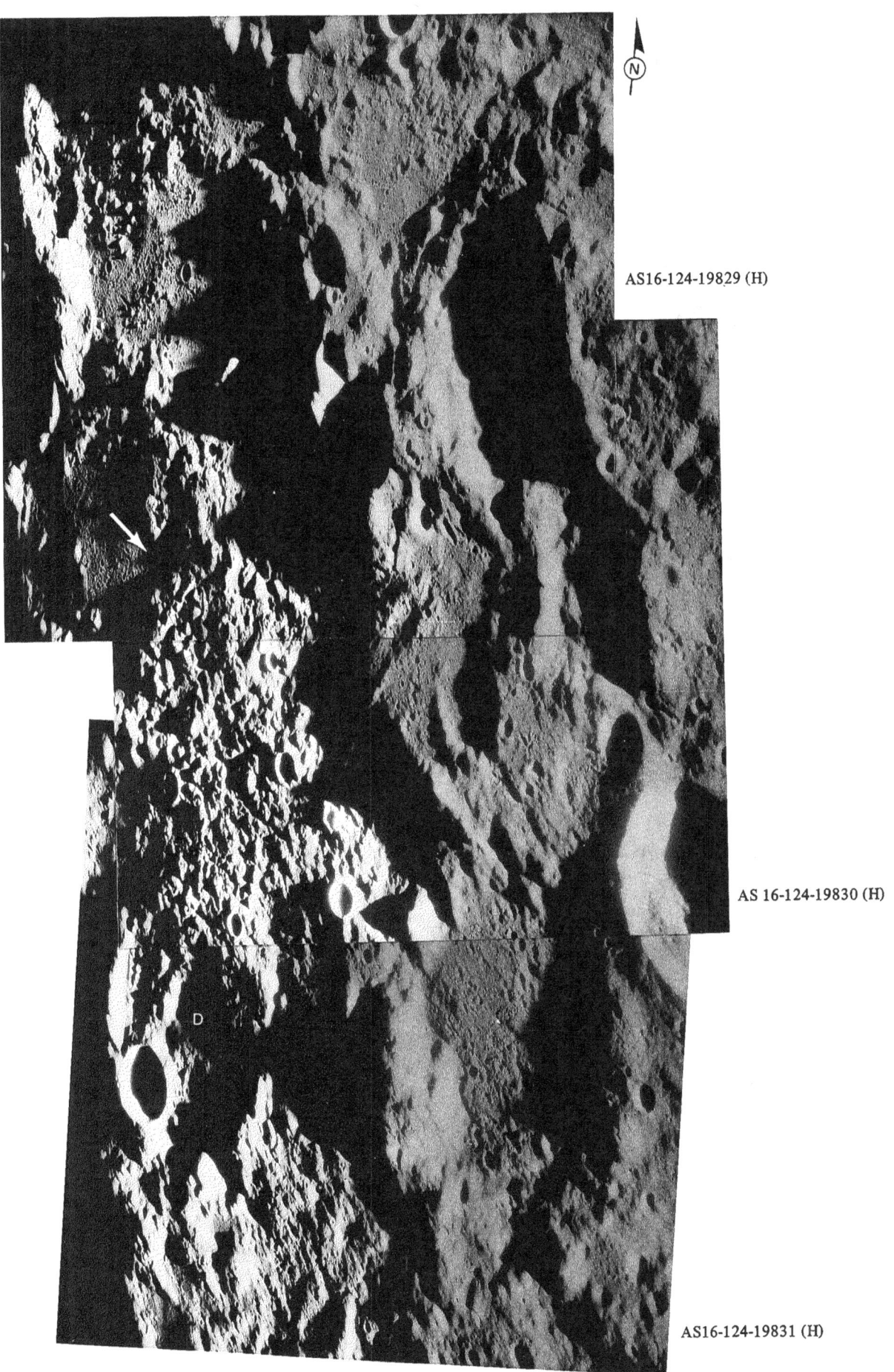

AS16-124-19829 (H)

AS 16-124-19830 (H)

AS16-124-19831 (H)

FIGURE 48.—A basin-forming event severely affects the terrain even beyond the outer margins of its ejecta deposits. This system of ridges and grooves known as Imbrium sculpture is radial to the Imbrium basin, located beyond the upper left horizon about 650 km from the center of this scene. The crater Herschel, 40 km in diameter, is at the center of the right edge. It is perched on the grooved rim of Ptolemaeus, part of whose floor and rim is visible in the lower right corner. Both secondary impact and faulting have been proposed as causes of the ridges and grooves, but in any case the sculpture must have been produced by the same event that produced the Imbrium basin.—D.E.W.

AS16-1411 (M)

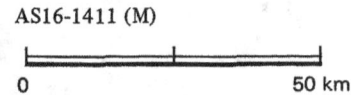

FIGURE 49.—An area centered about 900 km southeast of the Imbrium basin, illustrating again the radial fracturing and sculpturing of terra materials by the basin-forming event. The arrow points to a 120-km-long fracture that cuts the rims of the partly visible crater Albategnius in the lower left of the photograph and the crater Halley toward the upper left. It and similar trending fractures elsewhere in this picture are radial to the Imbrium basin and are related to its formation. The crater Hipparchus C (HC) is superposed on a fracture and, therefore, is younger than the Imbrium basin and the Imbrium sculpture. Light plains-forming materials (LP) are younger than the Imbrium event, as indicated by the absence of fractures and the scarcity of superposed craters. Light plains deposits are a major stratigraphic unit of the terra regions and will be illustrated in more detail beginning with figure 53.—M.W.

AS16-0982 (M)

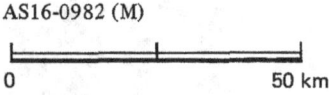

FIGURE 50.—Between Mare Crisium and Mare Serenitatis Apollo 17 approached its eventual landing site, which is in shadow at the left edge of the photograph (white arrow). The area consists of both mare and terrae. Maraldi, a crater described in figure 52, is shown by the black arrow. A markedly rectilinear pattern of major terra features and mare-terra contacts is characteristic of this area. Closely spaced intersecting structures also produce a finely textured pattern of equidimensional hills. The northeast and northwest directions of the large and small structural elements are less consistent with structural trends of the nearby Crisium and Serenitatis basins than with the more distant Imbrium basin. Therefore, Imbrium ejecta or seismic energy may have produced the structures. When viewed under conditions of high-angle lighting, the pattern of small hills is referred to as "corn on the cob" or as "sculptured hills." The appropriateness of these terms is demonstrated on the next figure, a stereogram enlargement of the rectangular area in the lower right corner of this picture.—D.E.W.

AS17-0304 (M)   0   50 km

FIGURE 51.—The southeastern corner of figure 50 is shown here as a stereogram of two Apollo 16 pictures taken when the Sun angle was much higher. The contrasting albedo (brightness) of the dark mare and bright terra is enhanced under these lighting conditions. When viewed stereoscopically, the "corn on the cob" texture of the terra is readily apparent. The rims and walls of the ancient craters Franz (F) and Lyell (L) have been severely degraded by erosion and show the same texture as the adjacent terra. These two craters contrast strongly with the much younger crater (C) whose original form has not been significantly degraded.—D.H.S.

0   25 km

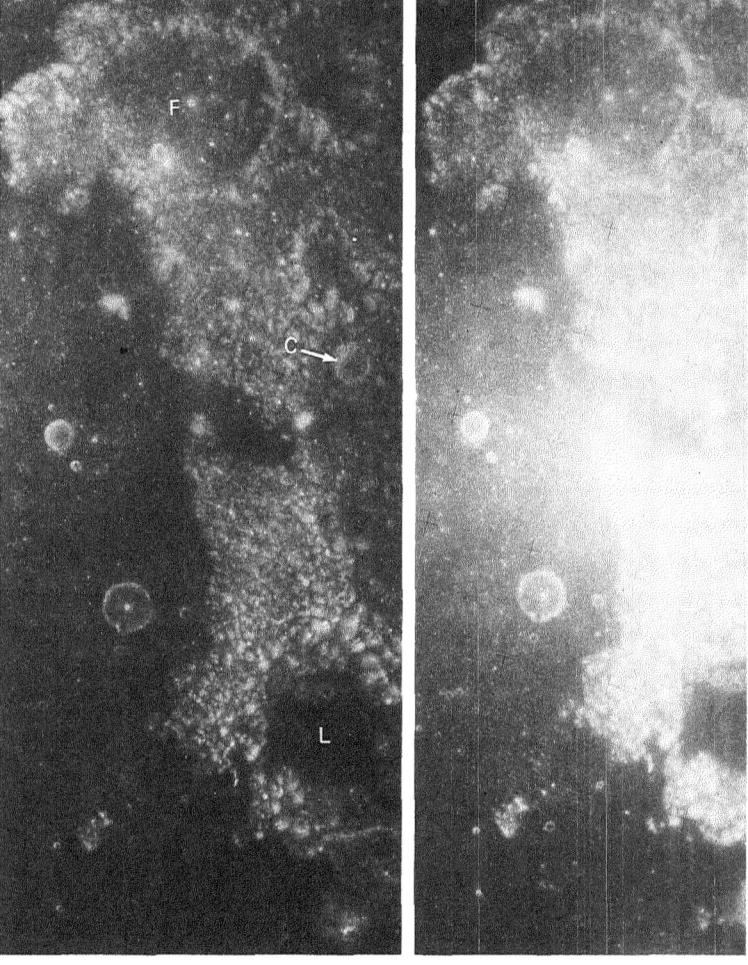

AS16-1648 (M)   AS16-1646 (M)

FIGURE 52.—Maraldi, a 45-km impact crater, was shown in its regional context in figure 50. Its rectilinear shape is in striking contrast to the circular or oval shape of most lunar impact craters. Faulting along northwest and northeast planes, probably generated by the Imbrium event, is the cause of the unusual configuration of its walls. Debris aprons form a narrow but continuous terrace along the base of the crater wall. The high rate of mass wasting on the steepest slopes is proven by the low density of craters superposed on the crater walls in contrast to that of the much younger mare surface in the floor of Maraldi. Ultimately, as depicted near the north edge of the picture, landforms evolve toward rounded forms partly buried under their own debris aprons.—M.J.G.

AS17-2302 (P)

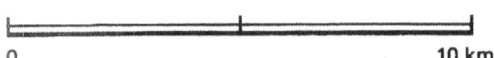

0   10 km

AS16-0708 (M)

FIGURE 53.—Light plains are a conspicuous feature of several lunar terra regions including the central near-side highlands. In this scene essentially all flat-lying areas are covered by plains deposits. No mare materials are visible. Light plains deposits are not completely planar but faintly reveal underlying relief. The subdued features that underlie the plains deposits in the floor of the large crater Albategnius (lower center) are craters and linear "Imbrium sculpture" troughs like those on the rim of Albategnius and in the adjacent highlands. The rim crest diameter of Albategnius is about 130 km.—D.E.W.

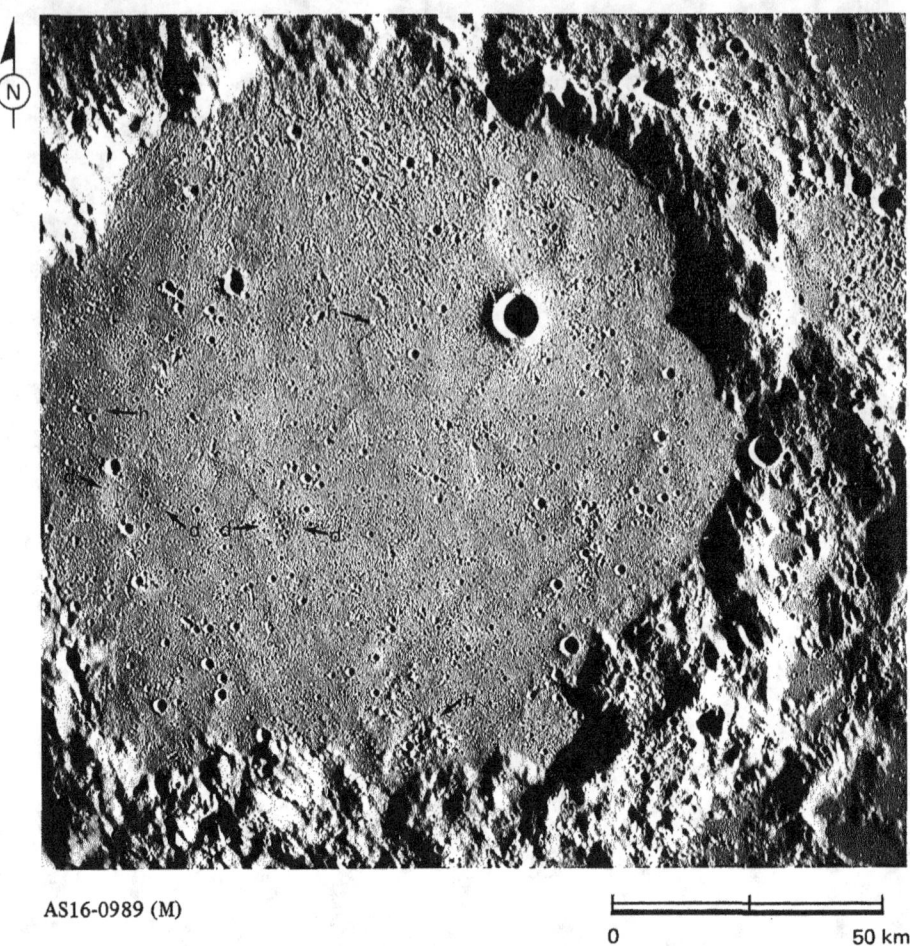

AS16-0989 (M)

0   50 km

FIGURE 54.—This picture shows the subdued, 150-km diameter, crater Ptolemaeus. The crater is filled to about half its original depth by the Cayley Formation, a unit with a gently undulating, nearly smooth surface. This unit forms similar smaller pools in numerous irregular depressions at various levels on the rim, flank, and wall of Ptolemaeus. The Cayley Formation consists of patches of light-colored plains materials that fill most depressions peripheral to the Fra Mauro Formation (fig. 44) in the central earthside lunar uplands. Undulating surface features on the Cayley include very subtle circular depressions (d) 10 to 15 km in diameter that are more than an order of magnitude larger than the craters superposed on the Cayley, and irregular swells, swales, and scarps. Other surficial features are small, equidimensional, steep-sided hills (h). The latter may have been formed on the surface of the Cayley by eruption of material from within the unit. In addition, more than 30 small craters on the Cayley have small central mounds (fig. 55). These mounds may represent relatively strong material that underlies a weak surficial layer of post-Cayley regolith, indicating that the regolith is thicker here than on mare surfaces.

The deposition of the Cayley in pools indicates that it moved partly as a fluid. The distribution of the pools peripherally to a deposit of basin ejecta, the Fra Mauro Formation, indicates a related origin. The large, subdued crater forms suggest that the Cayley material is a draped blanket of fragmental material. Therefore, my colleague G. G. Schaber and I have suggested that the Cayley is a deposit of basin ejecta that became segregated from the ballistically transported ejecta that formed the Fra Mauro Formation around the Imbrium basin. The Cayley was transported separately as a fluidlike cloud that flowed along the ground across the whole region; portions were left behind to accumulate in local depressions.—R.E.E.

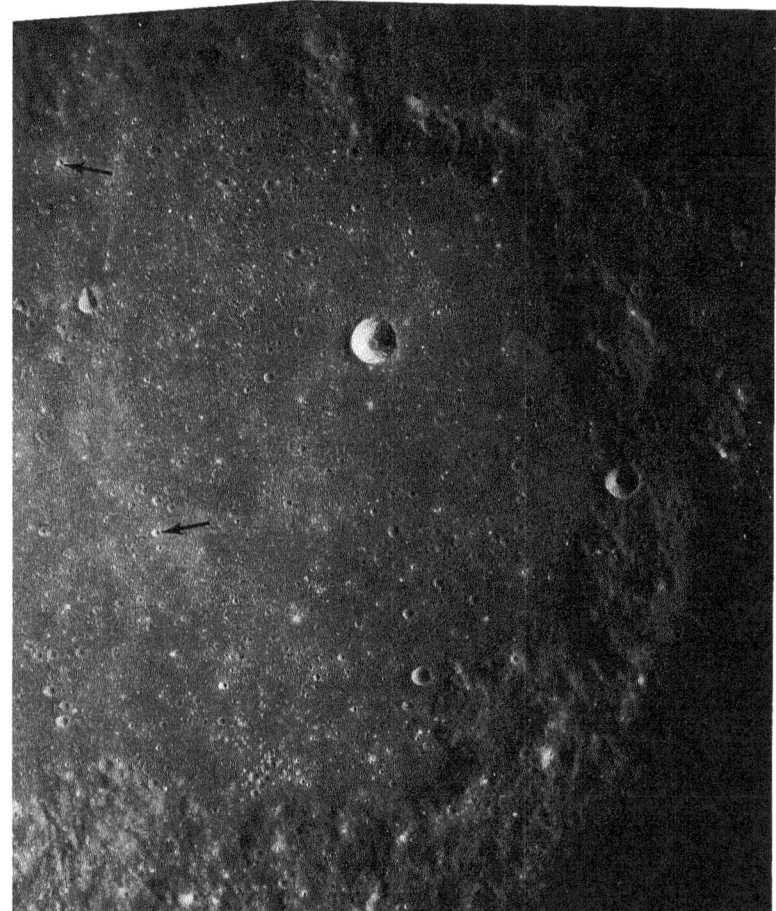

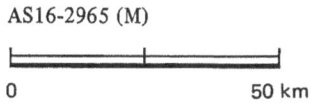

FIGURE 55.—This picture shows Ptolemaeus under a Sun elevation angle of 45°. The undulations on the Cayley Formation are so subtle that they disappear under the high Sun. The arrows point to examples of the small, relatively sharp, young craters with central mounds mentioned in figure 54. The mounds may be harder than the shallower materials and may indicate the depth of the regolith formed on the Cayley.—R.E.E.

AS16-2965 (M)

FIGURE 56.—The very large crater or small multi-ringed basin Mendeleev is shown here at the same scale as the preceding two pictures of Ptolemaeus. Like Ptolemaeus, Mendeleev is largely filled by plains material. In this case, however, subsequent cratering has been much more extensive, indicating that the Mendeleev plains are older than those in Ptolemaeus. The source of these plains materials on the far side of the Moon is unknown. The linear chain of elongate craters near the left side is probably of secondary origin, formed by the impact of fragments ejected from Tsiolkovsky (850 km to the southwest). —D.E.W. and C.A.H.

AS16-0875 (M)

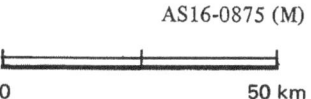

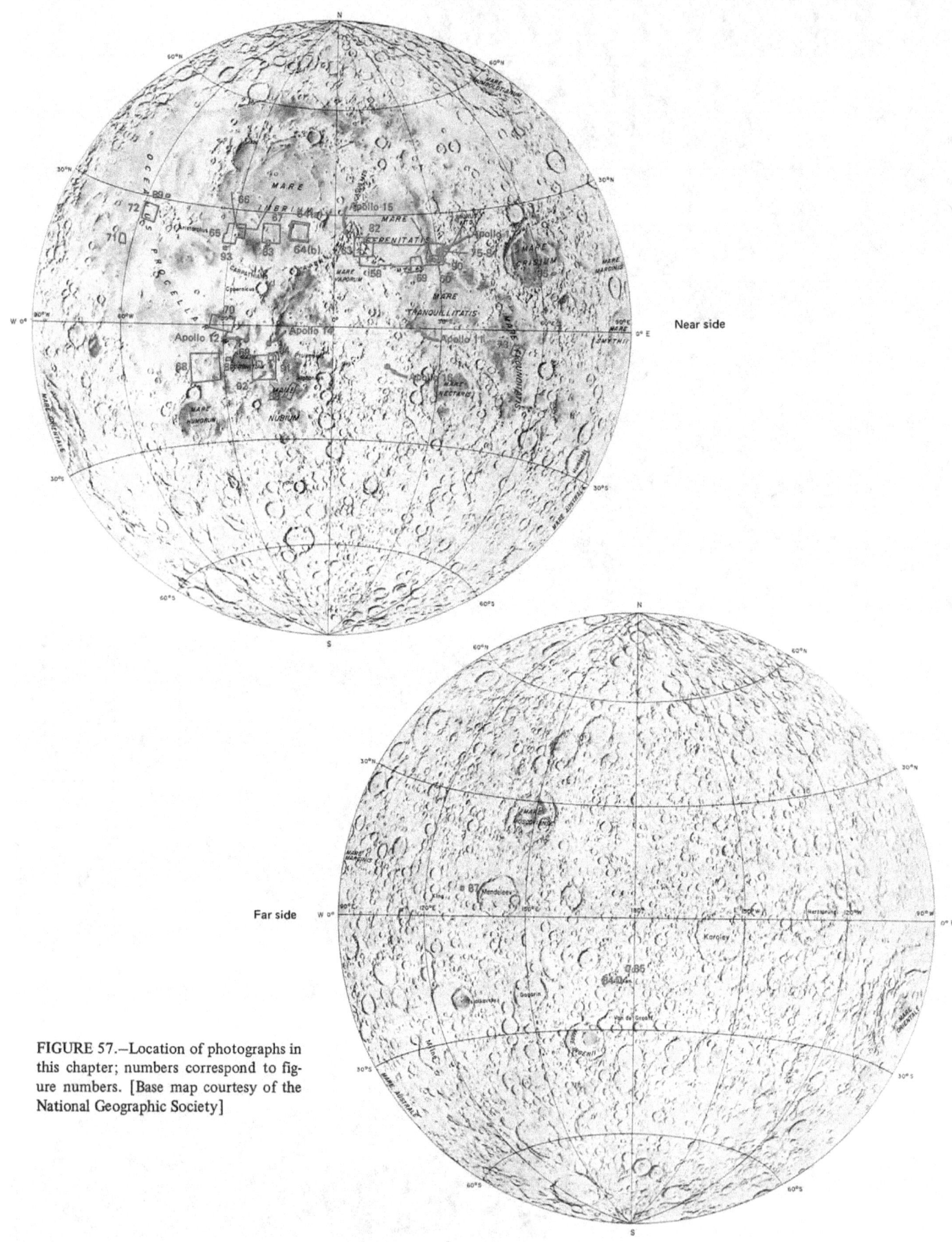

FIGURE 57.—Location of photographs in this chapter; numbers correspond to figure numbers. [Base map courtesy of the National Geographic Society]

4
The Maria

Areas of mare material occupy about 15 percent of the Moon's total surface. As shown in figure 14, most of them occur on the Earth-facing hemisphere. Mare areas are of two types, those that fill multiringed circular basins and those that fill irregular areas. The circular basins are believed to be impact features formed by the collision of giant meteoroids with the lunar surface; these were later filled to varying degrees by mare material. The basins lie at successively lower levels to the east, with Mare Smythii—the easternmost of the mare basins on the near side—lying almost 5 km below nominal lunar radius. The irregular maria lie in lowlands. The largest of these is Oceanus Procellarum, which lies on the west side of the Moon and is almost 2 km below nominal mean lunar radius.

Mare filling is characterized by several distinctive features that indicate a volcanic origin. These include many broad low domes with summit craters. Some of these domes closely resemble terrestrial basaltic shield volcanoes. In other areas, irregular and steep-sided volcanic piles dominate. Elsewhere, clusters of domes occur as in the Marius and Rümker Hills. Another type of feature is the broad lobate flow fronts that mark the edges of lava flows; these flow fronts extend several hundred kilometers in length and are as much as 100 m high. Other elongate flows closely resemble terrestrial flood basalts; samples returned by Apollos 11, 12, 15, and 17 confirmed this resemblance.

Other typical features on the maria are sinuous rilles and wrinkle ridges. Many sinuous rilles originate in craters near the higher margins of the mare basins and flow into the lowlands. Apollo 15 collected samples from the margins of Hadley Rille and confirmed the hypothesis that sinuous rilles are basaltic lava channels. Wrinkle ridges occur in all mare regions and form circumferential or medially transecting patterns.

Ages of the maria are being determined by two methods. Absolute ages are given by radiometric techniques. From these we know that the sampled lunar basalts are much older than their terrestrial counterparts. The basaltic lava flows range in age from 3.15 to 3.85 billion years, so the episode of lava filling on the Moon must have continued for at least 700 million years. Relative ages can be established by counting craters in mare surfaces. Comparison of crater counts on the lightly cratered lava flows in the northern part of Oceanus Procellarum with radiometric dates obtained for the basalts returned to Earth suggests that the Procellarum flows may be as young as 2 billion years. This date needs to be confirmed because it more than doubles the time of lava production.

Analyses of the returned samples show that the chemical composition of mare basalts varies across the Moon. These differences have also been correlated with the subtle color changes seen in spectral reflectance measurements; as a result, chemical variations can now be mapped far from the Apollo landing sites.—G.W.C. and H.M.

Rimae Sulpicius Gallus

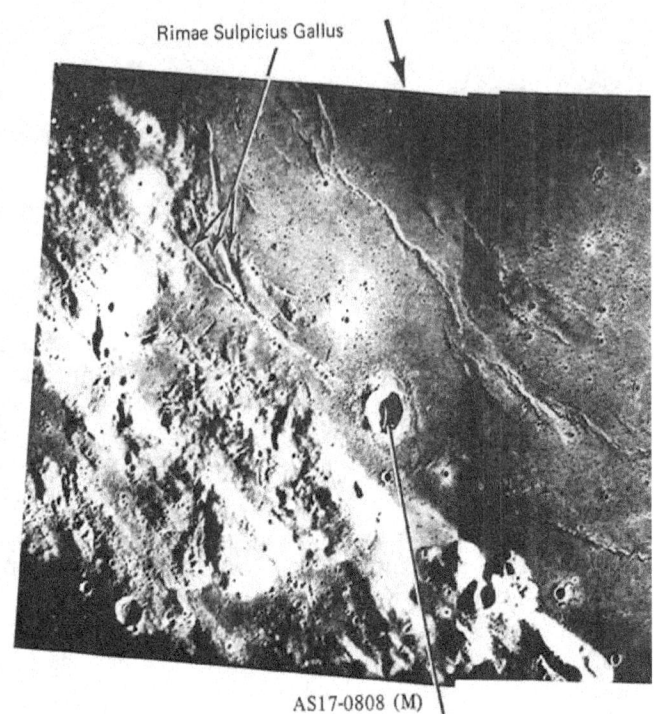

FIGURE 58.–This mosaic of Apollo 17 frames extends across southern Mare Serenitatis, one of the larger multiringed basins on the Moon's near side (fig. 14). The average diameter of the basin is about 680 km. Its generally circular outline is mimicked by the systems of arcuate rilles near the outer edge of the basin and also by the large system of mare ridges extending from arrow to arrow. Another feature of Mare Serenitatis is the nearly continuous ring of dark mare material that occupies the outer part of its floor. When these pictures were taken, the Sun angle was too low to show differences in albedo clearly. However, part of the ring of dark mare material is visible near the Plinius Rilles (Rimae Plinius) and the Littrow Rilles (Rimae Littrow). The stratigraphic relationships between the dark and light mare units are described in figure 59, which is an enlargement of the small area outlined in this figure.–G.W.C.

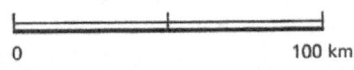

0   100 km

AS17-0808 (M)

Sulpicius Gallus

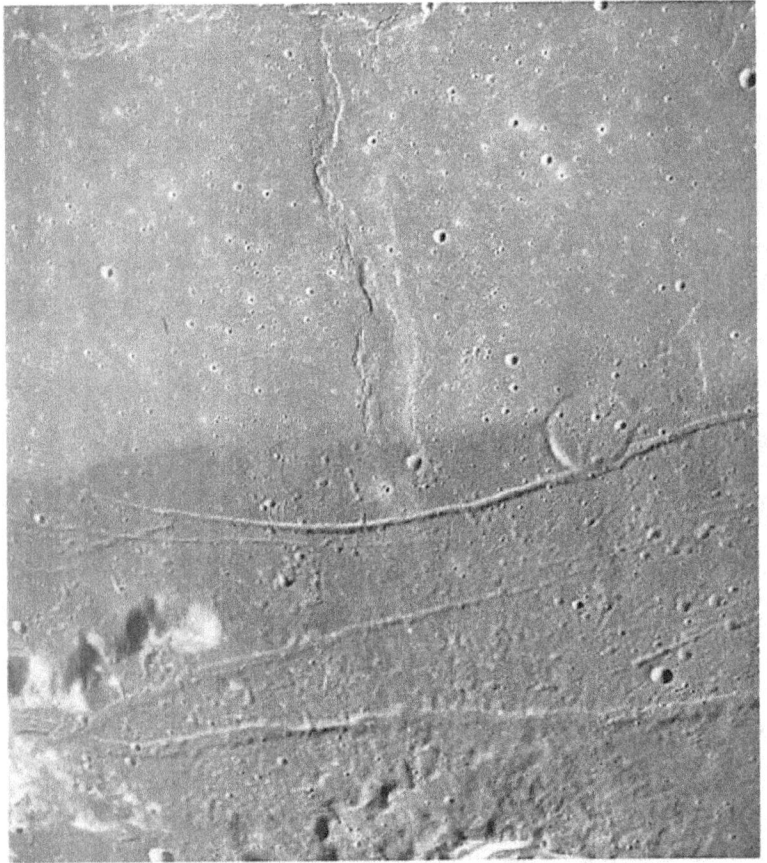

FIGURE 59.–Some of the strongest tonal, color, and structural contrasts among mare materials occur in Mare Serenitatis. Accordingly, it has become a classic area for studying the sequence (or stratigraphy) of mare rocks. Earlier studies of telescopic photographs seemed to provide evidence that the lighter materials in the center of the basin (top half of this view) were emplaced before the darker lavas erupted along the basin margin. However, pictures returned by Apollo 17 show that the opposite is true. The dark materials were emplaced first. They were then tilted northward and broken by faults, such as those that bound the Plinius Rilles, before the light lava flooded against them (Howard et al., 1973). The large mare ridge or wrinkle ridge deforms both light and dark mare units but is much more prominent in the lighter unit. Detailed spectral studies and visual observations by the Apollo 17 astronauts show that the lighter-toned mare is relatively browner and the darker mare is relatively bluer.–K.A.H.

AS17-150-23069 (H)

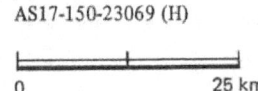

0   25 km

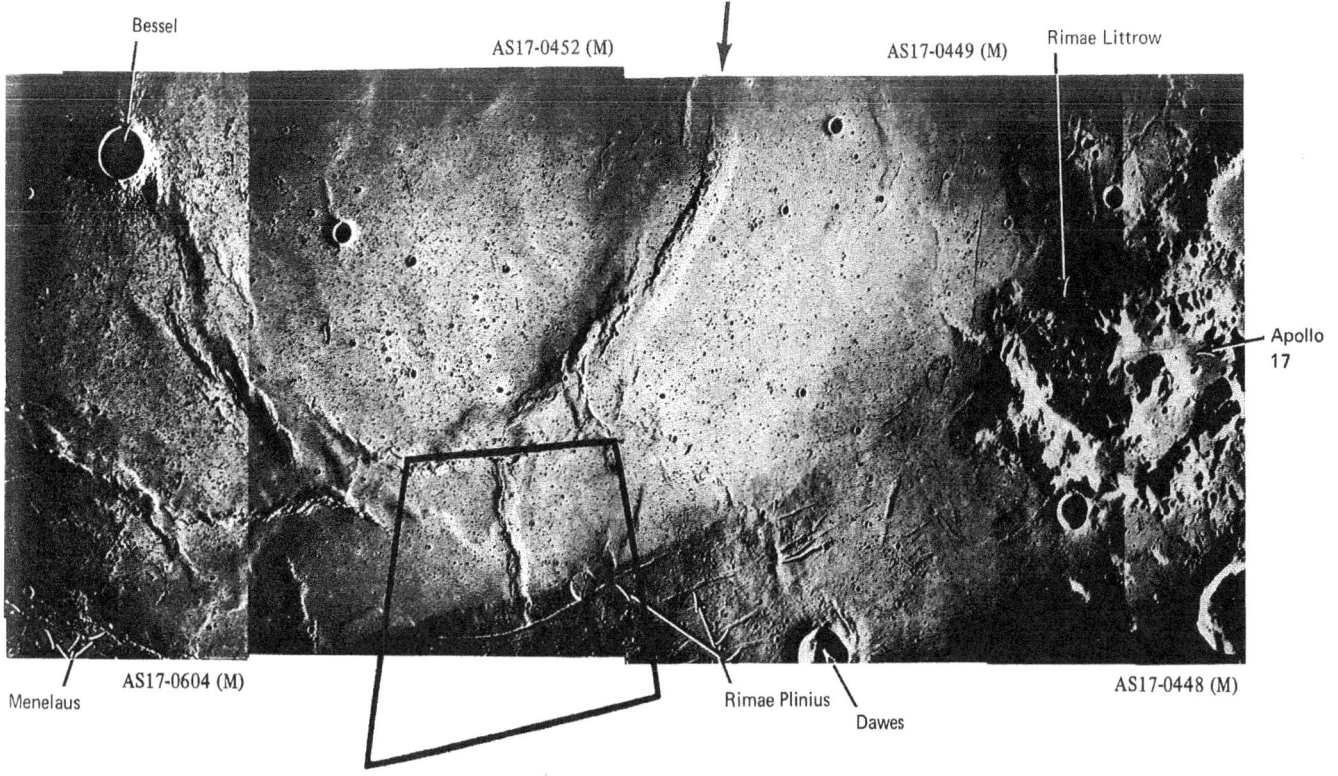

Bessel

AS17-0452 (M)

AS17-0449 (M)

Rimae Littrow

Apollo 17

AS17-0604 (M)

Menelaus

Rimae Plinius

Dawes

AS17-0448 (M)

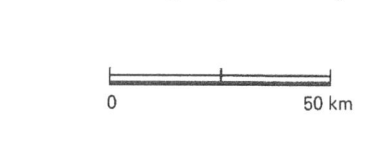

FIGURE 60.–The southeastern margin of Mare Serenitatis and the surrounding Taurus-Littrow highlands are shown in this high Sun angle photograph. Also shown is the Apollo 17 landing site (large arrow) in a dark-floored valley between bright mountain massifs. The rectangle surrounding the landing site outlines the area covered by the two maps that follow in figures 61 and 62. The boundary between light mare material in the central part of the basin and the very dark mantling material surrounding the landing site area is indicated by several smaller arrows. The difference in albedo is much more pronounced in this picture than in the mosaic (fig. 58) at the beginning of this chapter because this picture was taken when the Sun was at a higher angle above the surface. Before the Apollo 17 landing, the dark material was interpreted to be a blanket of pyroclastic debris (volcanic cinders and ash). It was thought to be as young as Copernican in age (see fig. 13), and hence younger than most other mare materials elsewhere on the Moon. Analysis of samples returned from the Taurus-Littrow area has shown that while the dark material may be predominantly volcanic in origin, its age is considerably greater than had been predicted. The dark mantling material most likely consists of black and orange glass beads that form a layer on top of the valley floor basalt and are reworked into the regolith, thus causing the low albedo.–B.K.L.

AS15-1115 (M)

0   50 km

N

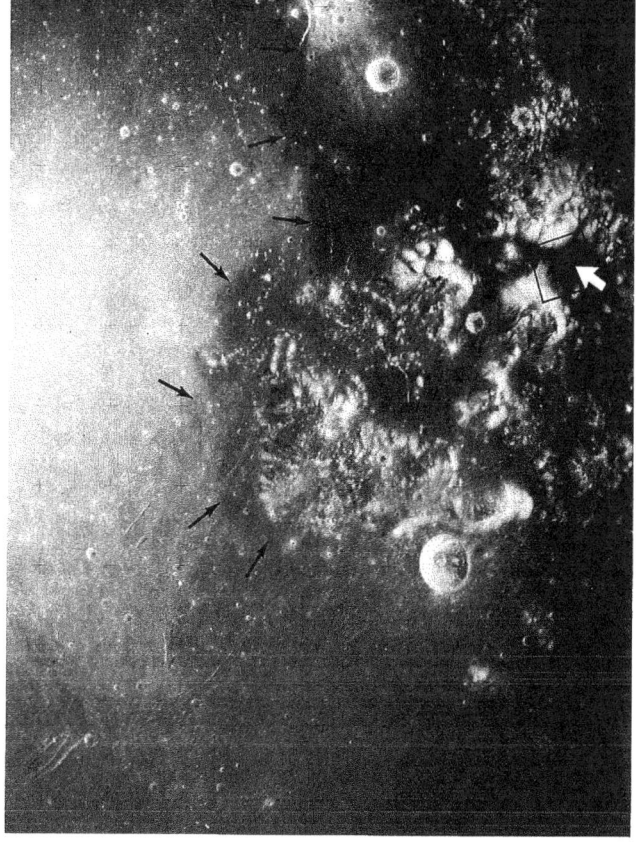

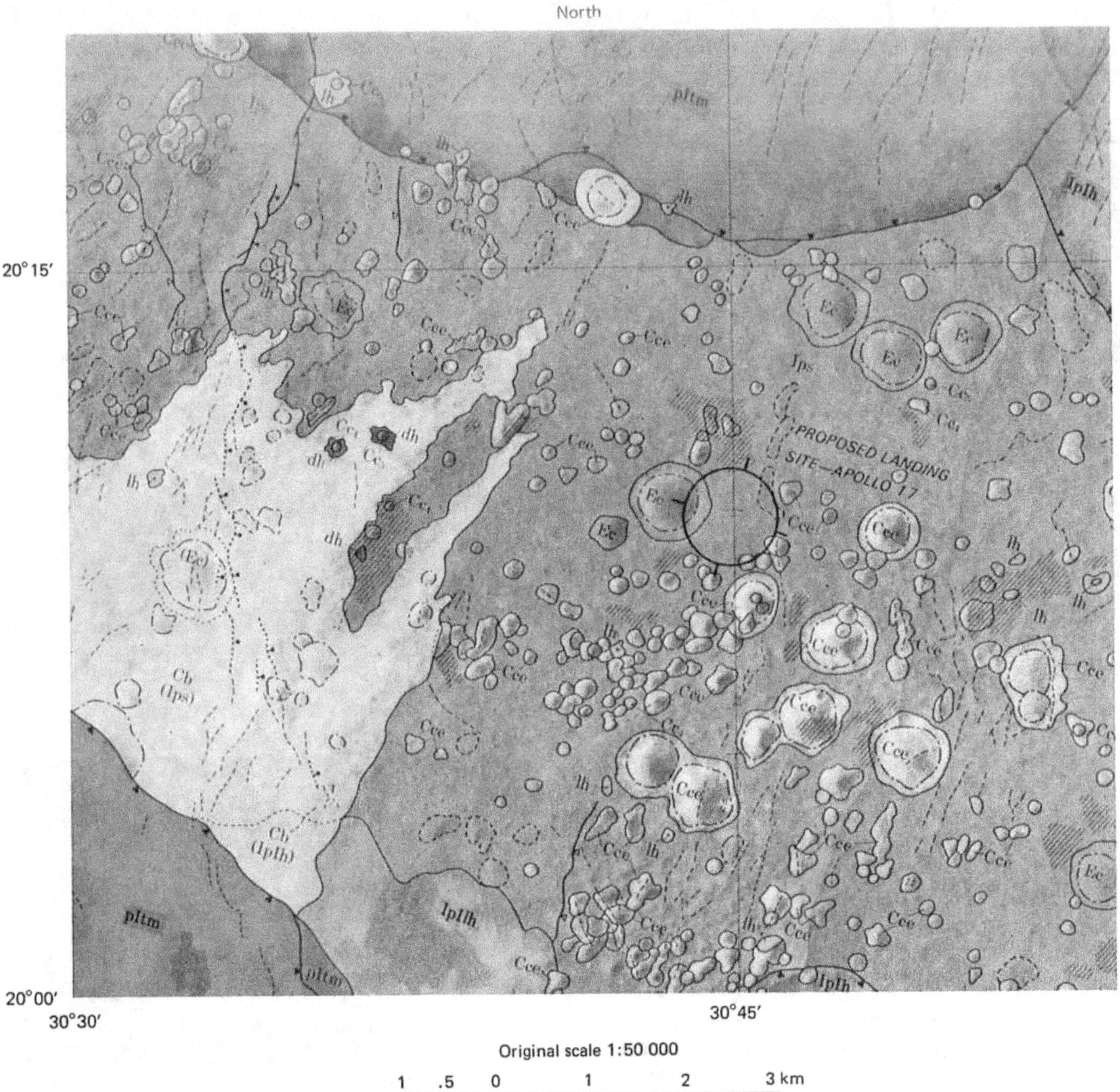

20° 15'

20° 00'
30° 30'

30° 45'

Original scale 1:50 000

1 .5 0   1   2   3 km

FIGURE 61.–This is part of a premission geologic map of the Taurus-Littrow area, compiled by B. K. Lucchitta (Scott, Lucchitta, and Carr, 1972) and published before Apollo 17 was launched. The actual landing point was very near the center of the large circle marking the proposed landing site. Letter symbols and colors designate the different types of rock materials and their relative ages as deduced from study of photographs available before the mission. Some refinements could now be made based on samples and data gathered by the astronauts on the surface and from orbital experiments. Apollo 15 panoramic camera photographs were the principal source of information for the original map, but mapping camera photographs, Orbiter pictures, and Earth-based telescopic pictures were also used.

On the explanation accompanying the map, each unit is identified and its relative position in the lunar time scale is shown. The explanation on the original map also included a description of the physical characteristics of each unit and a

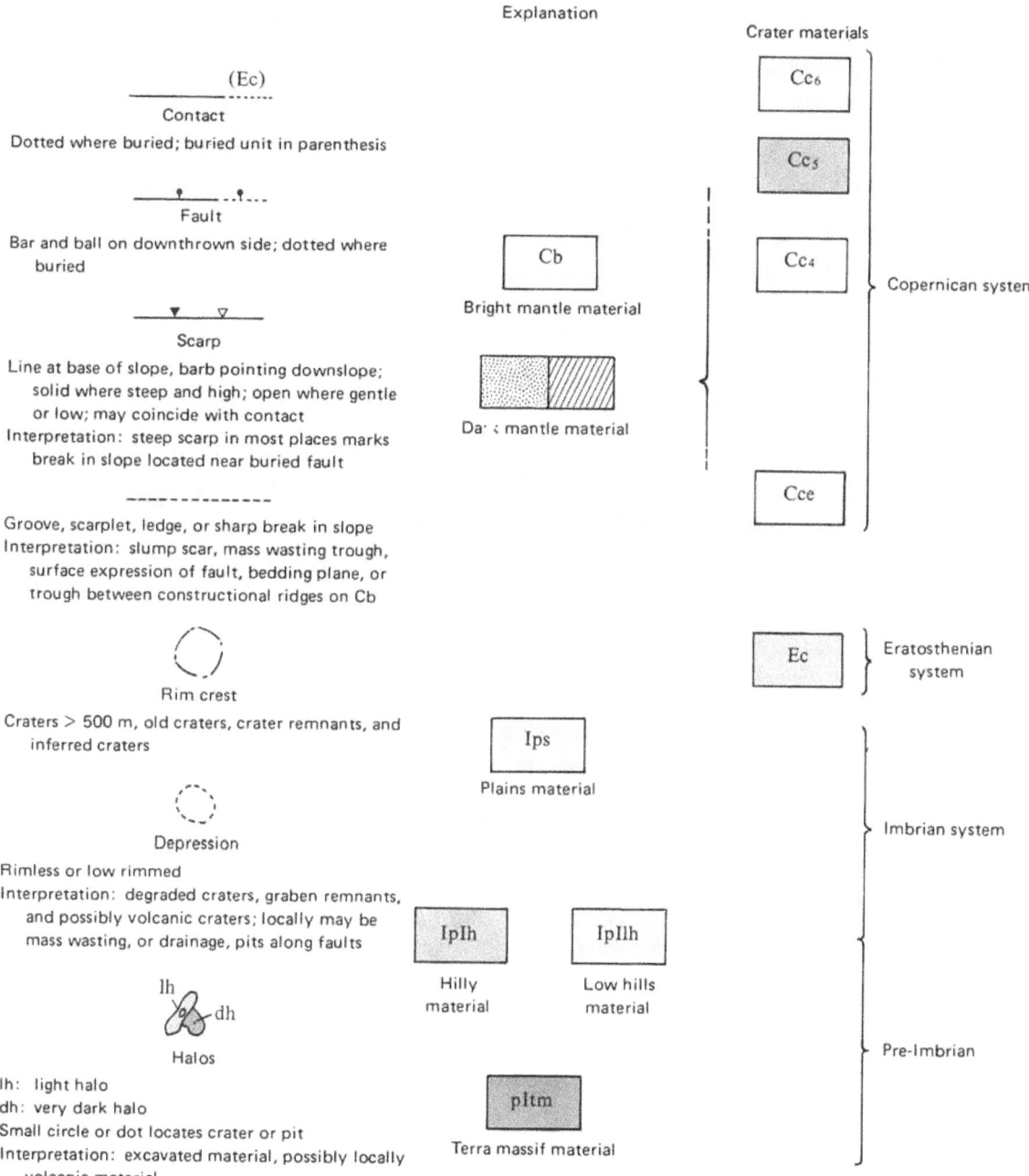

Explanation

Crater materials

Contact (Ec)
Dotted where buried; buried unit in parenthesis

Fault
Bar and ball on downthrown side; dotted where buried

Scarp
Line at base of slope, barb pointing downslope; solid where steep and high; open where gentle or low; may coincide with contact
Interpretation: steep scarp in most places marks break in slope located near buried fault

Groove, scarplet, ledge, or sharp break in slope
Interpretation: slump scar, mass wasting trough, surface expression of fault, bedding plane, or trough between constructional ridges on Cb

Rim crest
Craters > 500 m, old craters, crater remnants, and inferred craters

Depression
Rimless or low rimmed
Interpretation: degraded craters, graben remnants, and possibly volcanic craters; locally may be mass wasting, or drainage, pits along faults

Halos
lh: light halo
dh: very dark halo
Small circle or dot locates crater or pit
Interpretation: excavated material, possibly locally volcanic material

Cb
Bright mantle material

Dark mantle material

Ips
Plains material

IpIh
Hilly material

IpIlh
Low hills material

pItm
Terra massif material

Cc6
Cc5
Cc4
Cce
Copernican system

Ec
Eratosthenian system

Imbrian system

Pre-Imbrian

very brief interpretation of its origin and history. For example, unit pItm occurs on the steep hills north and southwest of the landing site and is interpreted to be composed of ancient rocks uplifted when the Serenitatis basin was formed. Unit Ips is a much younger, relatively smooth plains material that covers most of the Taurus-Littrow Valley. Before the mission it was interpreted as ejecta breccia or lava emplaced in a fluidized state; samples and other data gathered during the mission confirmed it was mare lava. Dark mantle material is shown by dot or line shading rather than by letter symbols and color. Throughout most of the valley it appears to be on top of (hence, younger than) unit Ips. It was interpreted as a blanket of pyroclastic debris. Unit Cb, bright mantle material, was interpreted as a deposit of avalanche debris derived from the steep mountain partly shown in the lower left corner of the map.—G.W.C.

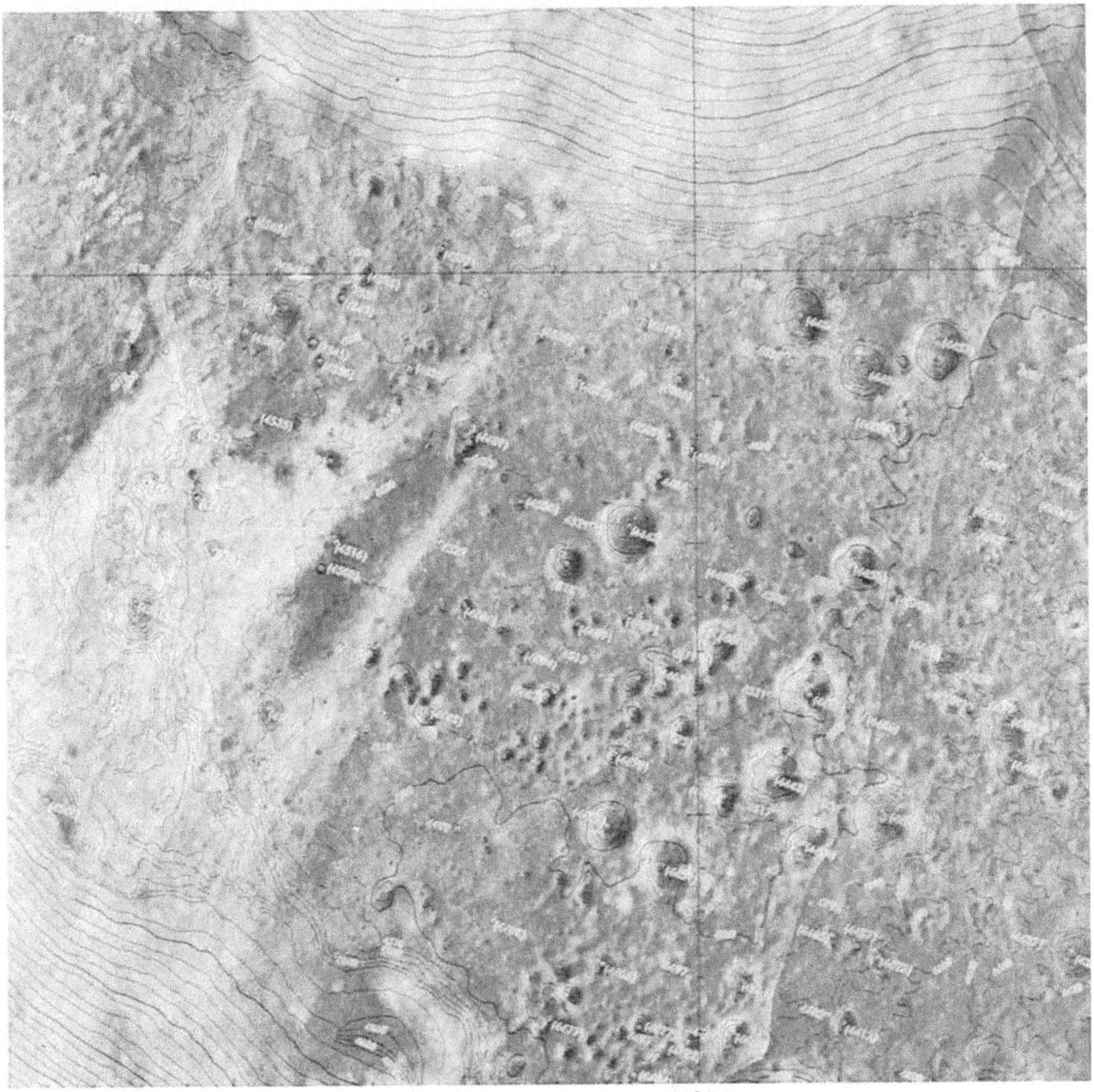

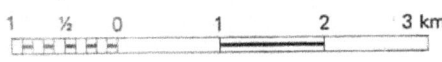

20°15′

20°00′
30°30′   30°45′

Original scale 1:50 000

1   ½   0   1   2   3 km

Contour interval: 50 m
Supplementary contours at 10-m intervals

Legend

Elevation (reference to datum)   . 5200
Elevation at bottom of crater   . (5180)
Slope ticks (downhill direction)   ∽
Index contour   ⌒⌒ 6000 ⌒⌒
Intermediate contour   ⌒⌒ 5950 ⌒⌒
Supplementary contour   --- 5910 ---

Prepared and published by the
Defense Mapping Agency Topographic Center,
Washington, D.C.

FIGURE 62.—This is a topographic contour map of the same area as the geologic map in figure 61. Topographic contour lines in red are superposed on an orthophoto base composed of rectified and mosaicked panoramic camera frames. The area shown is part of a larger map prepared by the Defense Mapping Agency Topographic Center and is included here to show the relationship between geology and topography. The steepness of the mountain slopes along the north edge and in the lower left corner is indicated by the closely spaced contours at 50-m intervals. These slopes are underlain by the very old rocks of unit pItm. The overall levelness of the valley floor—the area filled by younger rocks of unit Ips—is indicated by the widely spaced contours at 10-m intervals. An exception is the belt of closely spaced subparallel contour lines extending northward near the left edge of the map. These define an east-facing scarp or mare ridge interpreted on the geologic map as a fault. The average difference in elevation across the scarp is about 80 m, suggesting at least that much vertical displacement across the fault. The location and size of craters on the valley floor are shown by the many sets of circular contours.—G.W.C.

72

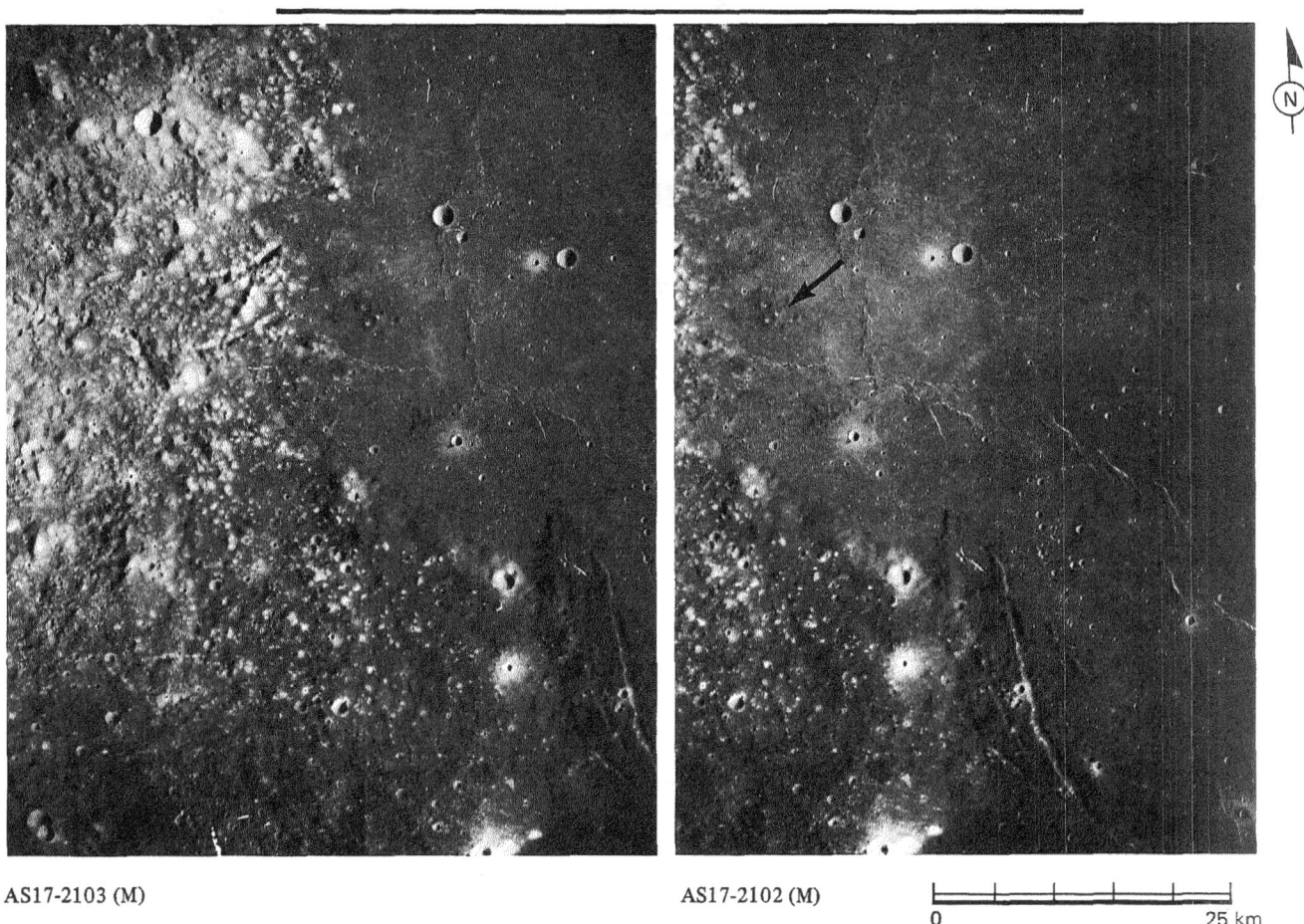

AS17-2103 (M) AS17-2102 (M)

0 25 km

FIGURE 63.—This stereoscopic view shows southwestern Mare Serenitatis "lapping against" its shore of ancient highlands or terrae. The highlands near the Sulpicius Gallus rilles in the lower part of the picture are unusually dark—darker even than the mare. M. H. Carr (1966) suggested from telescopic study that the darkness of the highlands is caused by a thin mantle of dark material, perhaps consisting of volcanic ash. The numerous small bright spots are knobs of highland material. They may have once been covered by the dark mantle but, if so, have since shed it. As elsewhere around the outer part of Mare Serenitatis, the rilles and the dark mantle in this area were originally thought to be younger than the lighter mare to the north. Apollo 17 photographs such as these have changed that concept. Now, the lighter mare is interpreted as embaying the faulted dark materials, just as in the Plinius rilles area (figs. 58 and 59). Isolated islands of dark mantled highlands that escaped inundation are shown by the arrow.—K.A.H.

73

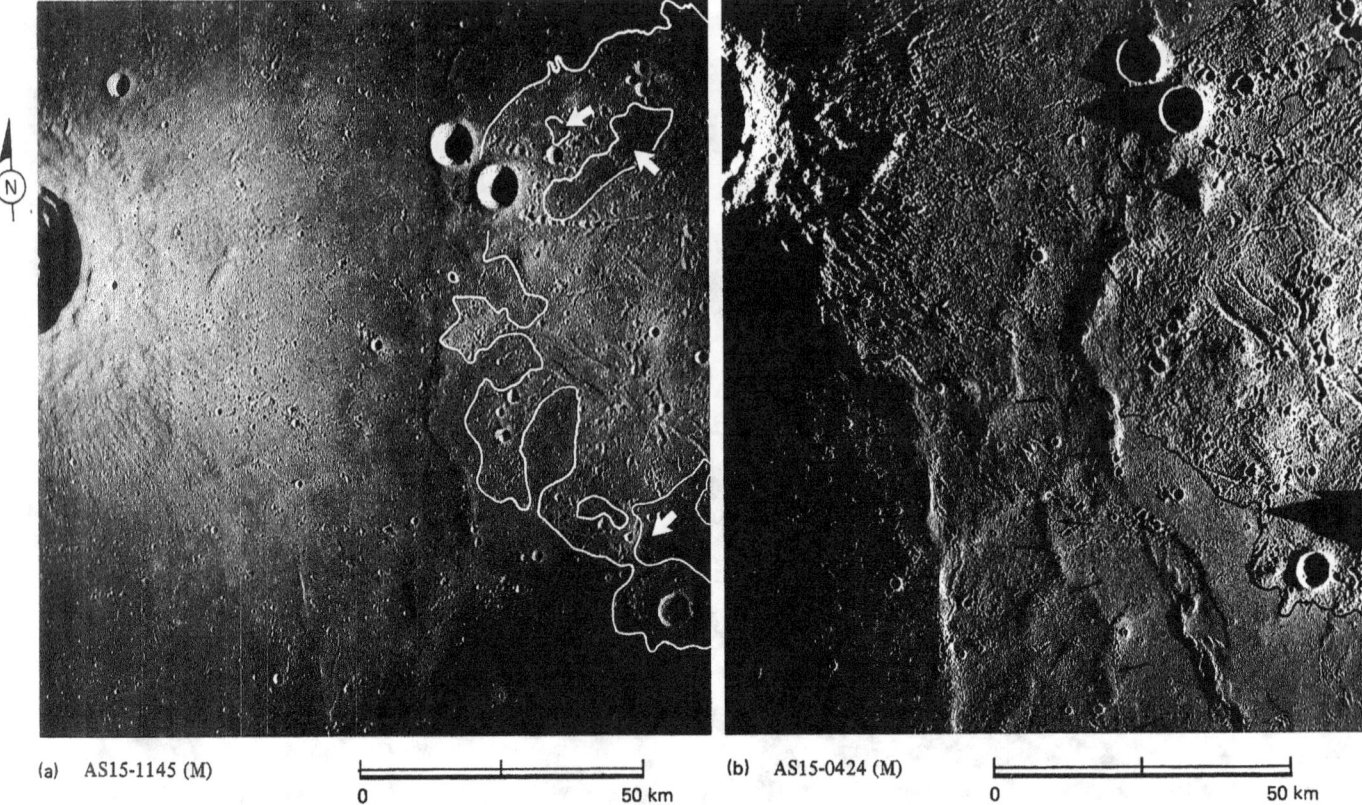

(a) AS15-1145 (M)

0   50 km

(b) AS15-0424 (M)

0   50 km

FIGURE 64.—These two contrasting pictures of the same area in southeastern Mare Imbrium were taken by Apollo 15, but on different revolutions under different lighting conditions. The picture on the left was taken when the Sun angle was 17°; the Sun angle was 2° when the picture on the right was taken. The large crater at the west edge is Timocharis. The area is dominated by three geologic units. The oldest is a fairly densely cratered fractured plains unit of moderate albedo that occupies the eastern part of the area. Next oldest is the mare unit in the central part, with its typically smooth, level surface and moderately low albedo. The youngest unit is the bright (high-albedo), highly textured ejecta surrounding Timocharis.

We have included the two pictures to illustrate the problems photogeologists sometimes face when drawing a contact line between units. The eastern edge of the mare is used as an example. Throughout most of the area shown the mare is in contact with the plains unit. Characteristically mare material is darker and smoother than plains material. Using the picture on the left in which albedo differences are enhanced because of the relatively high Sun angle, the contact might be drawn as shown. The line is equivocal in places, but, in general, it does satisfactorily separate darker areas from lighter areas. Using the picture on the right, in which surface relief is exaggerated because of very low Sun angle, the contact would be drawn as shown. Some areas dark enough to be mapped as mare in the first picture are here seen to be too roughly textured to be mare. As drawn, the line separates a unit that is both dark and smooth from a unit that is predominantly light and everywhere rugged.

Detailed stereoscopic examination of all available pictures of this area explains why some dark areas within the plains unit should not be classified as mare. In several of them there are structures resembling volcanic outlets (wide arrows on left photo). Similar structures were not found elsewhere within the plains unit. Therefore, it is likely that some if not all the darker areas of the plains are caused by veneers of dark volcanic ejecta so thin that the surface relief of the underlying plains is still visible.

An additional point of interest is the clearly defined sinuous rille (small arrows in right photo) that extends half the length of the picture; the same rille is almost invisible in the other picture.—G.W.C.

74

FIGURE 65.—This photomosaic of an area of relatively young mare lavas in southwestern Mare Imbrium shows a complex of overlapping lava flows. The complex has been traced to its apparent source northeast of the mountain mass Mons Euler (formerly called "Euler β") where the approximate location of a fissure has been deduced by detailed geologic mapping (Schaber, 1973). Individual flows are recognizable in this low-Sun (about 4°) picture as elongate lobes bounded by steep scarps. They are shown on the accompanying sketch map. Many contain one or more small rilles that are interpreted as flow channels. As individual flows are traced southward toward their source, they narrow and converge or terminate in the vicinity of the postulated fissure. South of the fissure distinct flow scarps are absent. A row of dark volcanic cinder cones along the southeast side of Mons Euler is alined with the postulated fissure, further strengthening the idea that this is an area of eruption. It is likely that the fissure is covered by its own lavas. The succession of geologic events in this area is easily decipherable. Secondary impact craters (as at S) from the large crater Copernicus overlie the lavas; hence the lavas are pre-Copernican, or Eratosthenian, in age. In turn the lavas have inundated part of the ejecta from the crater Euler; therefore, Euler is also pre-Copernican. Before Apollo pictures became available, it had been mapped as a Copernican crater. At the present time, only the lavas can be assigned an absolute age. About 2.5 billion years old (Schaber, 1973), they are older than all but a very few rock outcroppings in the entire United States. On the other hand, they are much younger than most, if not all, the samples collected on the Moon during the six lunar landings. Solid line shows position of lower left corner of figure 66.—G.G.S.

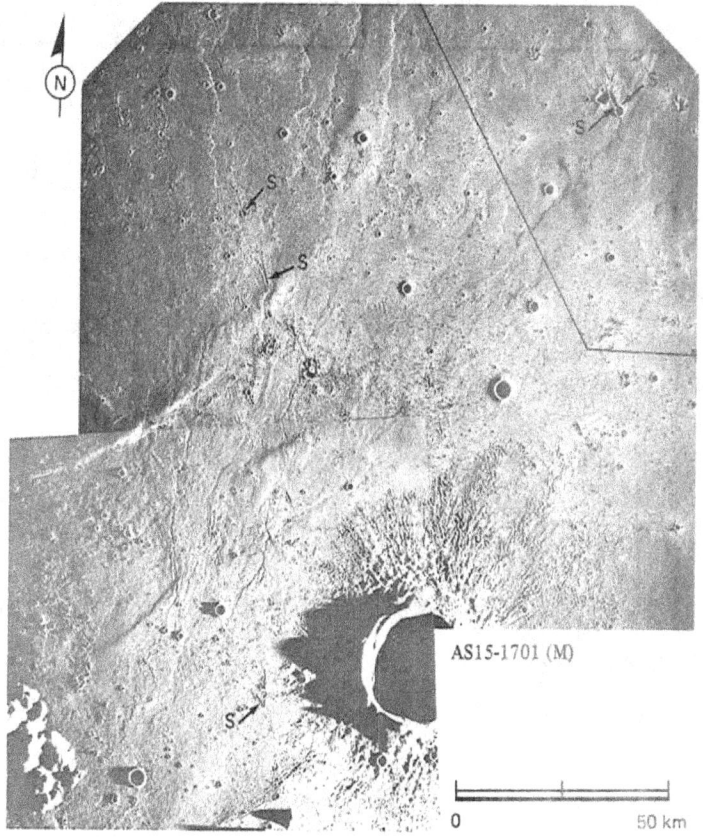

AS15-1701 (M)

0   50 km

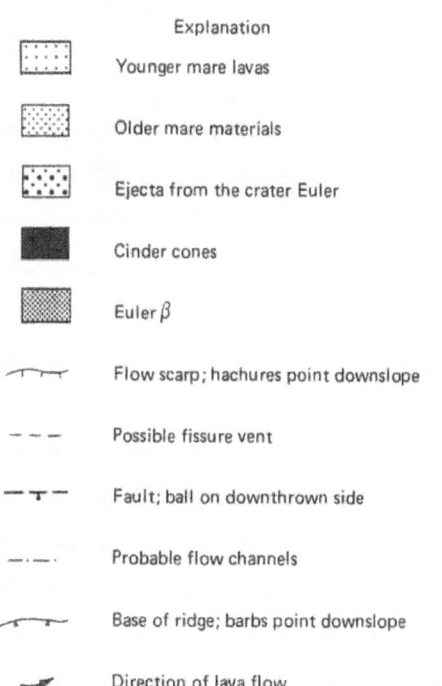

AS15-2295 (M)

Explanation

Younger mare lavas

Older mare materials

Ejecta from the crater Euler

Cinder cones

Euler β

⌐⊤⌐ Flow scarp; hachures point downslope

– – – Possible fissure vent

—⊤— Fault; ball on downthrown side

—·—·— Probable flow channels

⌐⊤⌐ Base of ridge; barbs point downslope

⟶ Direction of lava flow

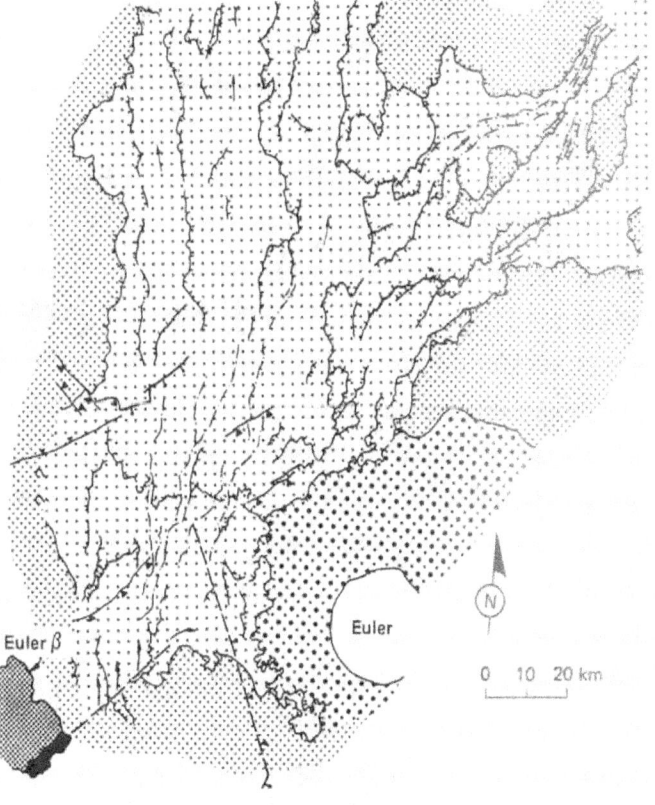

Euler β

Euler

N

0   10   20 km

FIGURE 66.—Two of the younger lava flows described in figure 65 are here followed northeastward to their terminations (see arrows) near the center of Mare Imbrium. The narrowness of the flows is somewhat accentuated by the obliquity of this view. From the postulated fissure vent near Euler β to the farthest terminal lobe is about 420 km. The mare ridge at A must have been nonexistent or at least much more subdued at the time the flow passed that point, otherwise the flow would have ponded behind the ridge. The spray of irregular craters in the center foreground was formed by material ejected from Copernicus, 460 km to the south.—G.G.S.

AS15-1556 (M)

FIGURE 67.—This area is east of that covered in figures 65 and 66; the large crater is Lambert. Here also lava flows of more than one age are present. A sinuous band of smooth, sparsely cratered mare extends northeastward through the center of the picture. It is most certainly a young lava flow and contrasts strongly with the much more densely cratered older mare southeast of the dashed line marking the contact between the two. The western boundary of the young lava flow clearly laps upon and embays the blanket of ejecta deposits surrounding Lambert. Many radial ridges of ejecta and radial grooves or chains of secondary craters radiating outward from Lambert are faintly visible beneath the younger flow near its western boundary. These relationships prove that the younger flow postdates the formation of the crater. Many clusters of secondary craters from craters other than Lambert are present. The shape, orientation, and freshness of some (indicated by arrows) lead us to believe they were probably formed by ejecta from Copernicus, which lies 360 km further south. They are present on the older mare, on the ejecta from Lambert, and elsewhere around this area. However, none is present on the younger flow. If this observation is supported by further study, the younger flow must postdate even the crater Copernicus, and thus be younger than any other extensive lava flow recognized to date.—H.M.

AS15-1010 (M)

AS16-2836 (M)

0   50 km

FIGURE 68.—Although this mosaic covers only a small part of Oceanus Procellarum, it shows a large number of different features that typify the mare surfaces of the Moon. All these features except the ubiquitous craters superposed on the mare are identified in the accompanying sketch map. The rugged terrae in the lower left corner mark the edge of Oceanus Procellarum. Similar terra materials project through the mare in many other places suggesting that the mare fill is thin here. Broad gentle arches, visible only in very low Sun pictures such as these, are numerous and seem to be independent of the even more numerous mare ridges or wrinkle ridges. The ridges are alined mostly northwest and to a lesser degree north-northeast. The alinement suggests—but does not prove—control by tectonic disturbance within the crust. A large forked rille emanates from three elongate depressions located on or

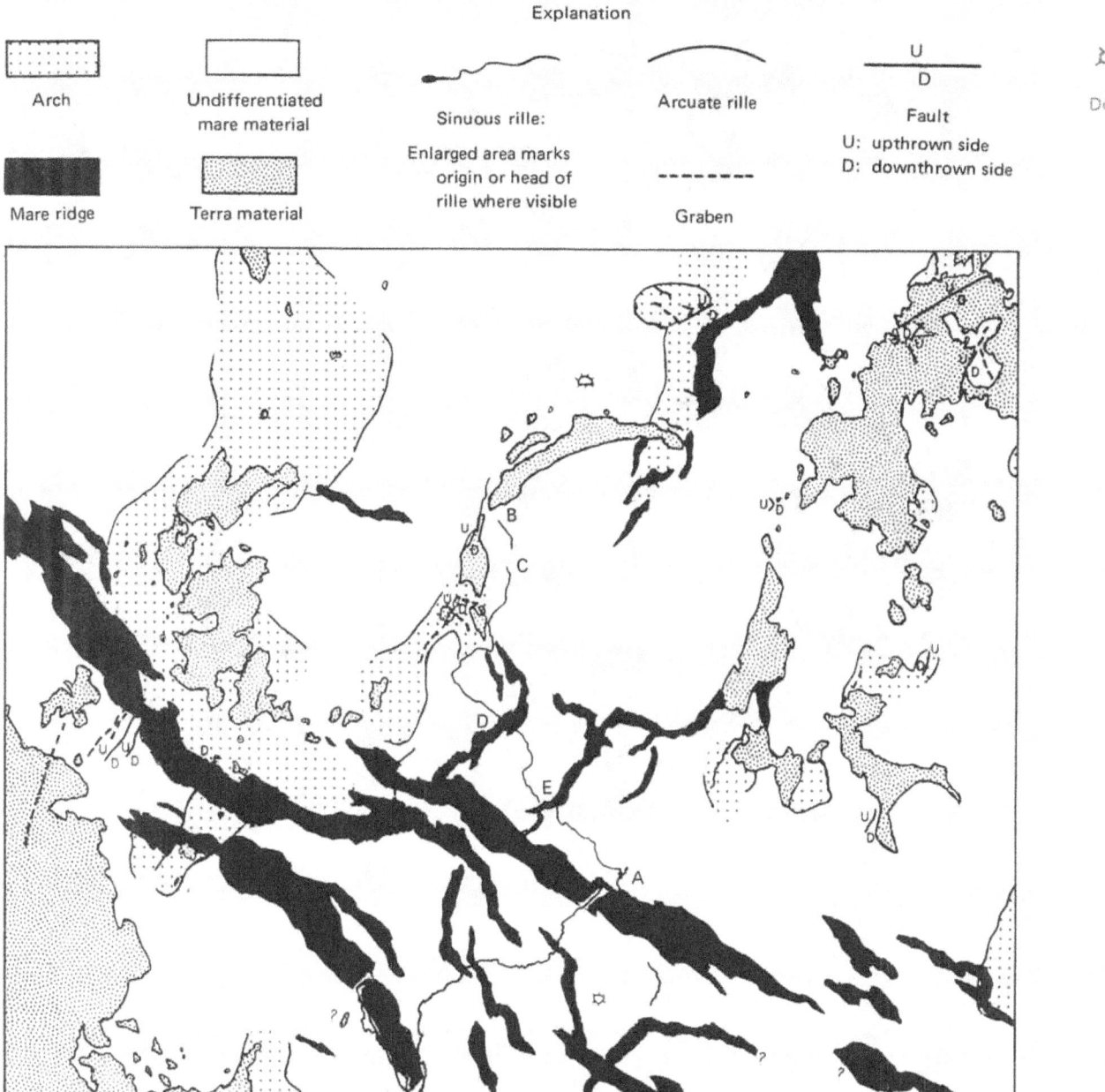

very near two of the largest ridges. This led Young (Young et al., 1972) to postulate that these rilles are lava channels or tunnels formed during "upwelling and outpouring of lava beneath a thin or viscous crust to form the mare ridges." The large rille extends southward for 135 km beyond the picture. From the depression at *A*, a smaller rille meanders about 100 km northward before disappearing at *B*. Note that this rille is interrupted by a small crater (*C*) and by the two mare ridges at *D* and *E*. The crater is obviously younger than the rille and, almost as certainly, so are the two ridges. Many small sinuous and arcuate rilles, small grabens, and faults are present. Although it is difficult to distinguish between some of them, most of these linear features are located on or near high areas of terra islands, mare arches, or mare ridges. Also identified on the map are the small possible volcanic domes.—G.W.C.

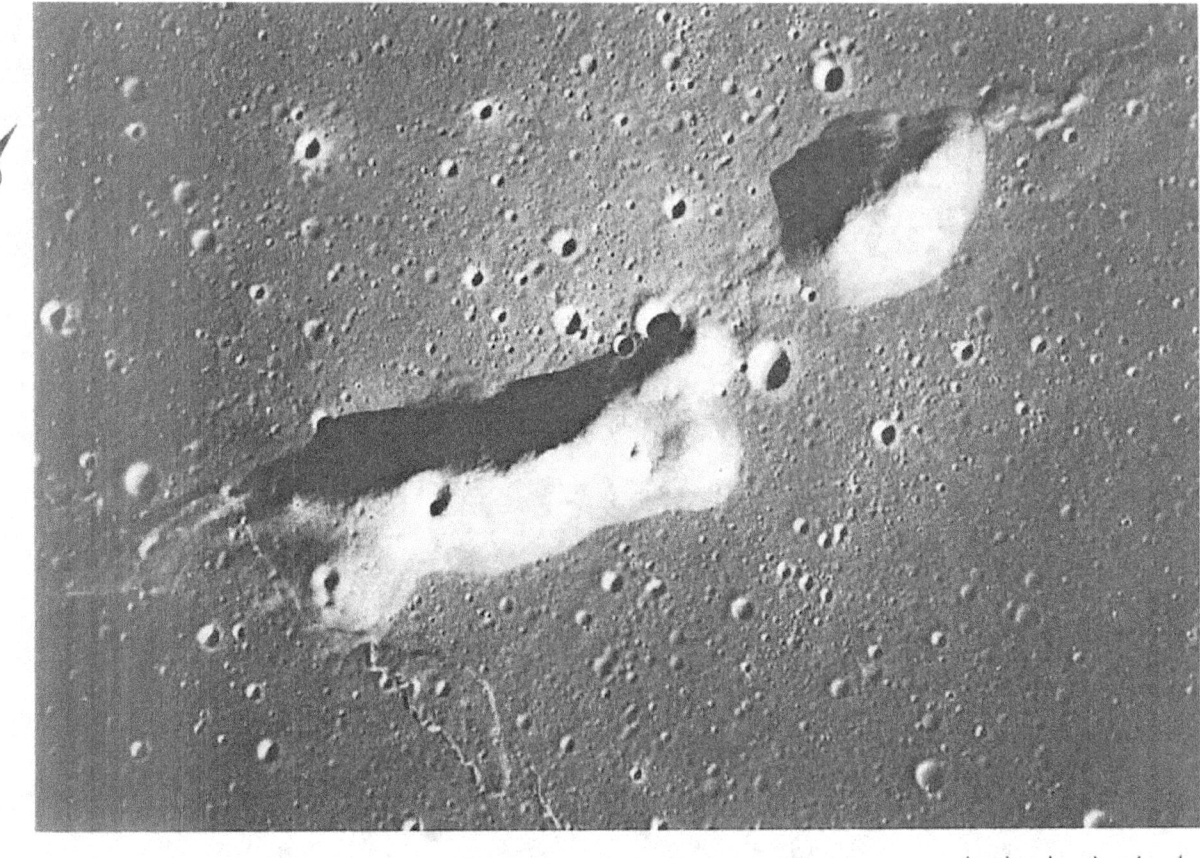

AS16-120-19244 (H)

FIGURE 69.—This cluster of features in Mare Cognitum was photographed in color with a handheld camera and 250-mm lens. Its location on a mare ridge is shown in figure 91. Thomas K. Mattingly, the astronaut who took the picture, noted the taluslike skirt of material around the base of this and many other positive relief areas in the western maria. In this photograph the apparent difference in tone (and in color on the original negative) between the hills and the slightly darker skirt of material around their base is largely, if not entirely, due to differences in slope—a more steeply sloping surface appears brighter, as in the walls of craters in the adjacent mare. However, a real color difference has been found at the base of other prominences. The suggestion has been made that in some mare basins the original level to which lava filled the basin has receded and that skirts such as these are, in effect, "bathtub rings" recording the higher level. The two hills pictured here are probably of volcanic origin.—M.C.M.

L.O. IV-1334

0   25 km

FIGURE 70.–These two pictures were taken of the same mare area near the southeastern edge of Oceanus Procellarum, south and east of the crater Kunowsky. The lower left (facing page) picture is part of a Lunar Orbiter 4 high-resolution photograph taken when the Sun was at a moderate elevation of 18° to 20°. The picture below is a mosaic of Apollo 14 frames taken when the Sun was exceptionally low–0° to 2°. Douglas Lloyd designed the special experiment by which these near-terminator photographs were obtained, using very high-speed film in the Hasselblad camera.

The density of craters (more properly, the number of craters per unit area of surface) has been used by geologists as a tool to determine the relative age of rock units on the Moon's surface. The method has been applied principally in mare areas because crater populations generated on the irregular highland surfaces cannot be accurately measured. These two pictures illustrate some of the problems encountered when applying the method.

The number of craters that can be seen and hence counted is affected by the Sun angle. For example, many more craters are visible in the mosaic of low Sun Apollo pictures than in the Lunar Orbiter picture, and a detailed count of all craters in each picture would result in two radically different relative ages for the same area. Further, comparison of the two views shows that the apparent discrepancy in abundance of craters exists only among the very small craters–those a few hundred meters or less in diameter. The number of craters a kilometer or more in diameter is the same in both pictures. The explanation is that most small craters can be recognized as such only by the shadows they cast. The materials on their walls and rims are commonly indistinguishable from those of the surrounding terrain. For each picture there is, depending on the angle of Sun elevation, a threshold value of slope below which no shadow is cast. In the picture at lower left that

value has been calculated to be about 5°. Consequently, those craters that have been degraded so that their slopes are less than 5° are not visible. In contrast, craters with slopes as gentle as 0.25° are visible in the picture below.

An observation immediately follows. Small craters have a relatively short lifespan. That is, once formed, they are rapidly degraded. Their rims are eroded, and their interiors are filled with debris from the continuing bombardment of the surface by other impacting bodies. By actual count about 80 percent of the small craters in this area have been so degraded that their slopes no longer exceed 5°.

It has also been shown (Soderblom and Lebofsky, 1972) that the small crater population here–and in most mare areas–is in a steady state. In other words, the rate of formation of new craters and the rate of destruction of existing craters (either by superposition of other craters, or by gradual erosion by much smaller craters) are balanced. It is fruitless, therefore, to count small craters because such counts will result in the same false age.

Fortunately, the crater counting method does yield satisfactory results when applied to larger craters. The same number of craters larger than about 1 km is visible in both pictures. This means that no craters this size or larger have been degraded to the extent that their slopes are less than 5°–as were 80 percent of the small craters. We may assume then that all the larger impact craters that ever formed on the upper part of the mare surface have been retained and that their relative abundances in different areas are a measure of the relative age of those areas.

With the careful application of this method, it has been possible to assign relative ages to most of the mare areas of the Moon. Using the absolute ages that have been determined for samples returned from the Apollo landing sites in the maria (Apollos 11, 12, 15, and 17), the relative time scale now has a quantitative base so that relative ages can be converted to absolute ages.–L.A.S.

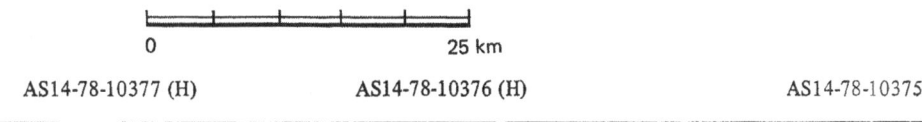

0   25 km

AS14-78-10378 (H)   AS14-78-10377 (H)   AS14-78-10376 (H)   AS14-78-10375 (H)

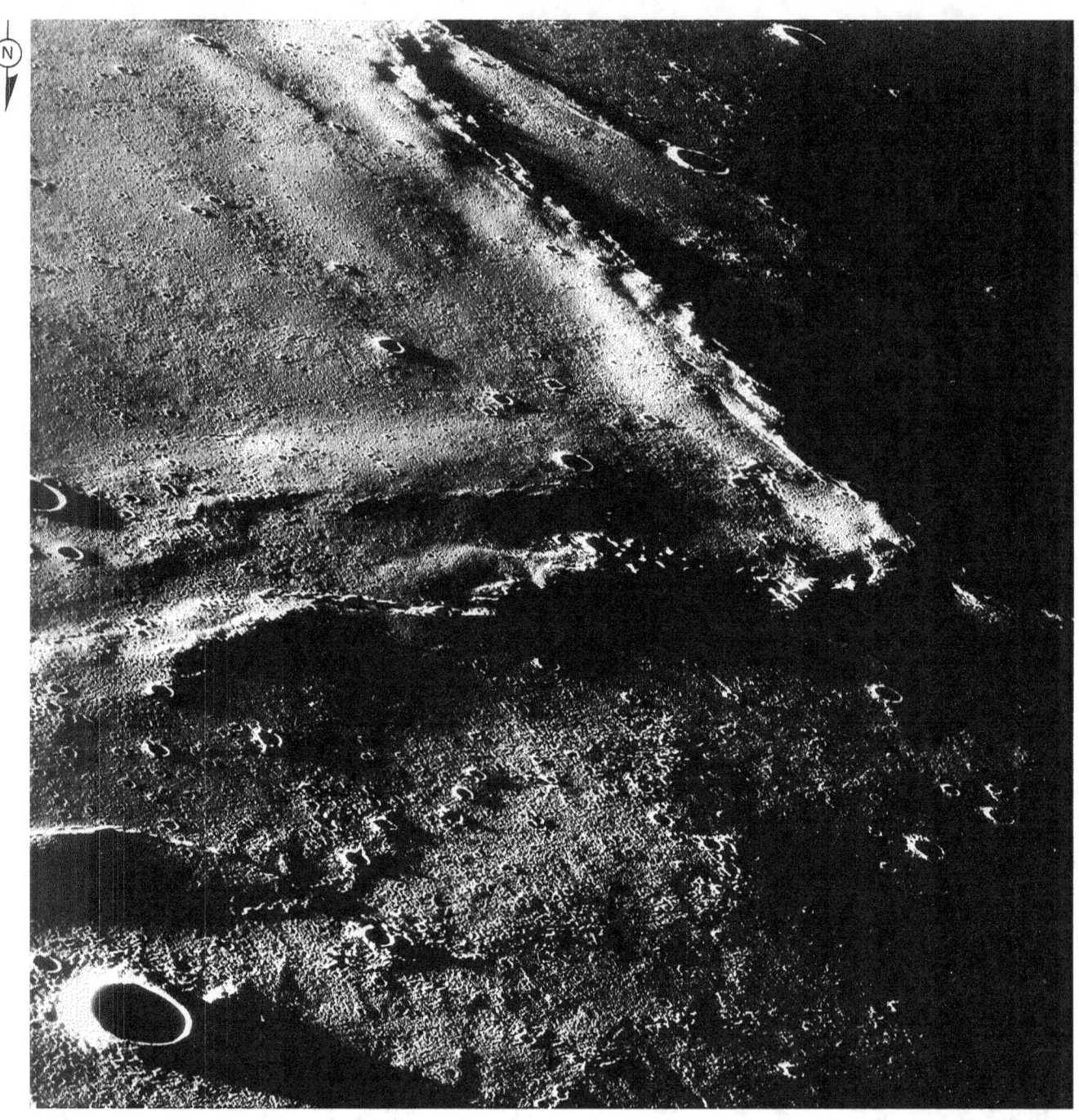

AS15-98-13361 (H)

FIGURE 71.—A very low Sun angle emphasizes detail in this south-looking oblique view of an area in western Oceanus Procellarum. The crater Seleucus is just outside the field of view to the right, and Seleucus E is in the lower left corner. The near-terminator view was exposed in a handheld Hasselblad camera with a 250-mm lens. Low Sun illumination enhances small surface features and subtle differences in slope so that the broad, gently sloping arches associated with the mare ridges are clearly delineated. The oblique viewing angle accentuates the angularity that characterizes the path of many ridges as they are traced across the mare surface. (See, for example, fig. 68.)—M.C.M.

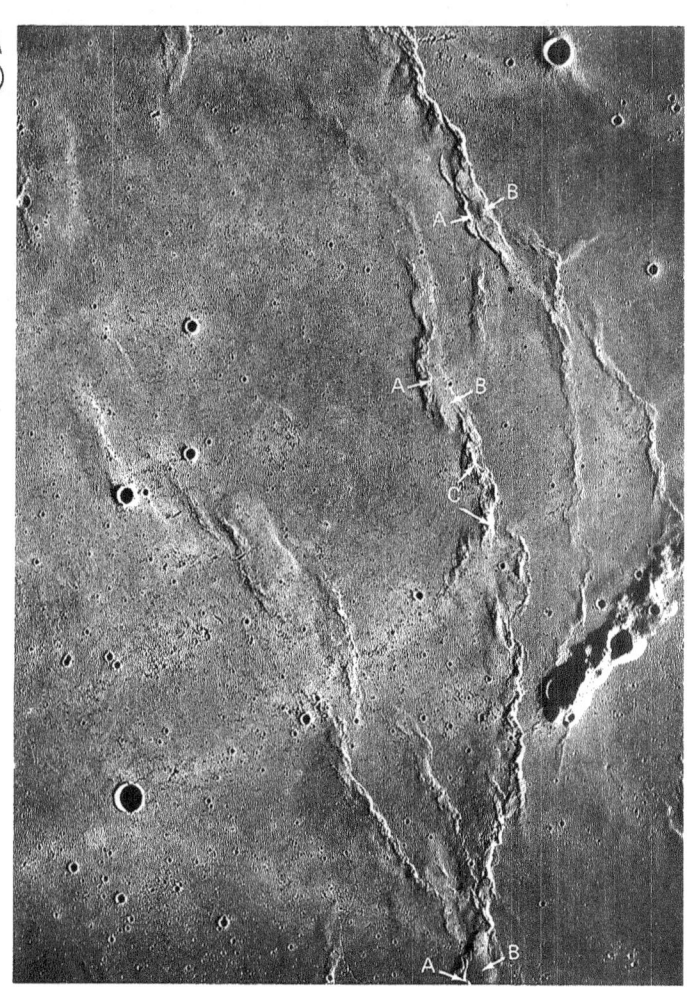

FIGURE 72.—Mare ridges in Oceanus Procellarum at the northwest tip of the Aristarchus plateau (lower right) are typical of mare ridges in many other areas. Among these are small crenulate wrinkle ridges (A) superposed on broad gentle arches (B), braided or en echelon patterns, a common tendency toward parallelism within limited areas, and a tendency to be deflected by obstacles such as the arcuate ridge belt of terra near the Aristarchus plateau. Measurements by photogrammetric methods have shown that mare ridges may project as much as 250 m (at C) above the surrounding mare surface. The origin of the ridges has yet to be explained satisfactorily by any single hypothesis. Among those proposed are (1) intrusion of deep-seated, postmare dikes and laccoliths, with some extrusion; (2) compressional buckling caused by sagging of domical mare surfaces; (3) buckling at the fronts of lava flows; (4) autointrusion of molten subcrustal mare lava forming laccoliths and squeeze-ups; (5) thrust faulting; and (6) drag folding along deep-seated transcurrent faults. The diversity in morphologies and patterns of ridges suggests that several hypotheses may be required to explain all of them.—C.A.H.

AS15-2487 (M)

0   50 km

FIGURE 73.—Stereoscopic viewing, made possible by the overlap of Apollo mapping camera frames, does wonders for some lunar scenes. When viewed monoscopically, these photos taken at high Sun illumination in Mare Fecunditatis show bright patterns of craters and narrow mare ridges. Viewed stereoscopically, they reveal the relative depths of the craters; the sharp relief of the narrow ridges; an unsuspected broad, high swell between the parallel "ranges" of narrow ridges; and the even more surprising differences in elevation between mare surfaces in the left (west) and right parts of the photograph. The elevations of the mare surfaces probably mimic the variable depth of the buried floor of the Fecunditatis basin. The center of the basin, presumably the deepest part, is located near the left of the picture, where the mare surface is lowest. The large crater Messier G is 14 km in diameter.—D.E.W.

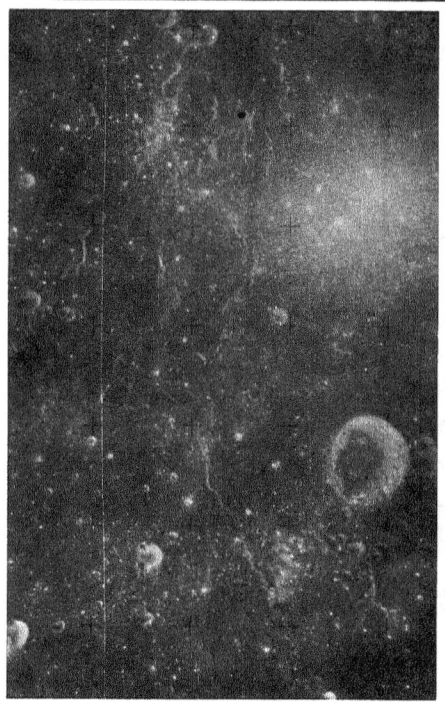
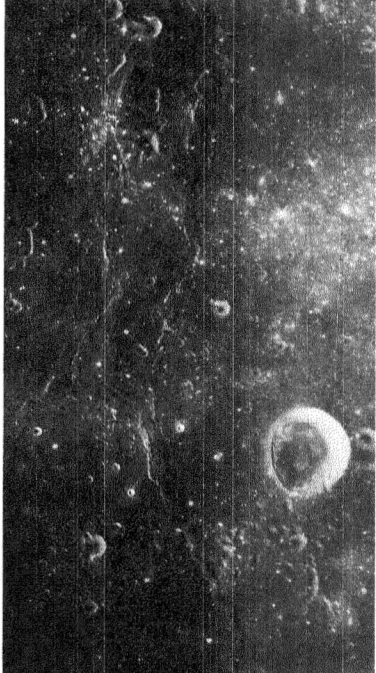

AS15-2401 (M)   AS15-2399 (M)

0   50 km

FIGURE 74.—An oblique view looking northward along the east side of Mare Serenitatis. The Apollo 17 landing site (arrow) is near the lower edge. A nearly identical view was used earlier (fig. 27), so the geology of the area will not be discussed again. We have included this picture here to serve as an index map for the next seven pictures, which are detailed views of parts of the mare ridges visible in this photograph.—G.W.C.

AS17-0939 (M)

AS17-2313 (P)

0   1 km

FIGURE 75.—This and the next six pictures are enlargements of specific areas to show the detailed form of the mare ridges seen at a much smaller scale in figure 74. Here can be seen a small segment of the most prominent ridge enlarged about 45 times. When seen in detail, the ridge is spectacularly crisp and well formed. The main ridge here is shaped like a string of sausages. Smaller wormy ridges appear on either side. Cracks in the top of the ridge probably formed by buckling of the mare lavas. As the following figures show, most other parts of the ridge are less symmetrical than this part.—K.A.H.

AS17-2313 (P)

0   1 km

FIGURE 76.–This part of the ridge appears to have flowed over the ground to the right. A careful search will reveal several craters overridden by the ridge along its right edge. A large oblong crater on the ridge near the bottom of the photograph has evidently been distorted by the ridge; in fact, the right edge of the crater seems to be cut off. (See arrow.) One can imagine that a sheet of soft putty might form a similar ridge if it were thrust over to the right. Model experiments and observations of deformed rocks on Earth have shown that large masses of rocks, even hard rocks like basalt, can behave like putty over sufficiently long intervals of time.–K.A.H.

FIGURE 77.—Here again the ridge has overridden craters along its right side. Many other mare ridges that are older may once have looked like this. The freshest examples of lunar features, like this one, are the best places to look for hints on origin.—K.A.H.

AS17-2313 (P)

FIGURE 78.—At its north end the ridge becomes a scarp that wraps around the base of the highlands like a shoved rug. Were the mare lavas thrust against the highlands? This might seem easiest to imagine if the lavas had been only partly solidified when they were deformed, but the lavas had had plenty of time to solidify. The ridge deforms numerous impact craters, such as the one near the bottom of the photograph. A long time—probably hundreds of millions of years—had to pass for these impact craters to form after the lavas crusted over and before the ridge formed.—K.A.H.

AS15-9298 (P)

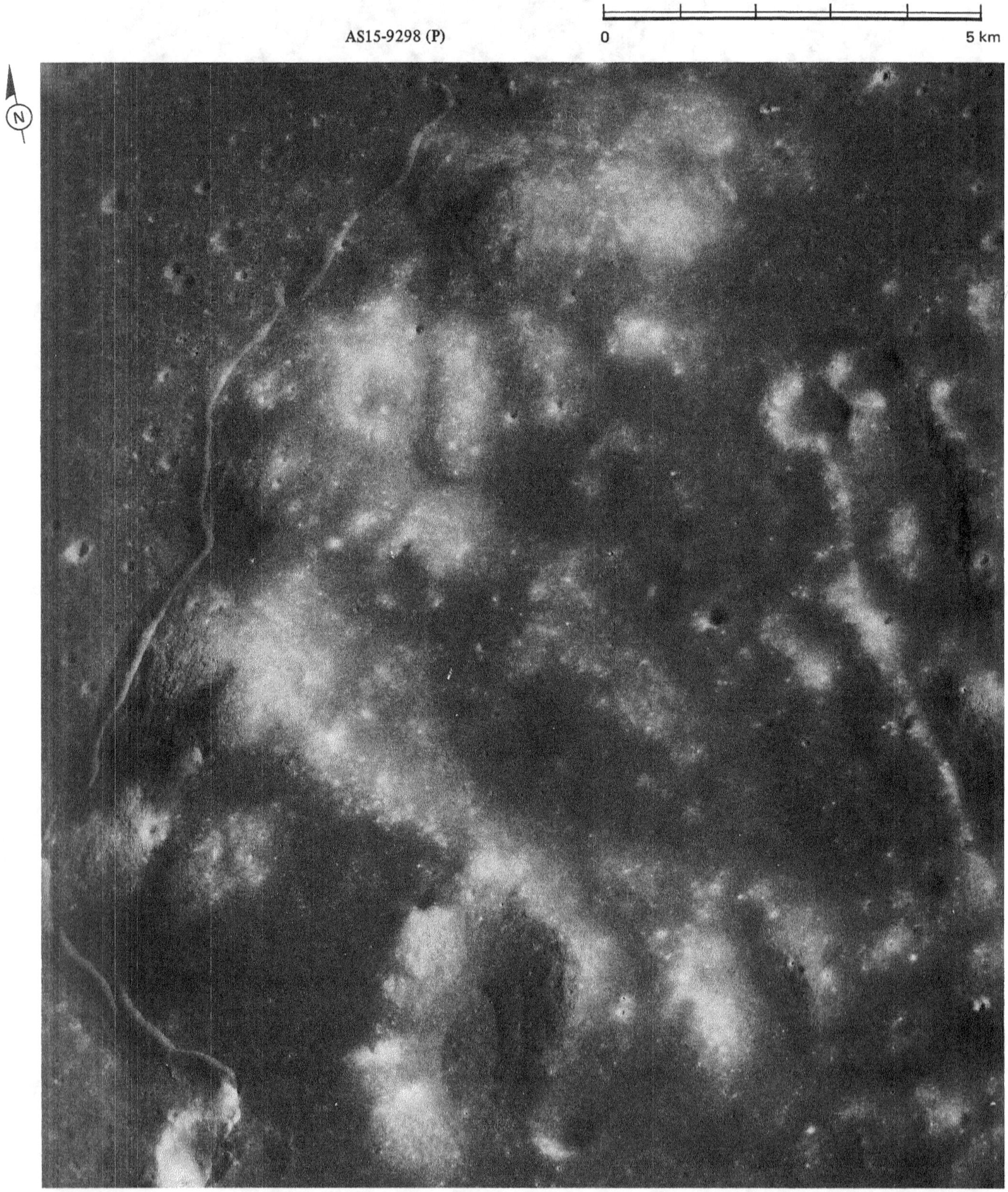

0   5 km

AS17-2313 (P)

0   5 km

FIGURE 79.—In the upper left corner of the photograph, a lunar ridge heads in a southeasterly direction toward the dark mantling unit that rings Mare Serenitatis. At its south end, one branch of the same ridge disappears in the dark mantling unit. The ridge boundaries are crisp and clear in the lighter mare materials in the left half of the photograph. However, in the much darker and topographically higher unit to the right, the ridge is subdued. (Owing to especially favorable lighting conditions, the true height of the east-facing scarp of the ridge is exaggerated.) The sequence of events worked out from the study of this area is as follows: (1) the Serenitatis basin was probably formed by the impact of a giant meteoroid; (2) the rim materials, and perhaps also the inner part of the basin, were flooded by dark basaltic lavas and associated volcanic (pyroclastic) debris; (3) the central part of the basin was filled by a lighter colored basaltic unit; and (4) gravitational adjustments to the enormous mass of volcanic fill probably caused the formation of the ridges in the light mare in the inner part of the basin and locally in the dark outer ring.—F.E.-B.

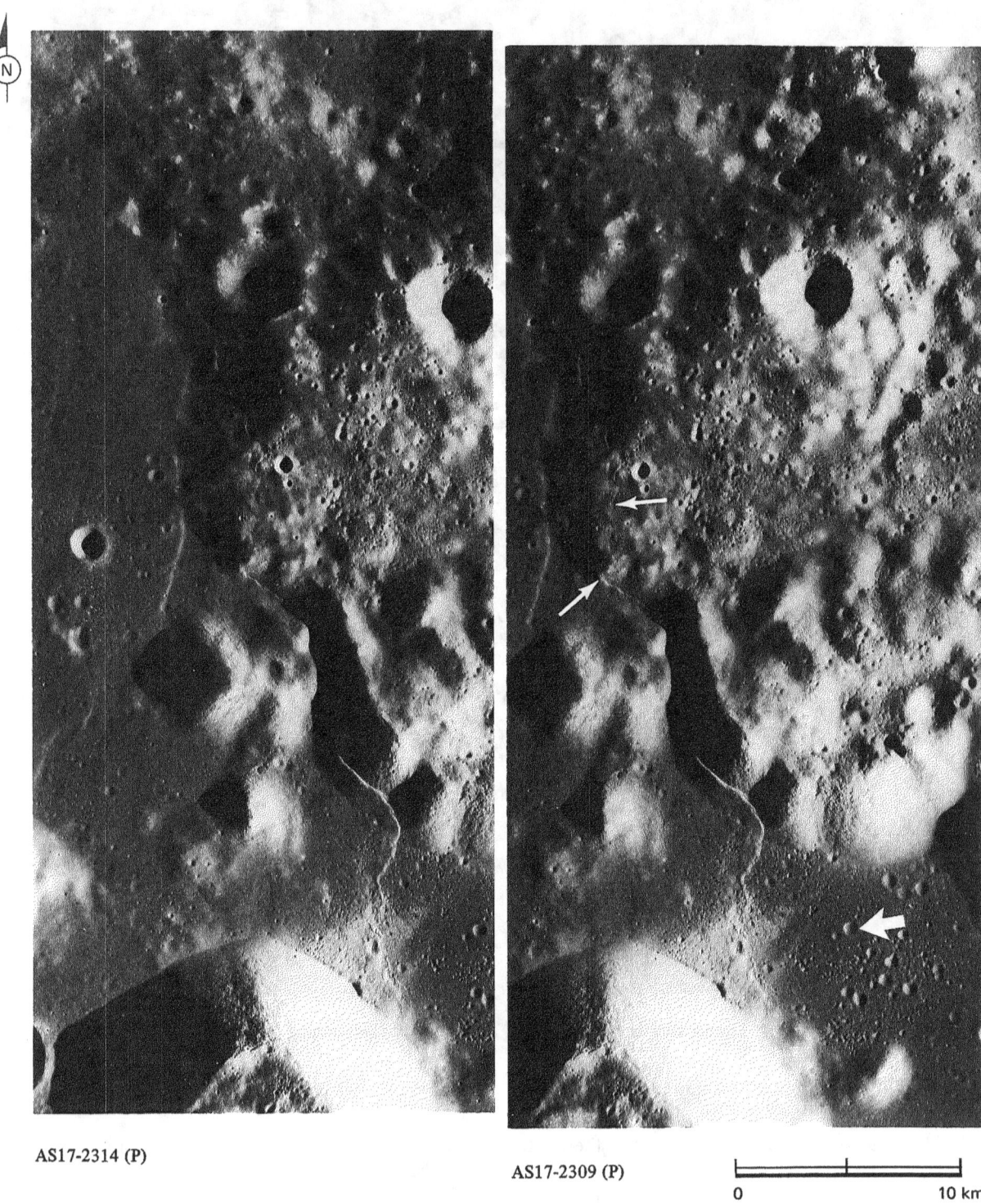

AS17-2314 (P)

AS17-2309 (P)

0   10 km

FIGURE 80.—A stereoscopic view of the only mare ridge climbed by man. The ridge crosses the Taurus-Littrow valley shortly west of the Apollo 17 landing site (large arrow) and was traversed by astronauts Gene Cernan and Jack Schmitt in the Apollo 17 lunar roving vehicle. Geologic and topographic maps of the area appeared earlier in this chapter (figs. 61 and 62). From the valley floor the ridge extends northward into the highlands. It is offset between the two small arrows and is partly buried by dark mantling material along its northern extension.—B.K.L.

FIGURE 81.—This enlargement provides a closeup view of part of the ridge. The fact that a fairly young impact crater (*A*) is deformed by the ridge and the generally fresh appearance of the ridge both suggest that it is fairly young. Within the valley, the ridge consists primarily of east-facing imbricate scarps, although west-facing lobate scarplets are also present. After crossing into the highlands (*B*), the ridge becomes a simple one-sided scarp resembling a fault scarp. If faulting was involved, movement was upward on the left side or, less probably, downward on the right. The Apollo 17 astronauts were unable to provide conclusive answers concerning the origin of the ridge owing to the thick cover of regolith and avalanche material. Photogeologic study indicates that part of the ridge may have formed by displacement along a fault or, less likely, by the upwelling of lava.—B.K.L.

AS17-2309 (P)

0   2 km

AS15-9361 (P)

0   5 km

FIGURE 82.—These intersecting sets of mare ridges are near the western edge of Mare Serenitatis. One set of generally small, sharply defined lobate scarps and ridges trends irregularly east to west. It cuts across a set of broader, more subdued ridges that trends north to south. On the basis of morphology and crosscutting relationships, the first set appears to be the younger of the two. Study of the better preserved younger set provides two clues to its origin. In several places (designated by A) incomplete craters are present on the flanks of ridges, but the other parts of the originally circular craters are absent on the adjacent mare floor. This suggests that some segments of the ridges are the frontal edges of thrust plates that have ridden over an already solidified and cratered surface. In several other places (B) craters on the mare floor appear to be partly overlapped by lobed projections of the ridges, suggesting inundation of the craters by a viscous material. The ridges in this area as in others may thus be the response of solidified rocks to faulting and possibly also to the movement of molten rocks.—B.K.L.

FIGURE 83.—The arcuate ridges shown here are part of the large crater Lambert R, which is mainly deeply buried and now outlined by mare ridges and a few isolated remnants of the original rim. This is one of many examples in which underlying topography is reflected by mare ridges, thereby suggesting a causal relationship. The ridge segments at *A* are probably caused by solidification and sagging of the mare lavas over the crater rim. Crustal buckling caused by sagging of the lava inside the crater rim may have formed the inner ridges *B* and *C*. The exact origin of mare ridges is still debatable.—C.A.H.

AS17-3075 (P)

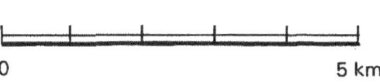

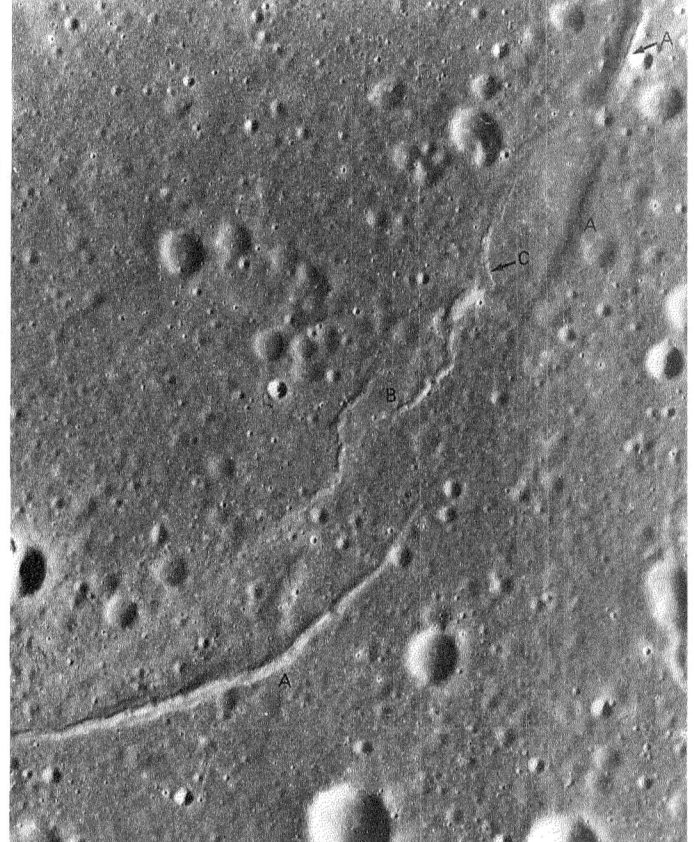

FIGURE 84.—This picture is included for several reasons. It is an example of a mare ridge on the far side of the Moon. It also shows that mare ridges occur in small bodies of mare. Finally, it serves as another example of a mare ridge transecting terra as well as mare. The scene covers part of the floor and eastern wall of Aitken, a large (135 km) crater near the center of the far side. The dark area is mare material, forming part of Aitken's floor. The ridge along the east edge of the mare fill extends southward (straight arrows) for a considerable distance beyond the edge of the picture, gradually ascending the east wall of Aitken. To the northwest (curved arrows) the same ridge extends for about 7 km diagonally along the north wall before disappearing. Whether in the mare or in the terra materials of the crater wall, the morphology and geometry of the ridge suggest that viscous material was squeezed out along a fault (?) plane that dips gently westward. The oval area patterned like a turtle shell is probably volcanic in origin. Other examples of this rather rare feature are shown in the chapter entitled "Unusual Features."—G.W.C.

AS17-1913 (P)

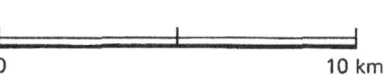

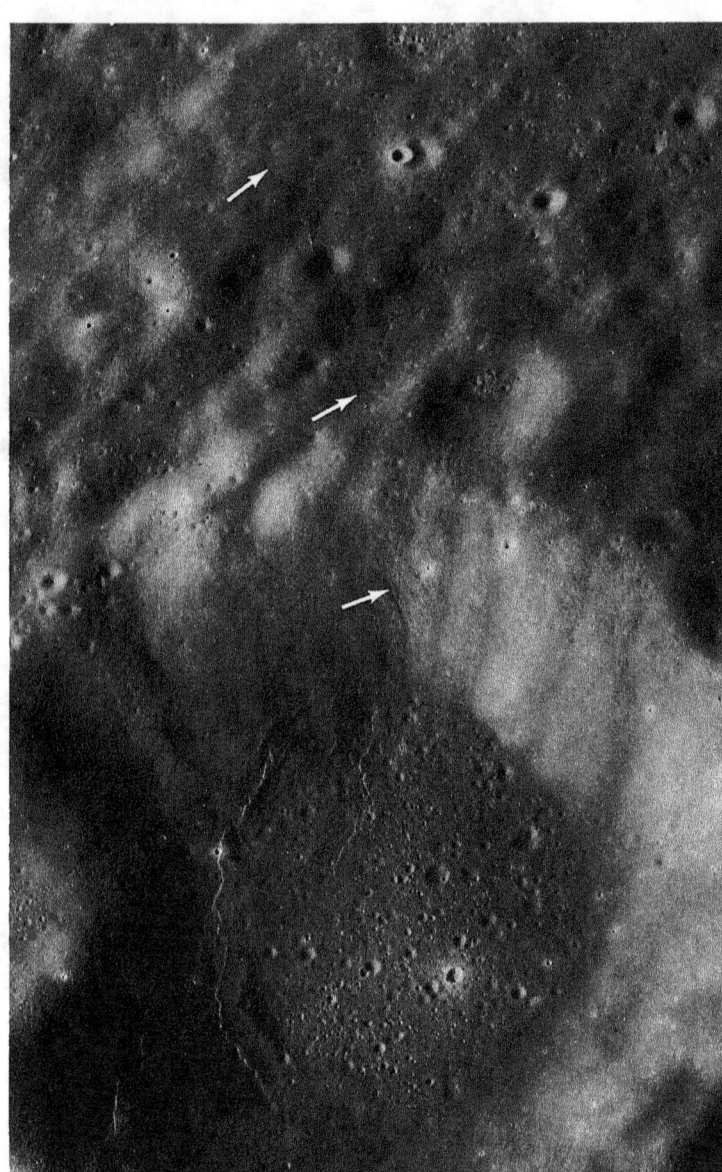

FIGURE 85.—In this unnamed crater on the far side we see ridges closely resembling those already described. Inasmuch as no mare materials are present here, the term "mare ridge" is not appropriate. The materials occupying most of the crater floor, while resembling mare, are plains deposits. Those along the western part of the floor and the lower part of the west wall probably consist of unconsolidated erosional debris from the wall. The crisp appearance of the ridges as they transect unconsolidated material proves that they are relatively young and that they are composed of resistant material. Here also the ridges may be related to faulting. The ridge nearest the center of the crater can be traced (arrows) part of the way up the north wall; interrupted segments are visible along the same trend northward to the edge of the picture. They may mark a steeply dipping fault plane.—G.W.C.

AS17-1676 (P)

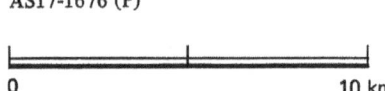

FIGURE 86.—In this picture we see another example of the close relationship between mare ridges and faults. The area is in the Montes Riphaeus between Mare Cognitum and Oceanus Procellarum. A small part of Oceanus Procellarum is shown in the upper left. A mare ridge (*MR*) extends across Procellarum to the edge of the Montes Riphaeus. In all visible respects it is typical of a great many other mare ridges. Across the Montes Riphaeus (arrow to arrow) the trend of the ridge becomes a fault or series of faults causing the valley. Narrower valleys or grooves in the lower part of the picture are parallel to the larger valley and they too probably originated by faulting. Near the east edge of the picture the Montes Riphaeus are bounded by mare (*M*). Parts of the contact are quite angular (heavy lines), suggesting a series of offset faults, possibly continuations of the main mare ridge-fault trend.—G.W.C.

AS16-5452 (P)

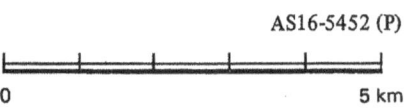

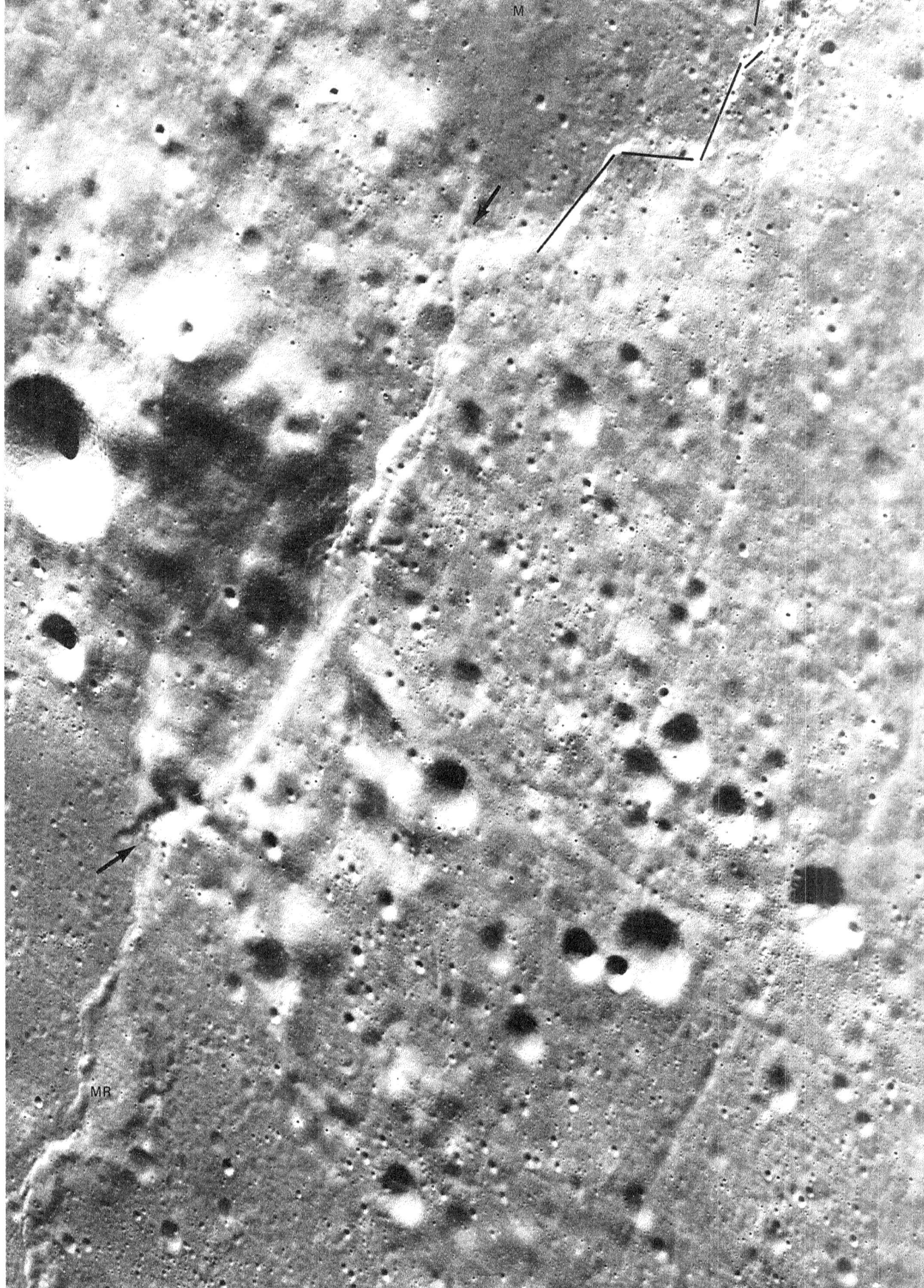

FIGURE 87.—This crenulate lobate scarp occurs within a small depression on the far side. It is not related to any obvious crater source, although it is similar to lobate flow fronts within the ejecta blankets of some young impact craters. Although it resembles some mare ridges, it is clearly confined to highlands terrane. Small scarps of this kind— unrelated to any apparent source of material or to any detectable tectonic control—are widely scattered in the highlands; their origin remains enigmatic. Mass wasting along a fault scarp might have produced the lobate configuration. A flow of locally derived surficial debris is also a plausible explanation; in this case, such a flow might have been triggered by small impacts on a ridge to the northwest.—C.A.H.

AS16-4970 (P)

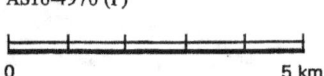

0   5 km

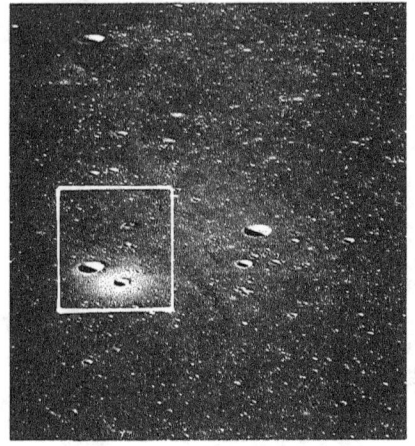

FIGURE 88.—This is an oblique view of an area in Mare Nubium west of the central lunar highlands taken with the Hasselblad camera on Apollo 16. The small picture to the left is the entire frame reproduced slightly larger (112 percent) than the size of the original negative, and the larger picture on the facing page is a 12.5X enlargement of the area outlined within the small picture. The view shows two distinct types of lunar craters in the mare. The two craters in the near field are typical of a great many small impact craters. The larger one has a raised hummocky rim but no bright ejecta blanket and no visible rays. The smaller and probably younger one to the right displays an extensive hummocky ejecta blanket and a system of bright, radial rays. The dark crater near the middle of the photograph has little in common with the first two. It has a prominent, smooth, convex rim. It is not surrounded by a hummocky or blocky ejecta blanket and has no rays. These characteristics suggest that this crater is volcanic in origin. Occurrences like this indicate that both impact and volcanic processes were active on the Moon, and that they are both responsible for crater formation. However, as one can see in this view, impact craters are more numerous.—F.E.-B.

AS16-120-19237 (H)

0   5 km   Approximate

97

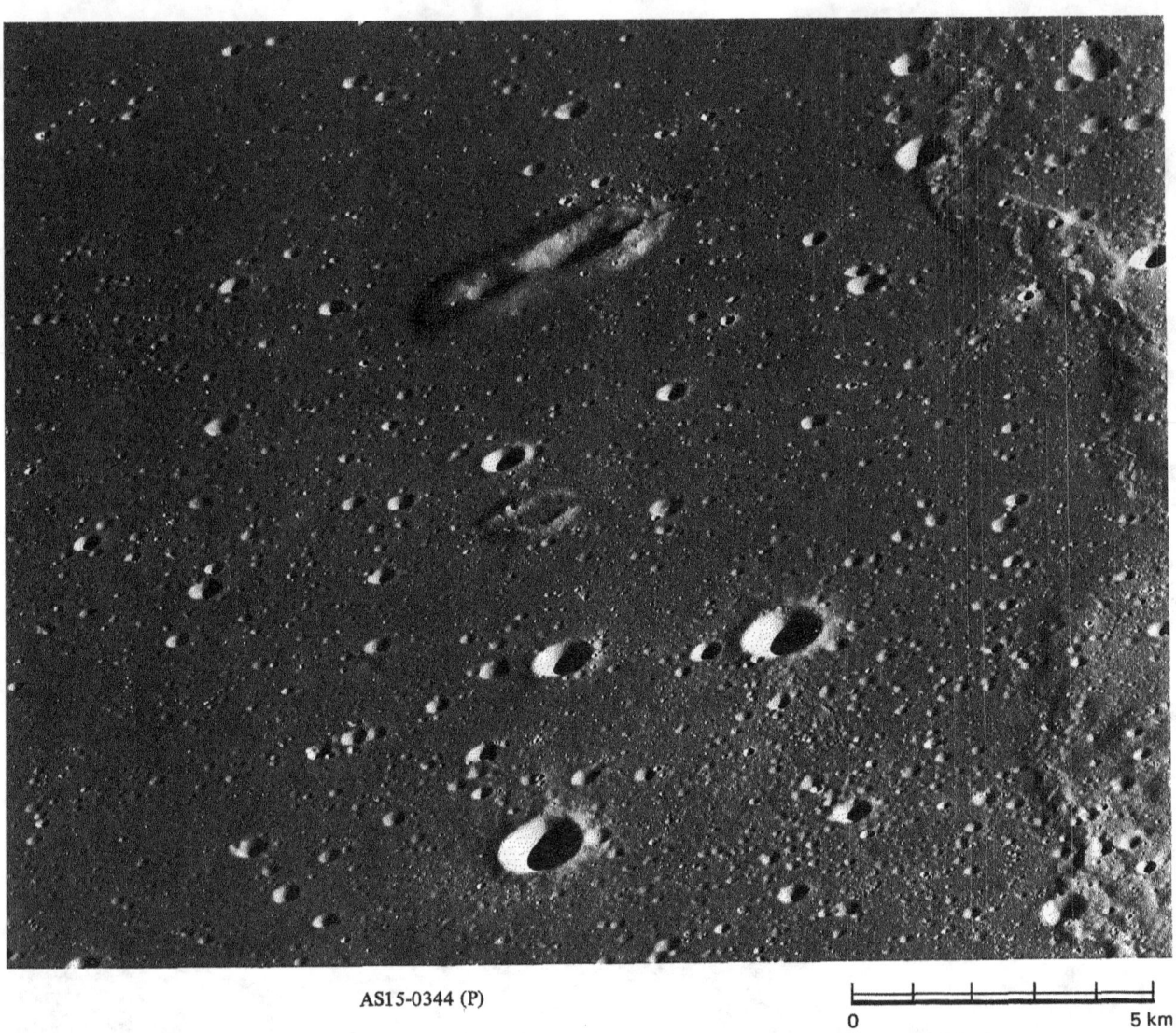

AS15-0344 (P)

0   5 km

FIGURE 89.—A few volcanic craters lie in a field of impact craters in Oceanus Procellarum, northwest of the Aristarchus plateau. Three in number, the volcanic structures are easily recognized by their high rims, planar slopes, and elongate shape.
They are clearly embayed and partly buried by mare basalt, and thus are older than
the mare plain surrounding them.—D.H.S.

FIGURE 90.—This panoramic camera enlargement shows two small (2 km across)
volcanic cones (A and B) near the southeastern margin of Mare Serenitatis. They are
intergradational with the surrounding mare basalt and probably are similar in age
(about 3 billion years). The same cones can be seen in their regional context in
figures 58 (lower right) and 60 (lower left). One cone (B) has an enclosed summit
crater and the other (A) has a breached crater leading to a rillelike valley. Both
conditions are common among volcanic cinder cones on Earth. The cones appear to
be alined along the buried extension of a straight rille (C), suggesting a relationship
between volcanism and tectonism that is common on Earth. In size, appearance,
and geologic setting these lunar cones are remarkably similar to terrestrial cinder
cones; by analogy they are probably composed of interlayered fragmental material
and lava flows. The elongate depressions marked D, as well as the straight and
arcuate rilles, are probably of structural origin resulting from faulting and collapse
along fractures.—D.H.S.

AS17-2317 (P)

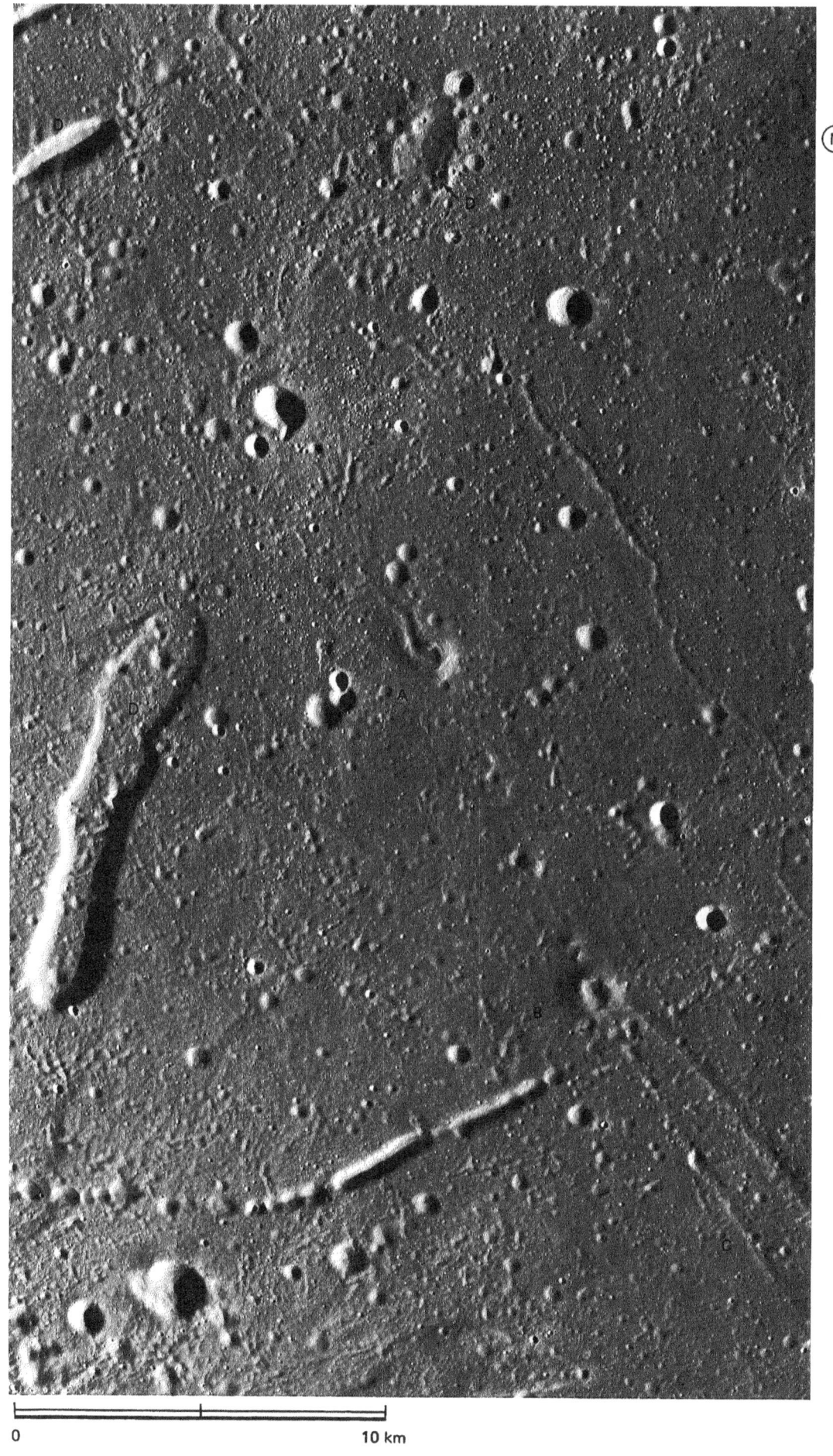

0   10 km

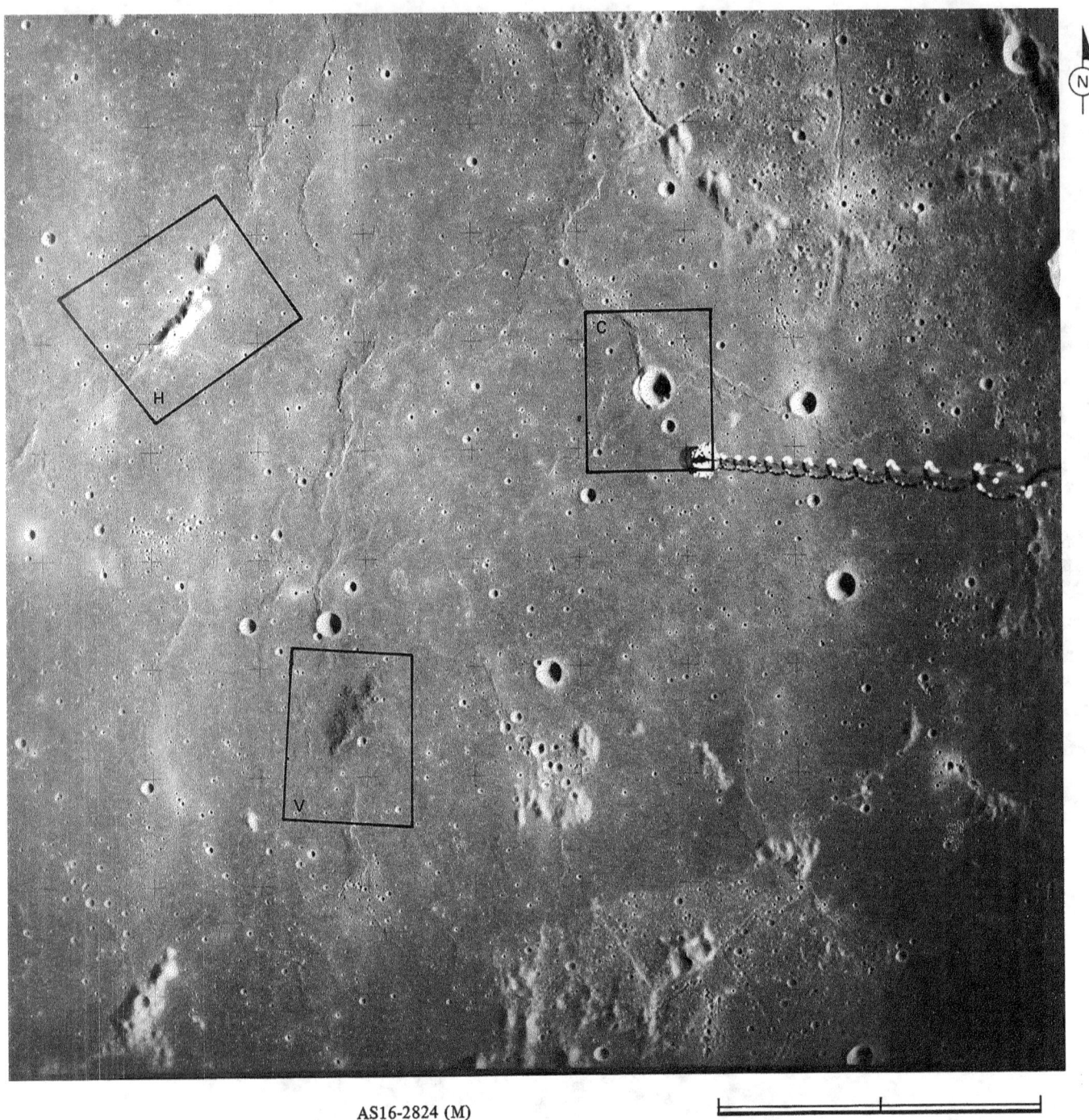

AS16-2824 (M)

0   50 km

FIGURE 91.—This picture of the eastern part of Mare Cognitum, while of interest in many respects, is used here largely to locate the dark hills (*V*) illustrated in detail in figure 92. Although small, the complex of dark hills is easily discernible because it is lower in albedo and is rougher than the surrounding mare lavas. Also included in this view are two bright hills (*H*) that were described earlier (fig. 69) as probably being of volcanic origin and an apparently faulted crater (*C*) that will be described later (fig. 246). Note the proximity of all three features to mare ridges. The dark hills lie within a westward bulge of one mare ridge, the two bright hills interrupt another ridge, and the crater is intersected by a third ridge. The association probably is not accidental.—M.C.M.

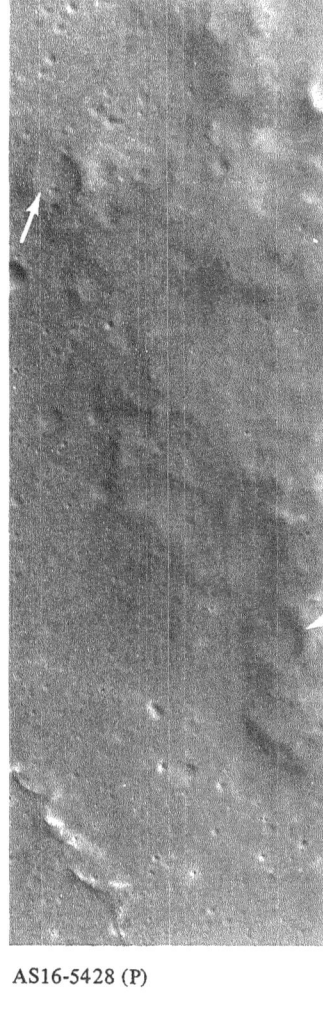

FIGURE 92.—The dark, rugged-surfaced complex of hills is enlarged many times in this stereogram. Without further detailed study, two explanations for its origin seem equally plausible. The complex may be a densely cratered block of terrae that was partly inundated by the lavas of Mare Cognitum and subsequently blanketed by dark volcanic ejecta. Within the darkened area, the concentration of fresh young craters is less than the surrounding mare surface, strongly suggesting that the dark blanketing was deposited appreciably later than the mare lavas.

Alternatively, the complex may be a pile of lava flows densely pockmarked by volcanic craters, and, as in the first case, subsequently covered by volcanic ejecta. The step-like, but discontinuous ledges along the east side of the complex probably represent successive flows of viscous lava. All the craters are shallow, probably because they have been filled by their own ejecta or by that from nearby craters. Nevertheless, several craters (see arrows) have steep raised rims, distinguishing them from impact craters. The difference is clear at A where a volcanic crater (left) can be readily compared to a normal impact crater (right).—G.W.C.

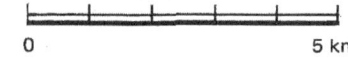

0   5 km   AS16-5433 (P)   AS16-5428 (P)

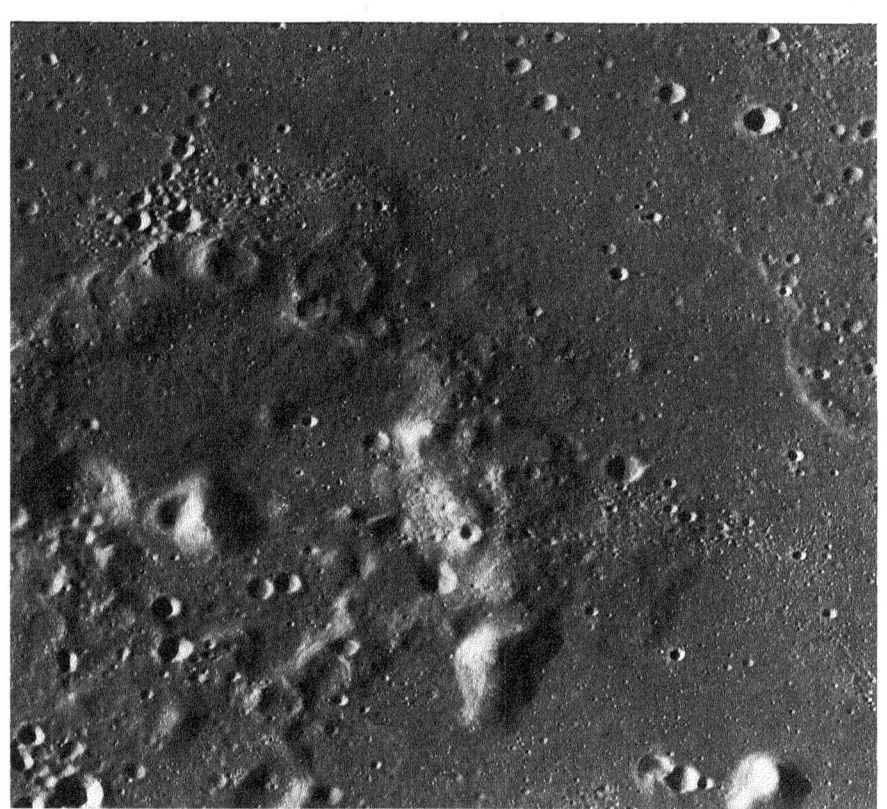

FIGURE 93.—Easily identifiable volcanic complexes like the one in figure 92 are not common, but some others have been recognized. This crudely circular belt of high-rimmed craters is located along the common boundary between Oceanus Procellarum and Mare Imbrium, about 130 km southwest of the crater Euler. Here also, a cover of dark material—presumably volcanic ejecta—is present. However, unlike the previous example, which was surrounded by a "sea" of unbroken mare, this complex is closely associated with many peaks of premare terra. They are distinguished not only by their appreciable relief, but also by their smooth surfaces and lighter tone.—G.W.C.

AS17-3114 (P)

0   10 km

AS16-5425 (P)

0   10 km

FIGURE 94.–Patches of dark material—presumably volcanic ejecta—are commonly associated with straight rilles or grabens. This area is near the Apollo 14 landing site. (See fig. 44.) The prominent ridge separates the large ancient crater Fra Mauro in the north half of the scene from the smaller craters Bonpland (*B*) and Parry (*P*). *The graben extending through the center of the picture cuts the floors of both Fra Mauro and Bonpland, as well as the intervening rim. Just north of the rim the graben has been completely filled and buried by dark materials. In appearance the low hills (see arrows) directly west of the graben resemble the volcanic complexes described in figures 92 and 93. The complexes are centered in the dark area and may well have been the source of the dark material. Starting with the formation of the three craters, the following events occurred: The craters were filled with mare lavas, the grabens were formed, and the dark volcanic materials were erupted. As everywhere else on the Moon, small bodies sporadically struck the surface throughout the entire interval of time.*–G.W.C.

FIGURE 95.—Resembling a dome whose top has collapsed, this unusual structure (see large arrow) may be just that. It is located in the southeastern part of Mare Crisium, a large mare-filled circular basin near the Moon's east limb. The mare ridges shown are part of an extensive ridge system that encircles the floor of the basin. The broad circular depression occupying the center of the dome may be analogous to the calderas that occupy the center of many volcanic edifices on Earth. It may have formed as the pressure exerted on a mass of molten rock being forced upward toward the surface was released, allowing the overlying rocks to settle. A similar appearing but smaller dome with a summit depression occurs on the northwest flank (lower right side as this picture is oriented) of the larger one. Another large circular structure (see small arrow) may have formed in the same manner, but the evidence is less compelling. It can just as easily be interpreted as being the remnant of a lava-flooded impact crater that was later blanketed by the dark material covering much of the area shown here.—G.W.C.

AS17-2224 (P)

0   5 km

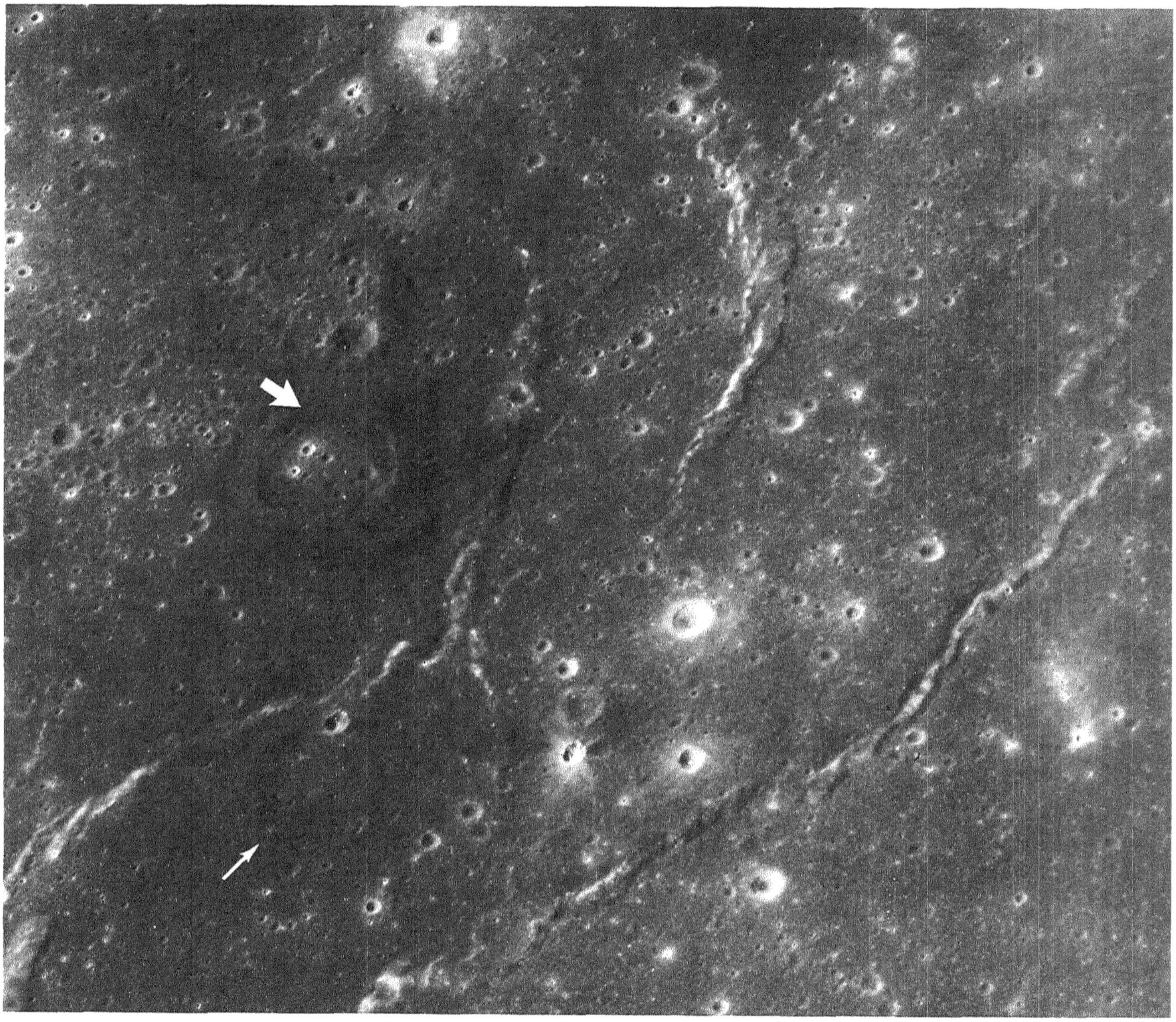

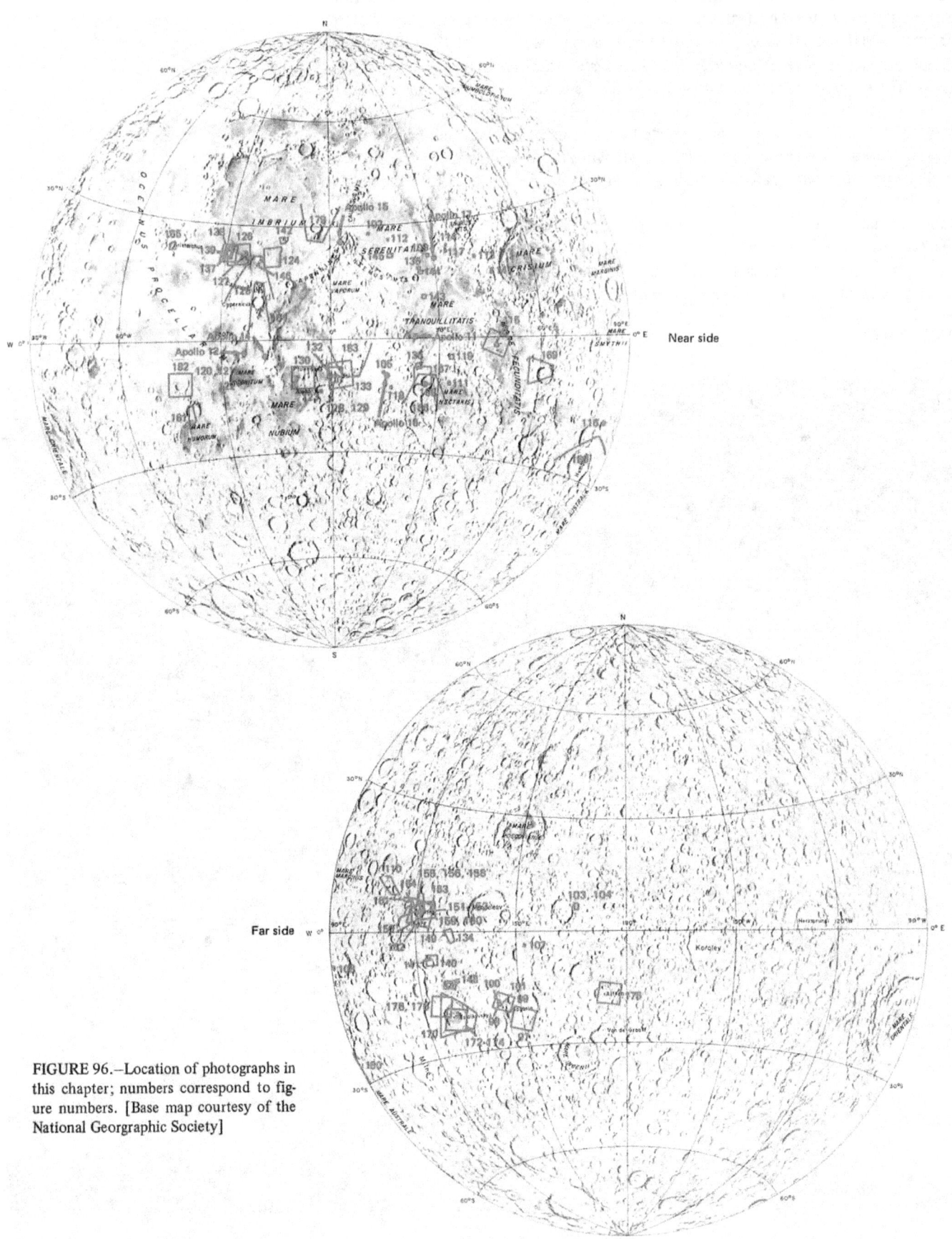

Near side

Far side

FIGURE 96.–Location of photographs in this chapter; numbers correspond to figure numbers. [Base map courtesy of the National Georgraphic Society]

5
Craters

The most common process that affects the lunar surface is impact cratering. Craters range in size from those that are 1000 km across and can be seen with the naked eye from Earth to those that are so tiny that the most powerful microscopes are required to study them. These tiny craters are abundant on the smallest grains in samples brought back from the Moon. In general, the size of the largest craters that pepper a particular surface is a measure of the age of that surface. Younger areas contain small craters; very old areas are covered with very large craters.

In addition to giving a measure of the age of the lunar surface, craters provide information about other processes that affect the Moon. Study of the shape of the craters and the distribution of the material ejected from the craters gives information on the nature of the projectile, its energy, and direction of impact. In turn, this information tells what kind of object caused the crater—whether it was a low density comet-like object or a high density asteroid.

Study of the material ejected from craters provides data on the temperature and pressures caused by the impact. Changes in the chemistry of the materials and in the mineral form—called impact metamorphism—also give clues to the nature of the projectiles. Of course, if pieces of the impacting object can be found, a direct determination can be made.

To better understand lunar impact structures, studies have been made of natural impact craters on the Earth as well as of manmade craters. Detailed studies of the craters made by explosives, ballistic missile impact, or by firing projectiles from high velocity guns at rock targets have resulted in some understanding of the process by which craters are made. As the high-velocity projectiles enter the target, a compressional shock wave spreads away from the entrance point, followed by a rarefaction wave that throws most of the ejecta out of the crater. Finally, the floor of the crater rebounds, forming the central peak. Three kinds of deposits can be found around a crater. One kind is the lines of ejecta thrown out along ballistic paths that are lines of secondary craters. On the Moon, these form the bright rays that extend for 10 to 30 crater diameters. A second type occurs when a continuous blanket of ejecta extends outward one to two crater diameters; such a covering formed by movement of the ejected material along the surface is the so-called "base surge." This flowing material may form ridges and dunes that vary according to the velocity of the flow and the pre-existing shape of the surface. Molten ejecta can flow down the interior walls of an old crater in its path and form a puddle on the crater floor or flow down the outside sloping rim of the crater. The third type of deposit is the lines of ejecta radiating from the central peak. These materials are the last to be ejected from the crater.

Craters fall into three groups. First, there are primary craters, which are generally randomly placed. Occasionally cosmic debris has struck the Earth in long lines, forming features such as the Campo de Cielo line of meteorite craters in South America; such an alinement of primary craters could occur on the Moon. Second, there are radiating and looped patterns of smaller craters surrounding the larger primary impact crater. These secondary craters in turn may have fans of ejecta thrown out from them in the direction away from the primary crater. Third, there are craters of internal (volcanic) origin. These have a different form, are often alined along fractures, and have ejecta blankets of different style and form from those of impact craters. Impacted ejecta has many blocks and forms very hummocky deposits; volcanic ejecta is fine grained and usually smooth.—H.M.

FIGURE 97.—Craters, craters, craters! The far side, even more than the near, presents a tortured record of the bombardment suffered by the Moon throughout its history. This scene exemplifies the relentless attack of impacting objects from space and from the lunar surface that has characterized most of lunar history.

The craters in this far-side area come in various shapes, sizes, and degrees of degradation attesting to a variety of formative processes, energies of formation, and ages. Each individual circular crater was probably produced by the impact of a body from interplanetary space—the larger the crater, the higher the energy; that is, the larger the body, or the greater its velocity upon impact. The first and largest such impact erased all earlier features and produced the crater that fills most of the scene, Gagarin, 265 km in diameter (rim crest outlined). A series of smaller craters followed, starting with crater A (46 km), itself heavily cratered, and ending with the sharp funnel-shaped craters and crater C (14 km). The young age of crater C is demonstrated by the sharpness of its rim crest and its halo of extremely fine, fresh ejecta and secondary craters. During the rain of objects from space, clots of lunar material ejected from impact craters outside this area landed here to form irregular secondary craters. Examples (D) are the elongate partly filled crater near the upper right corner and the elongate but deeper craters in the upper left. Conceivably, however, some irregular craters were formed by volcanism, a process that at one time was widely believed to be the cause of most irregular craters on the Moon. At some time late in the history of the region, an even more distant impact hurled a loose cluster of debris to form the group of sharp, circular (high-energy) craters in the left center of the picture.

In this section of the book other primary and secondary craters will be illustrated as will some possible volcanic craters and some craters whose properties are too obscured to reveal their origin—like crater B and its twin lying inside the older crater A in this picture.—D.E.W.

AS15-0293 (M)

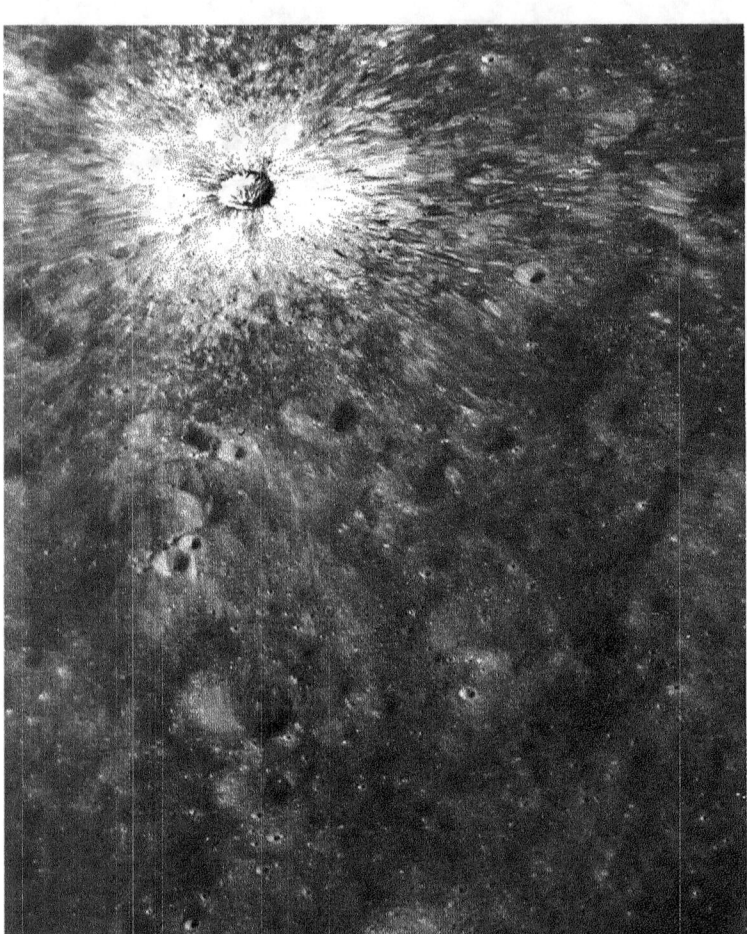

N Approximate

FIGURE 98.—This slightly oblique Hasselblad photograph, taken from Apollo 17, is an excellent view of a small, very young impact crater. Slightly less than 3 km in diameter, it is located on the west flank of the large crater Gagarin on the lunar far side. The youthfulness of this small crater is illustrated by its sharply defined rim crest; by its bright continuous blanket of ejecta extending outward for 1 to 2 crater diameters; and, most particularly, by its exceptionally well-developed radial pattern of bright rays. The rays consist of narrow, diffuse streaks and shorter bright spots. They are formed when ejected material from the impact site is redeposited as projectiles in the form of discrete blocks and clusters of disaggregated debris. The brightening is probably caused by the shock effects on the rock that was excavated from the impact site and deposited onto the older, darker surface materials.—F.E.-B.

AS17-150-23102 (H)

FIGURE 99.—This is the same small crater as appears in figure 98, oriented in the same way, but shown in a vertical view taken with the panoramic camera on Apollo 15. The western rim of Gagarin is marked by a dashed line. For easier comparison with figure 98, north is at the bottom. The Sun angle is very nearly the same in both pictures, but the oblique view recorded by the Hasselblad camera is certainly more spectacular. However, in many respects this vertical view is better suited for detailed study. The symmetries of the crater proper, its ejecta blanket, and its pattern of filamentous rays are even more apparent. Some of the rays extend to the edge of the frame, a maximum distance of slightly more than 50 km, or about 15 times the diameter of the crater. Figures 100 and 101 are enlargements of other photographs of the small areas indicated by rectangles.—G.W.C.

AS15-0102 (M)

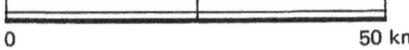
0   50 km

AS15-8936 (P)

0   2 km

FIGURE 100.—This is an enlargement of a small part of a panoramic camera frame showing the same crater in much more detail. The high resolution inherent in the camera system shows that the bright radially grooved ejecta contains many large blocks near the rim and that dark fine-textured material overlies the light coarse ejecta. Notice how the dark material occupies depressions on the rim, slumps down the walls, and forms a smooth-surfaced pool at the bottom of the crater. Also visible are ridges of bedrock in the upper walls and lobes of light-colored debris extending down onto the floor. Most aspects of the crater indicate that it was formed by impact. The dark material probably was penetrated at depth and ejected late in the explosive stage to accumulate in the vicinity of the rim. A volcanic origin for the dark material is possible but less likely.—B.K.L.

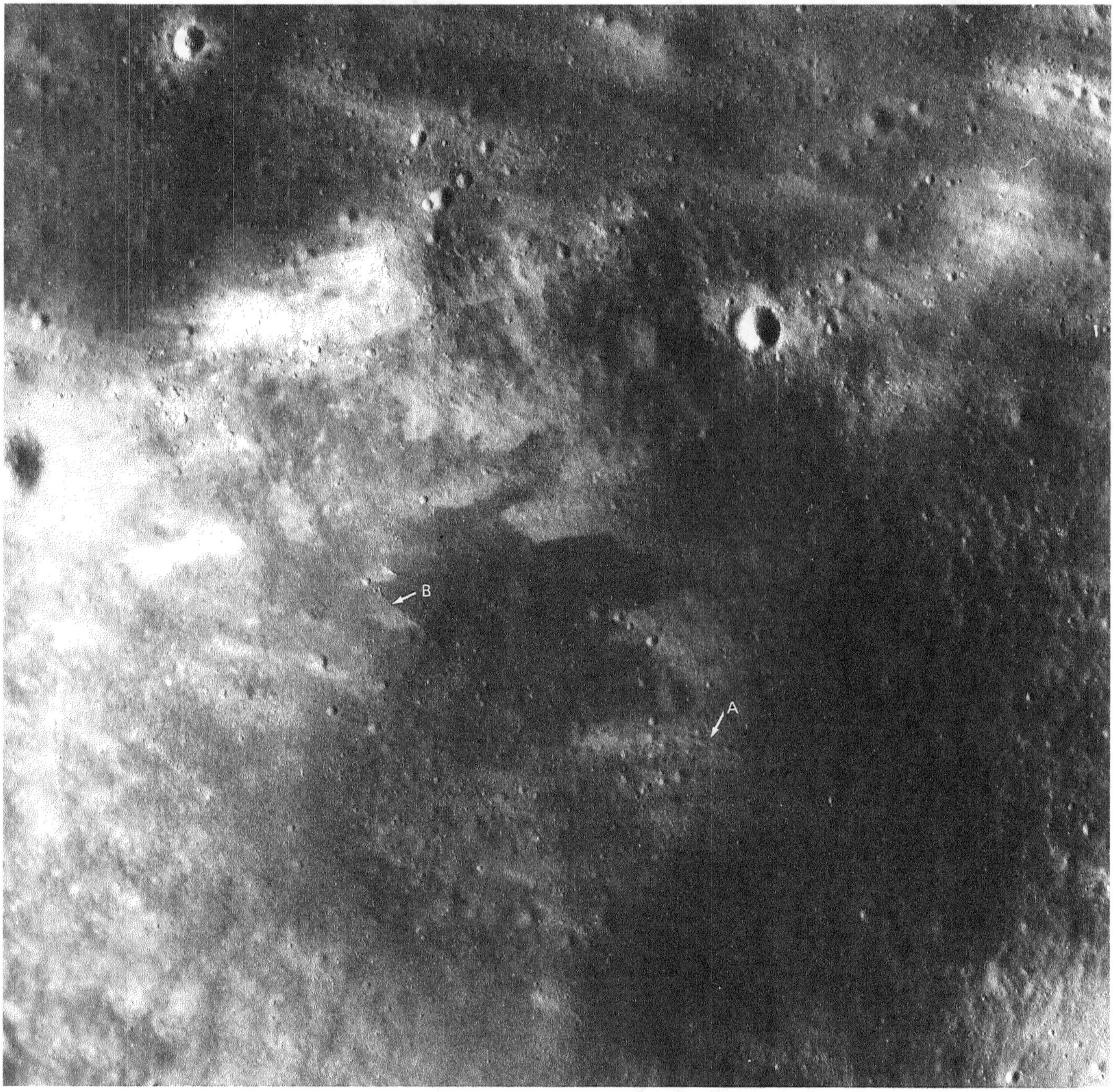

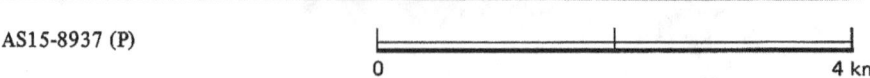

AS15-8937 (P)

0   4 km

N

FIGURE 101.—Ray material from the fresh crater described in figures 98 to 100 covers an old, degraded 12-km crater located a few tens of kilometers farther east (fig. 99). The light linear streaks and pointed markings are the rays from the fresh crater. In this picture the light material overlies dark regolith. Fine linear depressions (A) in the light patches are chains of secondary impact craters. They are associated with most of the light patches, and in some places (as at B), lie at their origin, thus showing that much of the ray material may be composed of local material disturbed by secondary impacts. In some places no such association is obvious, and the rays may consist of material ejected from the primary crater.—B.K.L.

FIGURE 102.—The characteristics of an extremely youthful small crater are illustrated in this Apollo 15 panoramic camera photograph of the crater Linné as viewed from the south. The 4-km-diameter crater is located in the western part of Mare Serenitatis. The rim of the crater slopes steeply up from the mare surface, and the rim crest is sharp and even. Boulders are abundant on and near the rim. Concentric dunelike features occupy the inner part of the ejecta blanket and, with increasing distance from the crater, give way to irregular arcuate clusters of satellitic craters. A more subtle radial pattern in the ejecta can be seen where the Sun's rays shine across it, to the north and south of the crater. The bright ejecta blanket and ray pattern displayed here are typical of very youthful craters of all sizes.—M.C.M.

AS15-9348 (P)   Approximate 0   10 km

FIGURE 103—This view looks southward near Mandel'shtam on the lunar far side. Most young lunar craters wider than about 40 km have flows on their rims that resemble lava flows or mud flows on Earth. The unnamed crater near the top is about 14 km wide and was recognized by H. J. Moore (1972) as being the smallest crater known to have such flows. Flows in the middle of the picture surged downhill off the high rim of the crater making lobes and tongues and leaving behind drained channels with levees. In the area to the right of the crater, enlarged in figure 104, are some thin lobate flows that apparently rode over small hills, as if these flows were propelled outward from the crater with sufficient velocity to climb the hills. Ejecta deposits farther than about 1 km from the rim are radially lineated and are smoother than the ground immediately surrounding the crater. The crisp, blocky zone around the crater is typical of many fresh craters.—K.A.H.

AS16-4136 (P)

0   10 km

AS16-4136 (P)

0   2 km

FIGURE 104.—This enlarged view of part of figure 103 shows some of the smooth flows that originate near the crest of the crater rim at the left side of photograph. Arrows point to the lower ends of two flows. The origin of the flow material is controversial. It was probably molten material generated by shock-wave compression of lunar rocks and ejected at relatively low velocities during the late stages of the formation of the impact crater; or it may have resulted from the flow of rock debris mixed with a fluidizing agent such as gas or water; or it may have been volcanically generated lava.—H.J.M.

113

FIGURE 105.—The two bright-rayed craters in this picture have been examined more closely from the lunar surface than those illustrated earlier in this chapter. The larger one, South Ray (*S*), is located 6.2 km southwest of the Apollo 16 landing point, indicated by an arrow at the top edge. At sampling station 8, 3.4 km from the edge of South Ray, astronauts John Young and Charles Duke sampled some rocks from South Ray that had been deposited as ejecta along one of its numerous rays. Studies of these rocks indicate that South Ray formed about 2.5 million years ago bringing rocks 3.8 billion years old to the surface. The impact that created South Ray crater occurred when Australopithecus lived in Africa—well before contemporary man evolved. The much smaller rayed crater is Baby Ray (*B*). Its rim crest is much sharper than South Ray's and its rays overlie those of South Ray. Consequently it is younger, possibly having formed within modern man's time span.—H.J.M.

AS16-4558 (P)

FIGURE 106.—Another view of the crater South Ray. This detailed topographic map was prepared under the supervision of S. S. C. Wu of the U.S. Geological Survey by photogrammetric techniques. Prepared from a stereo pair of panoramic camera photographs, it is one of the largest scale maps ever done with Apollo orbital images. The area shown here is a small part of a map that was used for compiling geologic data from the Apollo 16 mission. Dashed lines have been added to the original map to show the crest of South Ray's rim. From this type of topographic representation, the width, depth, slope of the walls, and other parameters can be measured. For example, the maximum depth of the crater, measured from the high point on the rim (7771 m) to the low point in the floor (7628 m), is 143 m.—G.W.C.

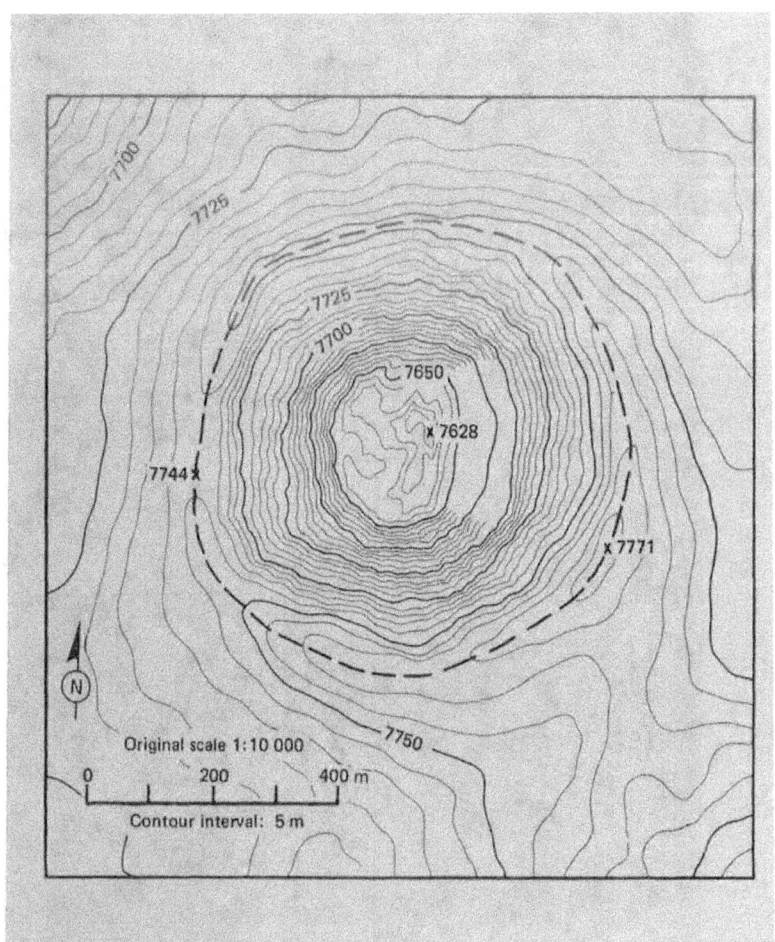

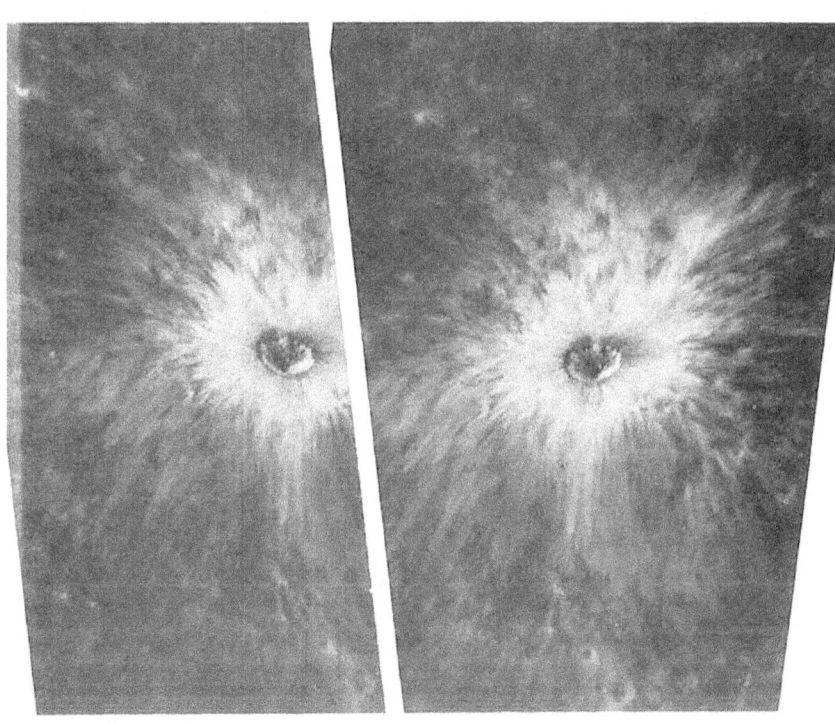

AS17-1769 (P)   AS17-1764 (P)

FIGURE 107.—The bright rays around this fresh 1.5-km crater on the Moon's far side show the effects of ejecta thrown out on rugged highlands. The Sun is nearly overhead; therefore, the relief is not conspicuous. When viewed stereoscopically, the crater is seen to be perched on the top of a high bluff that slopes steeply toward the foreground. On top of the bluff, the ray pattern resembles splashes, but on the steep slope the rays cascade in lobes like streamers in a waterfall. The lobes may be avalanches triggered by ejecta striking down on the slope. Dark materials on the rim and wall of the crater and pooled in the crater floor may be glassy rocks melted by the impact. (*Note:* This pair demonstrates the problem caused when areas near the ends of unrectified panoramic camera frames are used for stereoscopic viewing. Geometric distortion makes it impossible to view all the overlapped area stereoscopically.)—K.A.H.

115

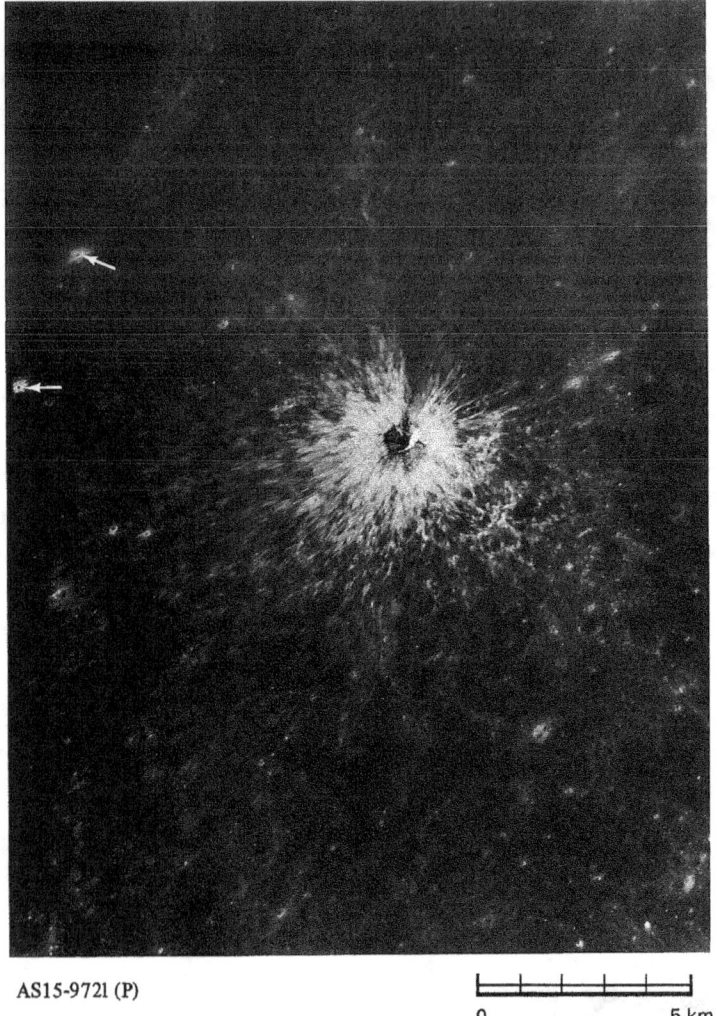

AS15-9721 (P)

0   5 km

FIGURE 108.—Another fine example of a small crater with extensive bright rays. It presumably was formed by meteoroid impact. Both light and dark rocks are visible in the walls, indicating that more than one layer of rock was penetrated. Rays formed by secondary craters, blocks, and pulverized ejecta extend more than 13 crater diameters beyond the crater proper. The marked irregularity of the pattern southeast of the crater probably was caused by topographic irregularities that deflected the radially outward movement of the ejecta. Arrows indicate smaller bright-rayed craters.—M.W.

FIGURE 109.—The configuration of this small crater has been affected by the topography of the impact site. It is located near the eastern edge of Mare Serenitatis, north of the Apollo 17 landing area. The body that excavated the crater impacted at the foot of a westward-facing fault scarp that bounds one side of a graben, or fault trough. The fault scarp is coincident with the diffuse narrow band of shadow extending between the arrows; the area west of the scarp, the floor of the graben, is noticeably lower than the area to the east. The flow of ejected debris toward the east was apparently blocked by the scarp, resulting in an abbreviated bright halo and ray pattern in that direction.—G.G.S.

AS15-9301 (P)

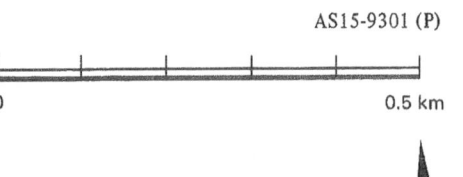

0   0.5 km

116

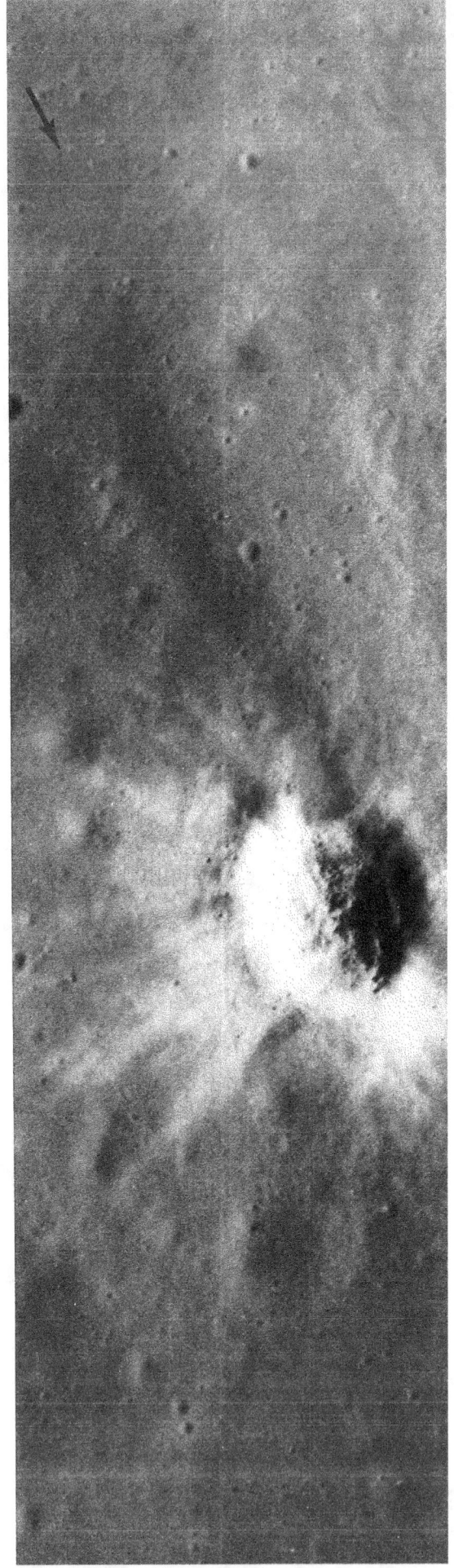

FIGURE 110.—An oblique view looking northwest at part of the wall of the crater Lobachevsky on the lunar far side. It shows a small crater on Lobachevsky's wall with unusual streaks of dark material that appear to have originated from the lower rim of the structure and to have moved down toward the floor of Lobachevsky. This feature was first noticed by T. K. Mattingly, the Apollo 16 CMP, who described the darker streaks as probable lava flows (Mattingly, El-Baz, and Laidley, 1972). However, the streaks can also be explained by the downslope movement of dark fragmental debris excavated from Lobachevsky's wall by the small crater. Closer to the lower border of the photograph is a bright area extending across Lobachevsky's rim. This area and other sinuous light-colored markings in the upper half of the photograph are on the periphery of an enormous field of light-colored swirls in this part of the far side (El-Baz, 1972a). The origin of the swirls is not well understood.—F.E.-B.

AS16-121-19407 (H)

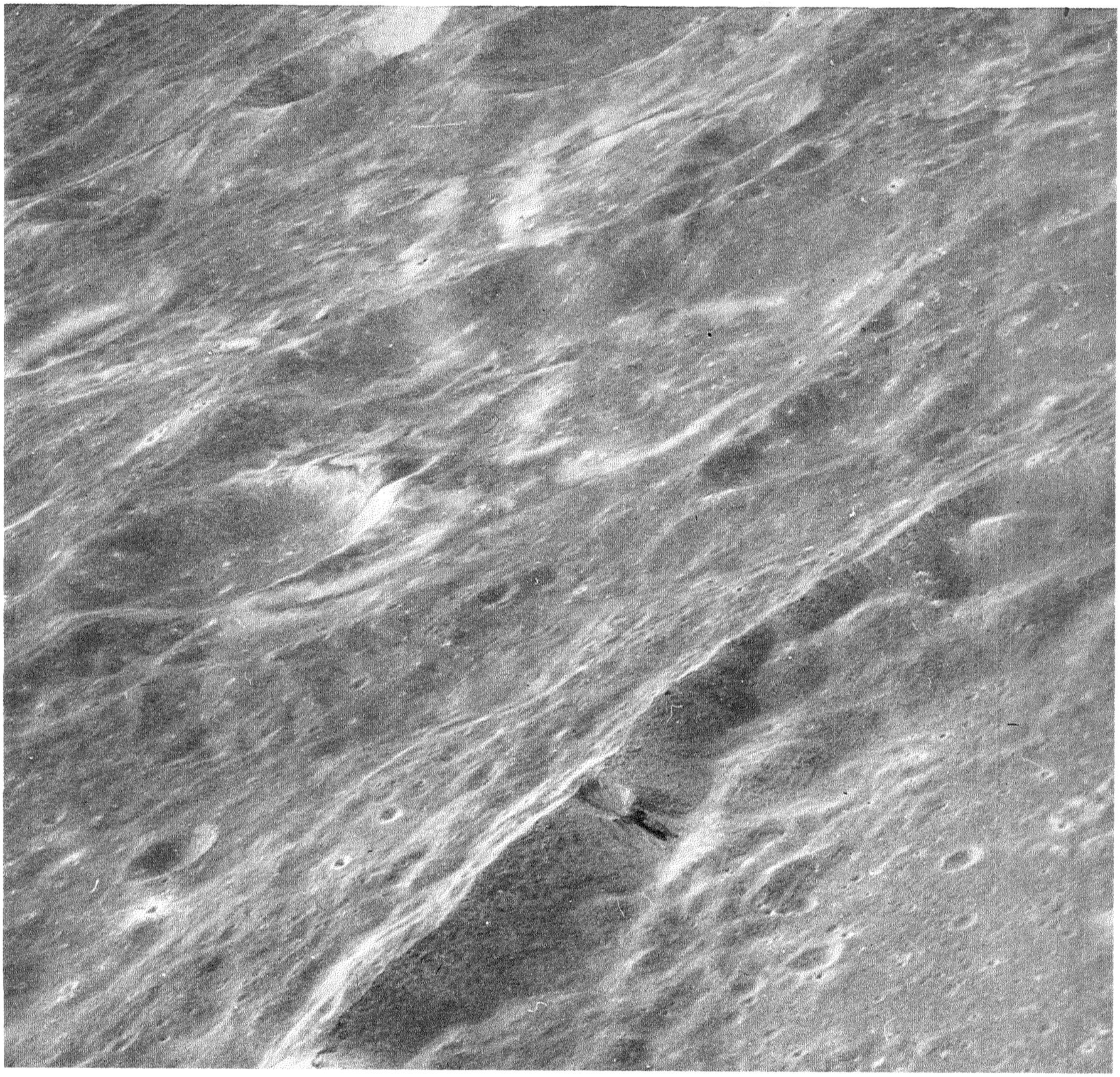

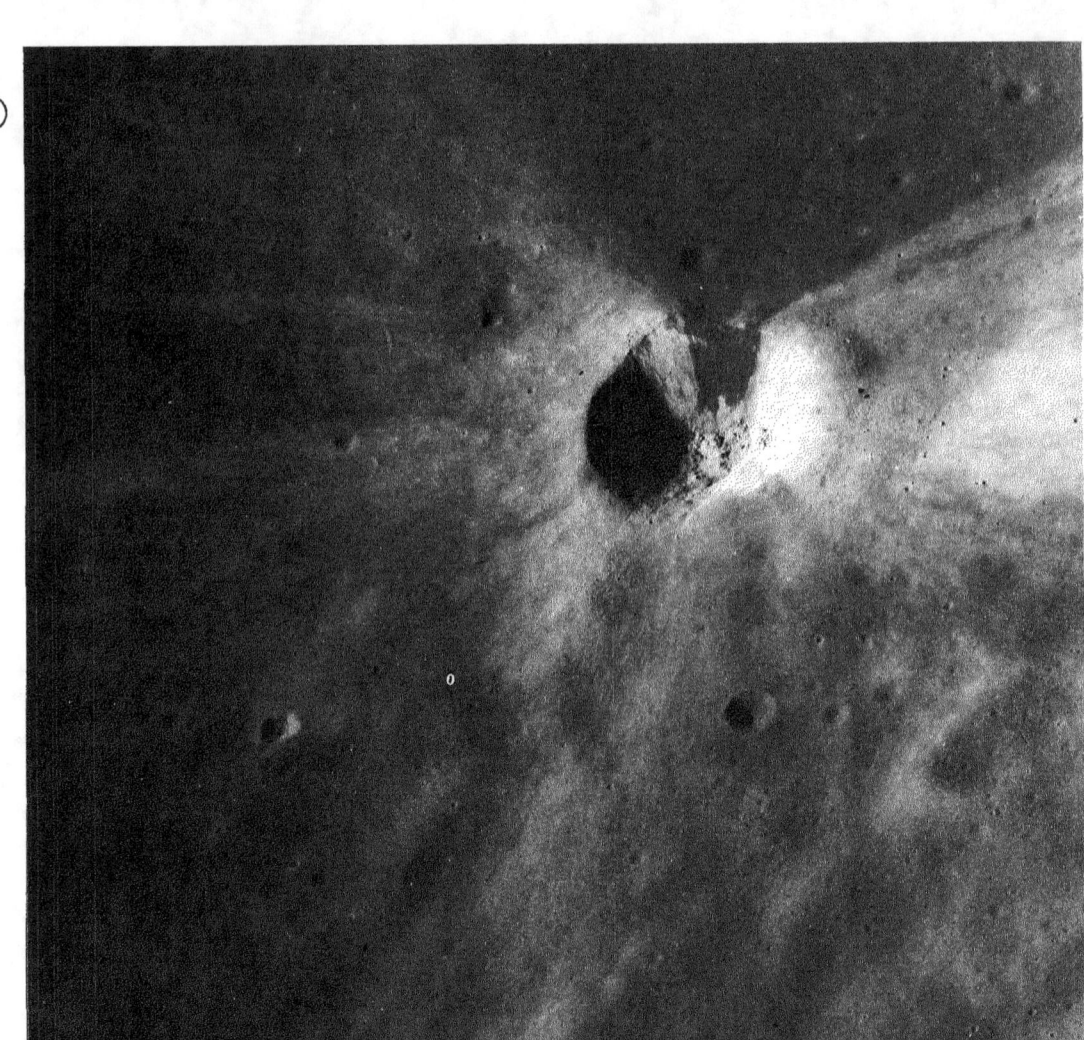

AS16-4511 (P)

0   4 km

FIGURE 111.—For easier viewing, this picture is oriented with north at the bottom of the page. It shows the striking bilateral symmetry of the rays of a small (2-km-diameter) crater in the floor of the large crater Daguerre in Mare Nectaris. Continuous areas and narrow filaments of light-gray ejecta extend from the crater across the dark mare surface through 270°, but are entirely absent in the southern 90° sector. Within the crater, dark material occurs on the southern crater wall while the remaining walls are bright. (The reader may wonder about the material whose reflectivity cannot be observed because it lies in shadow on the east wall of this crater. Until the area is observed under high Sun conditions, we are forced to make the simplifying assumption that it is bright because most of the materials visible elsewhere in the walls are bright.) This crater probably resulted from the impact of a projectile traveling from south to north along an oblique trajectory. Its pattern of ejecta distribution is similar to that of small craters produced by the impact of missiles along oblique trajectories at the White Sands Missile Range, N. Mex. Some observers postulate that the dark material is a talus deposit of mare material that has fallen into the crater.—H.J.M.

Another geological explanation is that the unusual pattern may be due to an intrinsic characteristic of the local terrain, probably an abrupt lateral change in the composition of the bedrock within the area that was excavated.—F.E.-B.

FIGURE 112.—This small young crater has an incomplete
ray pattern. It is located in central Mare Serenitatis. The
small raised features on the flanks of the crater are blocks
of ejected bedrock, probably mare basalt. The largest is
about 30 m across. In this picture, where sunlight is from
the east (right), raised features appear as bright spots cast-
ing a shadow to the west. Note the relative scarcity of
blocks on the side lacking rays. The absence of rayed ejecta
and of discrete blocks to the west suggest that the impacting body was traveling from west to east along an oblique
trajectory. The roughly concentric ledges in the rayed
crater result from the interaction of shock and stress waves
with the layered substrata.—H.J.M.

AS15-9337 (P)

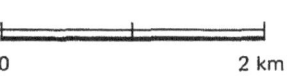

0   2 km

FIGURE 113.—This elliptical crater is 1 km long with an
unusual, winglike pattern of rays. This ejecta pattern is
similar to those around some small experimental impact
craters produced by missiles traveling along low-angle trajectories at White Sands Missile Range, N. Mex. From the
shape of the crater and the distribution of the rays, it is
difficult to tell whether the meteoroid was traveling from
north to south or south to north. The higher albedo (brightness) of the north wall and the concentration of high-
albedo ejecta on the northwest and northeast flanks suggest
that it traveled from south to north.—H.J.M.

AS15-9254 (P)

0   2 km

AS17-2744 (P)

0   2 km

FIGURE 114.—This is an oblique view of another crater that probably was formed by a meteoroid following a relatively low-angle trajectory. This crater, 4 km in diameter, is located in the highlands east of Mare Serenitatis. Compared to the crater just described (see fig. 113), this one is less elliptical and its bilobate ray pattern is much less pronounced. The differences may be attributed to a higher trajectory angle of the impacting body that formed this crater as it struck the surface. H. J. Moore (1976), in his study of craters formed by impacting missiles at White Sands Missile Range, recognized a characteristic asymmetric profile along the axis of trajectory for craters formed in this manner. The wall beneath the missile trajectory is typically less steep than the opposite or down-trajectory wall, and its rim crest is lower and more rounded. These observations, when applied to the lunar crater in this photograph, indicate that the impacting body was traveling toward the east when it struck the Moon.—H.M.

AS15-2405 (M)

0   50 km

FIGURE 115.—Messier (*1*) and Messier A (*2*) are a pair of unusual craters in northwestern Mare Fecunditatis. Messier is elliptical and has bright walls and light rays of ejecta extending at right angles to its long axis (approximately 16.5 km). Messier A is a doublet crater having two very long rays or filaments of ejecta extending westward from it. The east part of the doublet has steep, bright walls, whereas the west part is dark and appears mantled. Differences between the two parts are more clearly shown in this oblique view of Messier A composed as a stereogram. Both Messier and Messier A resemble some small experimental impact craters produced in sand by projectiles following shallow trajectories (4° or less from the horizontal) at velocities of approximately 1.7 km/s. In separate experiments using single projectiles, both elliptical craters with lateral ejecta lobes and doublet craters have been produced. Thus, it can be inferred that these lunar craters were produced by high-velocity projectiles following shallow trajectories. By further analogy with the experiments, the projectiles that formed Messier and Messier A apparently traveled from east to west.—H.J.M.

AS16-4471 (P)   AS16-4469 (P)

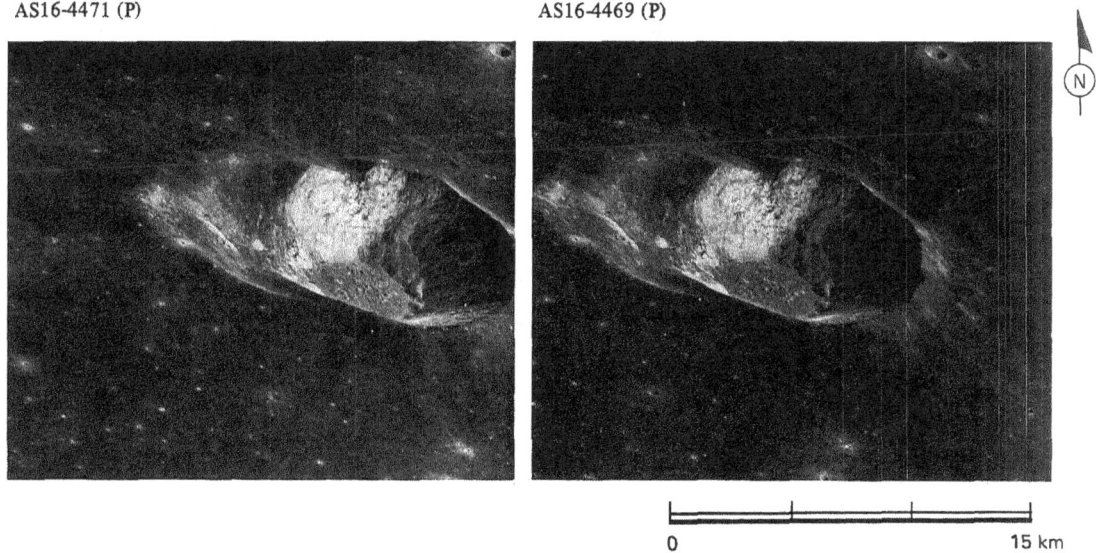

0   15 km

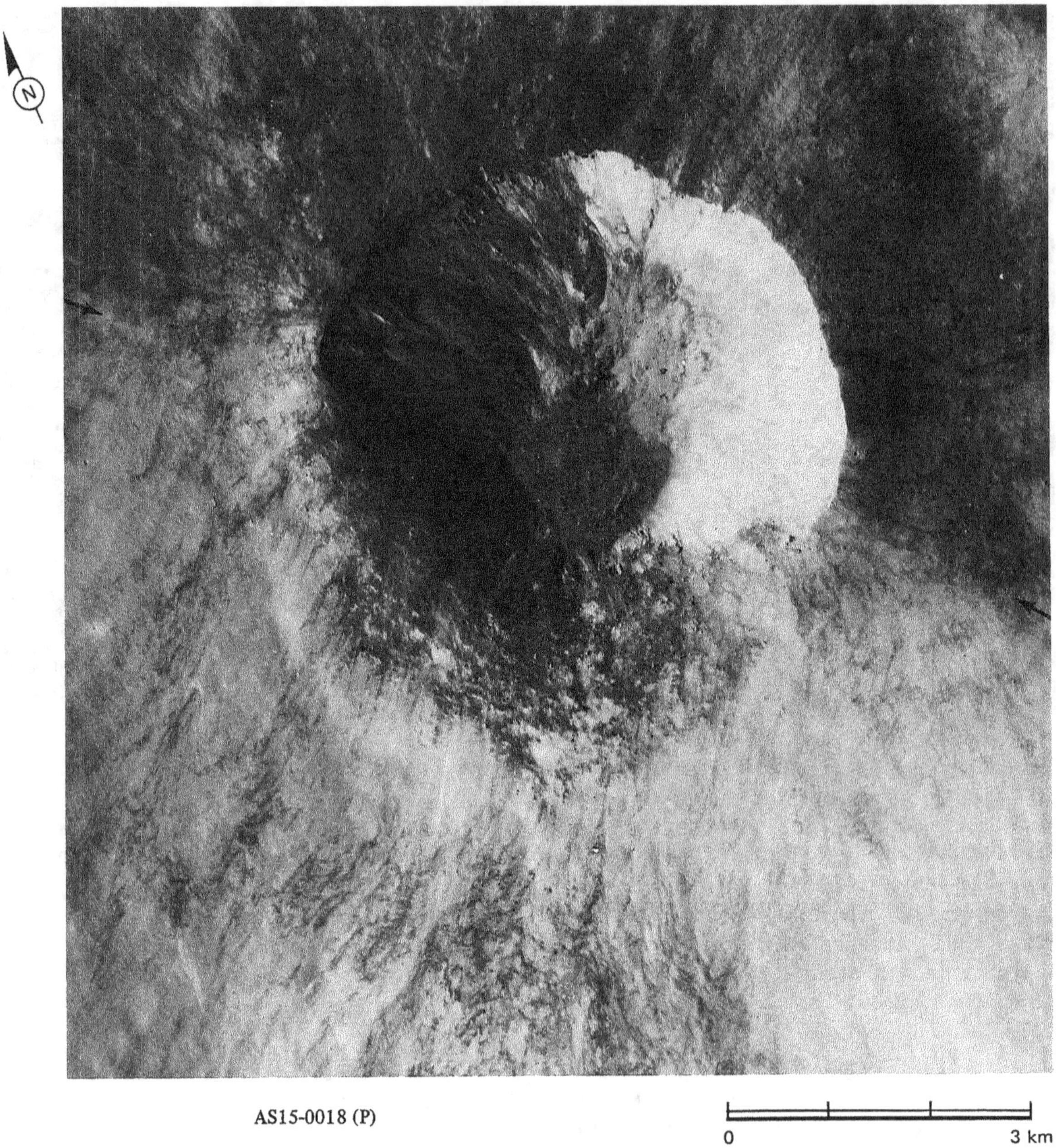

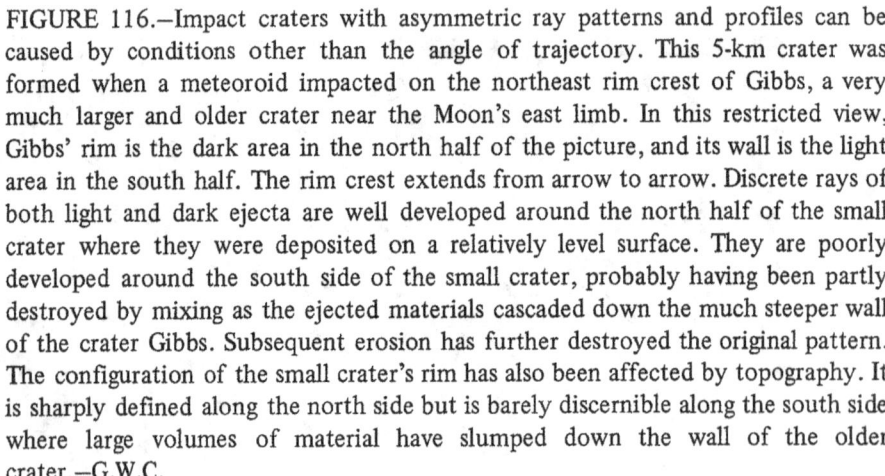

AS15-0018 (P)

0   3 km

FIGURE 116.—Impact craters with asymmetric ray patterns and profiles can be caused by conditions other than the angle of trajectory. This 5-km crater was formed when a meteoroid impacted on the northeast rim crest of Gibbs, a very much larger and older crater near the Moon's east limb. In this restricted view, Gibbs' rim is the dark area in the north half of the picture, and its wall is the light area in the south half. The rim crest extends from arrow to arrow. Discrete rays of both light and dark ejecta are well developed around the north half of the small crater where they were deposited on a relatively level surface. They are poorly developed around the south side of the small crater, probably having been partly destroyed by mixing as the ejected materials cascaded down the much steeper wall of the crater Gibbs. Subsequent erosion has further destroyed the original pattern. The configuration of the small crater's rim has also been affected by topography. It is sharply defined along the north side but is barely discernible along the south side where large volumes of material have slumped down the wall of the older crater.—G.W.C.

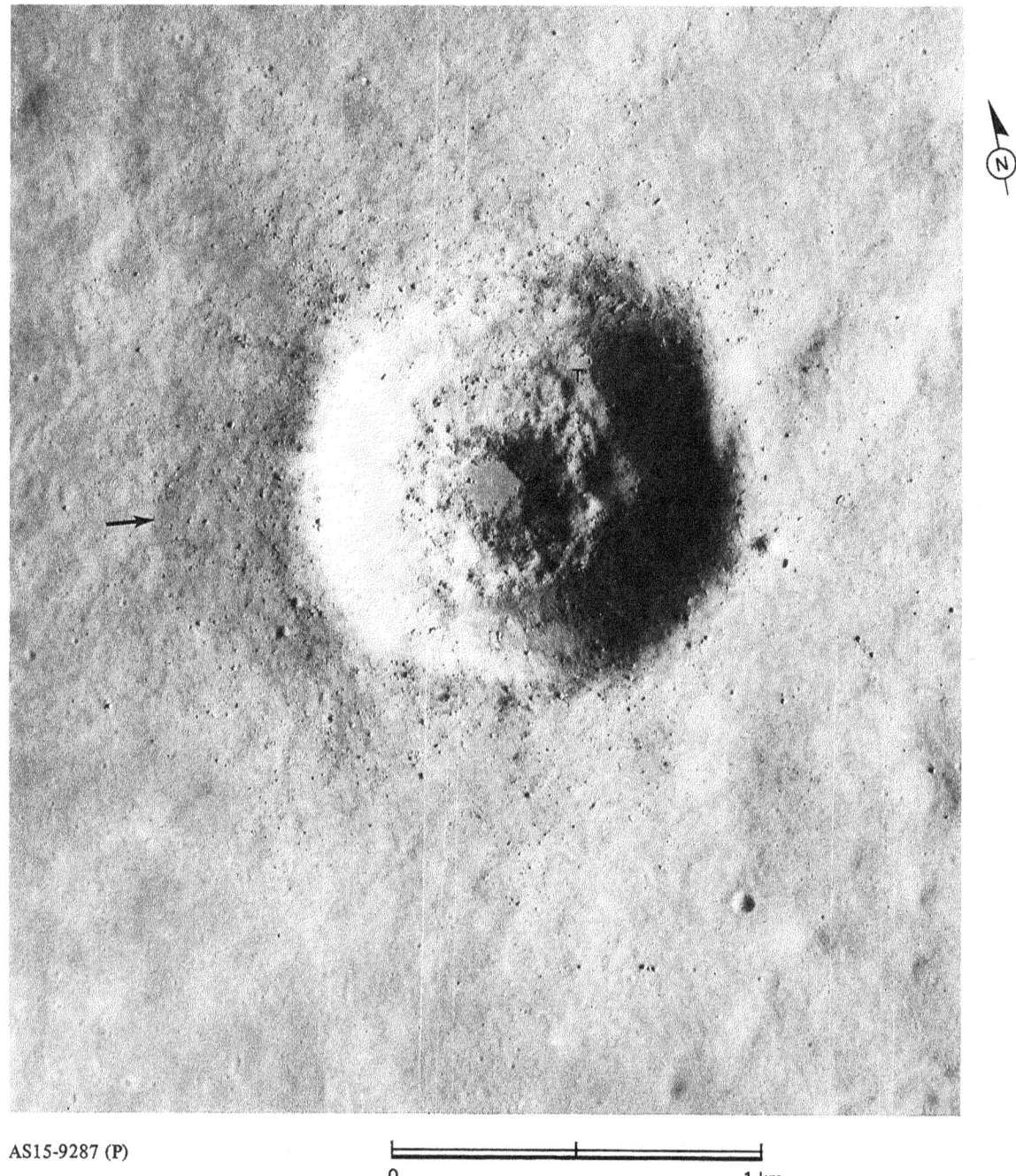

AS15-9287 (P)

0   1 km

FIGURE 117.—Remarkable detail is shown in this enlargement of a small part of a panoramic camera frame. In most respects, the crater itself is typical of a great many craters its size—about 1.2 km. Because it does not have rays, it is believed to be older than most other craters discussed previously in this chapter. Its rounded rim crest and slightly raised rim (extending outward to the arrow on the west side) also point to its greater age. On the other hand, it is young enough that some of the original dunelike texture of the ejecta blanket is preserved (especially to the west), a great many large blocks of ejecta are still visible, and the original depth of the crater has not been greatly lessened by infalling debris. The largest blocks, which are about 30 m in size, occur near the rim. The terrace (T) extending partly around the wall about 100 m below the surface probably marks the top of a resistant rock layer. However, if there were other signs of bedrock stratification within this crater, they have been obscured by the movement of debris down the walls. The very smooth floor is the only unusual feature of this crater. It may consist of a solidified pool of rock melted by heat generated from the impact.—H.J.M.

AS16-4559 (P)

0   2 km

FIGURE 118.—Kant P is the larger of these two craters in the central highlands on the Moon's near side. About 5.5 km in diameter, its overall shape is not in the least unusual. However, the younger, small pear-shaped crater on Kant P's north wall is an excellent example of the controlling effect that topographic relief plays on the shape of an impact crater. Because the small crater was formed on a steeply sloping surface, its ejecta was deposited chiefly downslope and formed a broad rim. The original rim and wall on the upslope side have been obliterated by slumping. The slumping has left a landslide scar and has caused talus and scree to be deposited in the lower part of the crater.—H.J.M.

124

FIGURE 119.—This oblique view of the crater Isidorus D was taken with the panoramic camera on Apollo 16. Isidorus D is about 15 km in diameter and is located in the highlands between Mare Tranquillitatis and Mare Nectaris. Evidence of avalanching (Howard, 1973) and of other types of downslope movement of material are clearly visible on the inner walls of the crater. The streaks resembling shooting stars on the left wall appear to be avalanche scars. The avalanches probably were spearheaded by large blocks followed by fine-grained material. On the near wall (arrow) a larger landslide terminates in a straight line against the relatively flat crater floor. In the shadowed part of the crater wall many short irregular benches or narrow terraces mark the tops of masses of slumped material. The brightness of the avalanche scars is an indication of their freshness; in general, freshly exposed lunar materials are brighter than undisturbed materials nearby.—F.E.-B.

AS16-4502 (P)

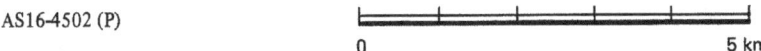

0   5 km

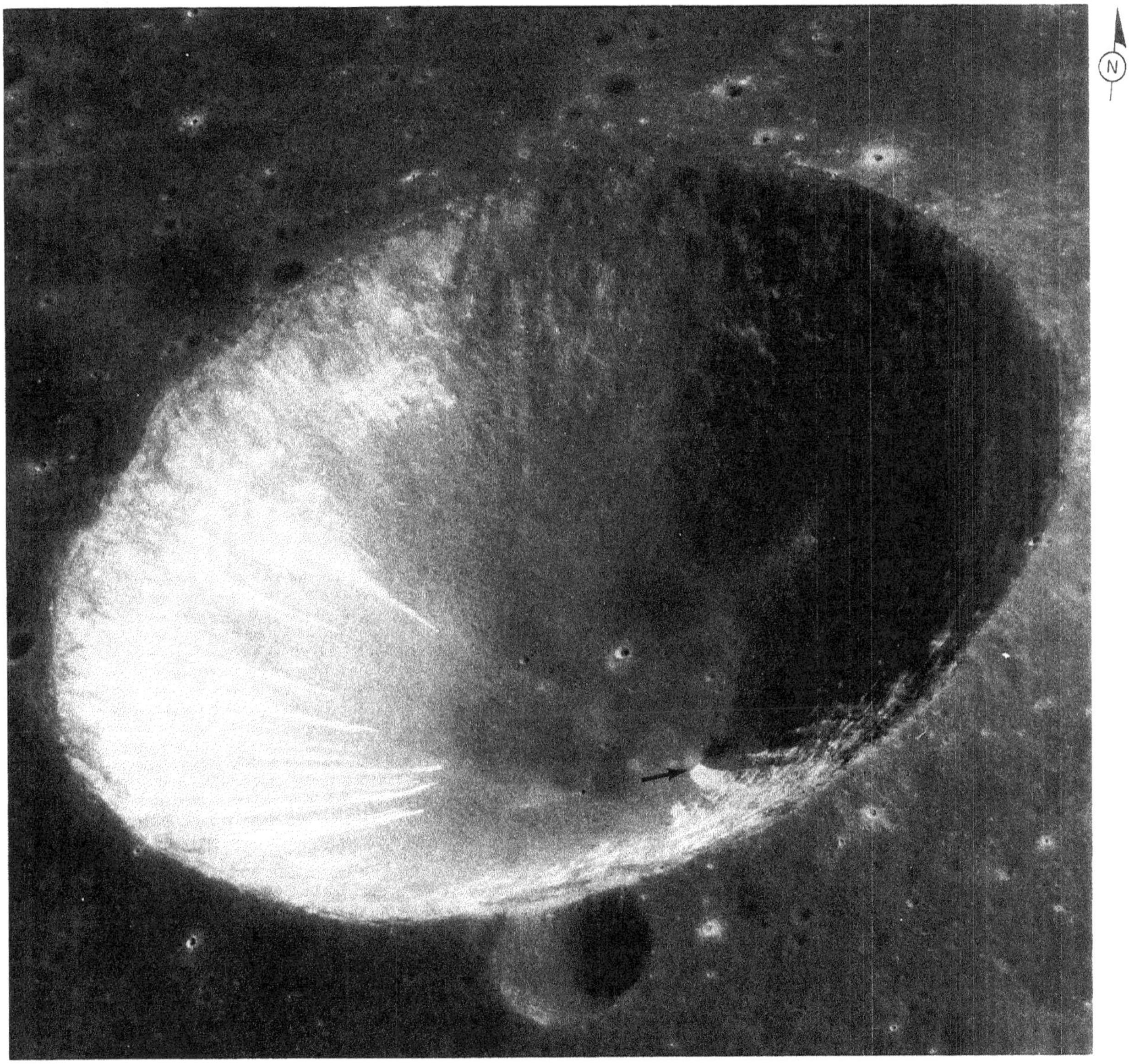

125

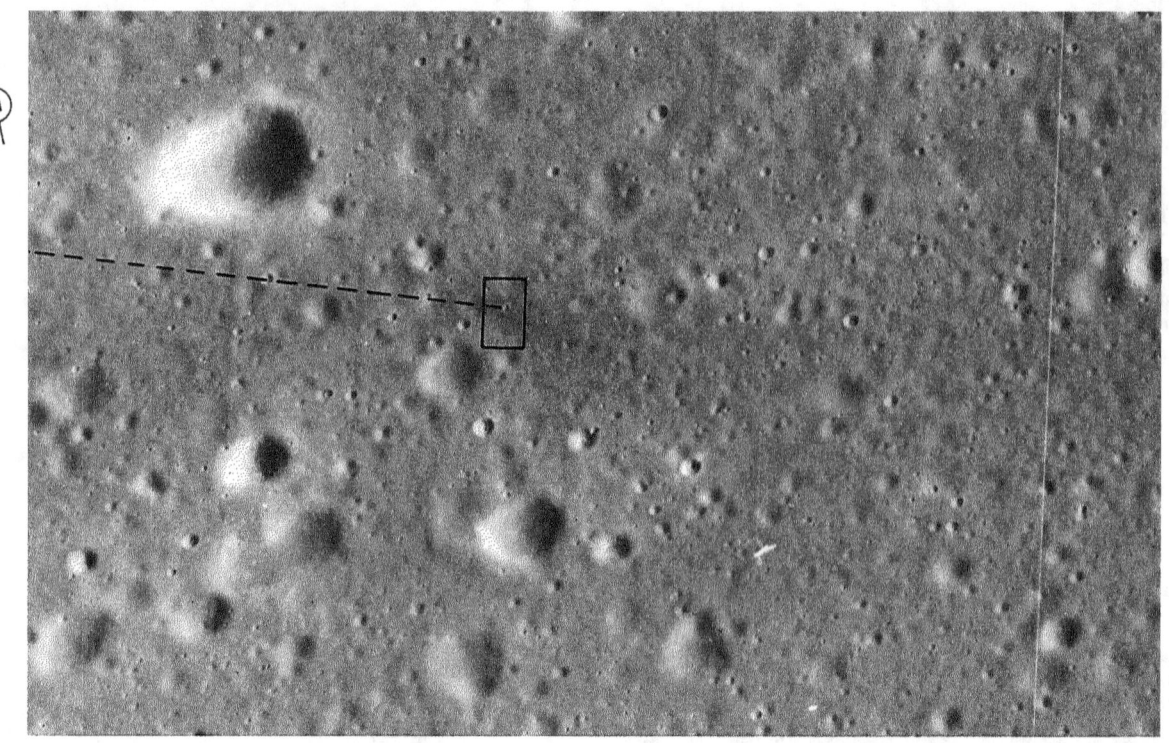

AS16-5444 (P)

0   2 km

FIGURE 120.—Not quite all the craters on the Moon are natural features. About 28, and possibly as many as 36, according to Ewen A. Whitaker of the Lunar and Planetary Laboratory, University of Arizona, have been caused by man. Only about five of these artificial craters have been visually identified. One of them (centered in the small rectangle) is shown in this enlarged portion of a panoramic camera frame of 24 km² in western Mare Cognitum. The more than 2400 craters visible in this small area are typical of the density of craters in many areas of the Moon. First located by Whitaker (1972), the manmade crater is easily identified by its pattern of dark rays interspersed with a few light rays. It was created in February 1971 by the SIVB stage of the Apollo 14 launch vehicle, which was directed to crash onto the lunar surface to generate seismic waves for detection by seismometers set up at the Apollo landing sites. The surface trace of the SIVB's trajectory is shown by a dashed line.—H.J.M.

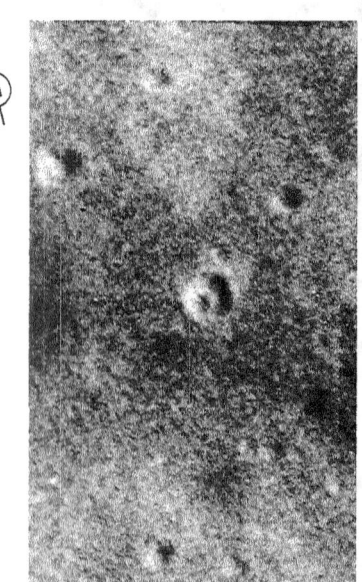

FIGURE 121.—The same crater of figure 120 is shown in more detail in this photograph, an enlargement of the area indicated by the rectangle in figure 120. About 40 m in diameter, the crater differs from most natural impact craters in this size range by having a prominent central peak.—H.J.M.

AS16-5444 (P)

0   200 m

126

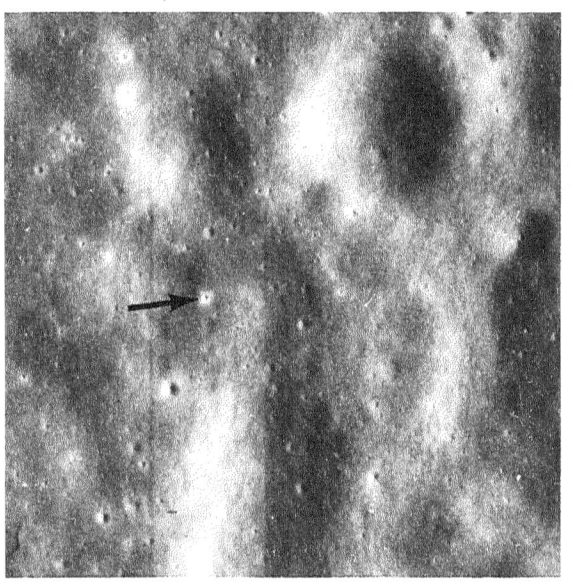

AS16-5430 (P)

FIGURE 122.—An enlargement (approximately 15X) of a panoramic camera photograph showing a crater (arrow) formed in Mare Cognitum by the impact of the Ranger 7 spacecraft. Following an oblique trajectory, Ranger 7 struck the lunar surface on July 31, 1964, while traveling at a velocity of 2.65 km/s. The resulting crater is about 14 m across. Its bright ejecta blanket contrasts with the generally dark blanket around the SIVB crater and is more typical of natural lunar impact craters.—H.J.M.

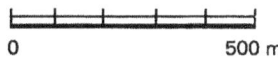

0   500 m

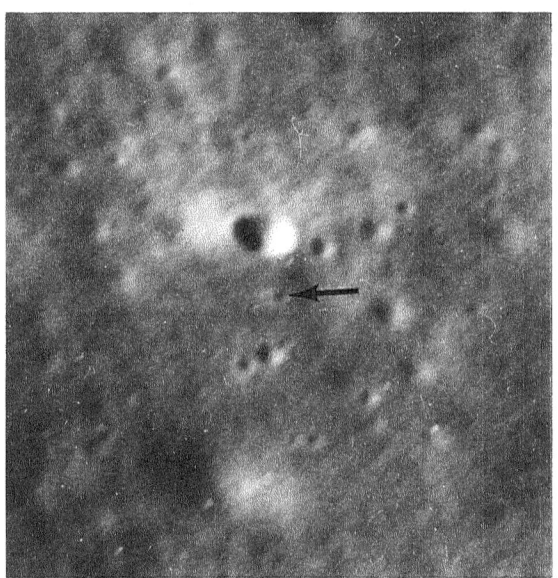

AS16-4658 (P)

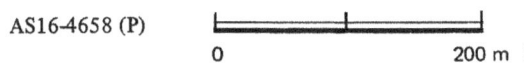

0   200 m

FIGURE 123.—An enlargement of approximately 40X showing another manmade crater. This crater was excavated in the floor of the large near-side crater Alphonsus on March 24, 1965, by the Ranger 9 spacecraft. Prior to its crash, Ranger 9 had transmitted the highest resolution imagery obtained to that date from any spacecraft. Both Ranger vehicles impacted on mare surfaces while traveling at the same velocity along trajectories with the same inclination. It is, therefore, not surprising that the Ranger 7 and 9 craters have, within half a meter, the same diameter.—H.J.M.

127

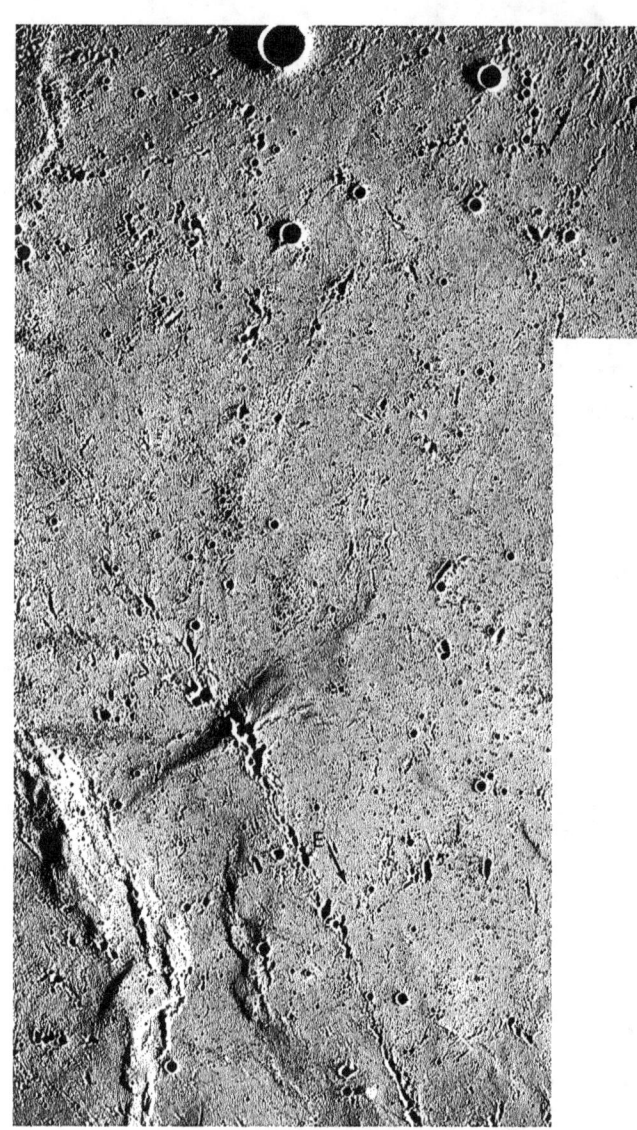

FIGURE 124.—This photo taken at low Sun angle emphasizes how common secondary impact craters are on the Moon. The primary craters whose ejecta formed most of the secondaries in this part of southeastern Mare Imbrium can be identified by observing the orientation of the secondary crater chains and of the "herringbone" ridges that splay outward from individual secondary craters. The chains are radial to the primary or parent crater and the apexes of the "herringbone" ridges point toward it. The most conspicuous chain and the chain at the left center of the picture are secondary to Eratosthenes, 250 km to the southeast in the direction of the arrows labeled *E*. Lying athwart these chains is a large younger cluster (arrow *C*) secondary to Copernicus, 400 km to the southwest. The chains in the extreme upper right corner (arrow *T*) are secondary to Timocharis whose rim is only 35 km northeast of the pictured area. Most of the other, smaller chains, clusters, and "bird's-foot" gouges can also be traced to Eratosthenes, Copernicus, or Timocharis, but some probably were created by fragments from more distant sources. The largest crater in the scene is Timocharis A, 8 km in diameter.—D.E.W.

AS17-2120 (M)

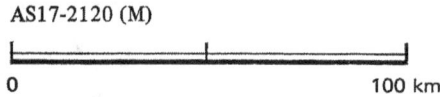

0   100 km

FIGURE 125.—This oblique view across southern Mare Imbrium looks toward Copernicus, the large crater near the horizon. The distance from the lower edge of the picture to the center of Copernicus is 400 km. The mountains at the edge of Mare Imbrium are the Montes Carpatus, and the large crater near the center of the picture is Pytheas, almost 19 km in diameter. Copernicus is one of the youngest of the Moon's large craters. It is visible from Earth, even without the aid of a telescope because of its bright ejecta blanket and its extensive bright rays. The many chains and clusters of small irregular craters and the many bright streaks or rays extending across Mare Imbrium are caused by the secondary impact of debris ejected from Copernicus. The viewing angle accentuates the radial pattern of the secondary impact features. The Sun angle is sufficiently low to show their relief, but high enough to show the contrast between the bright streaks and the normal dark mare surface. As in figure 124, herringbone ridges point toward the primary crater, and the flaring sides of the secondary craters point away from it. The arrow midway between Copernicus and the left edge of the photograph points to a less common pattern of secondary craters; these are concentric to Copernicus.—M.C.M.

AS17-2444 (M)

FIGURE 126.—Here is another area in southern Mare Imbrium that shows a profusion of bright rays and chains of secondary craters from Copernicus, which is located about 380 km southeast of the center of the picture. The large crater near the left (west) side is Euler. (See figs. 64 and 137 to 139.) Here lighting conditions are ideal for showing how the material ejected from Copernicus secondaries is splashed downrange (away from Copernicus) to form the herringbone ridges and the bright patches or streaks. Individual craters and their associated ridges sometimes combine to form a pattern that resembles the imprint of a bird's foot. The area within the small rectangle is shown in much more detail in figure 127.—H.M.

AS17-2291 (M)

0   100 km

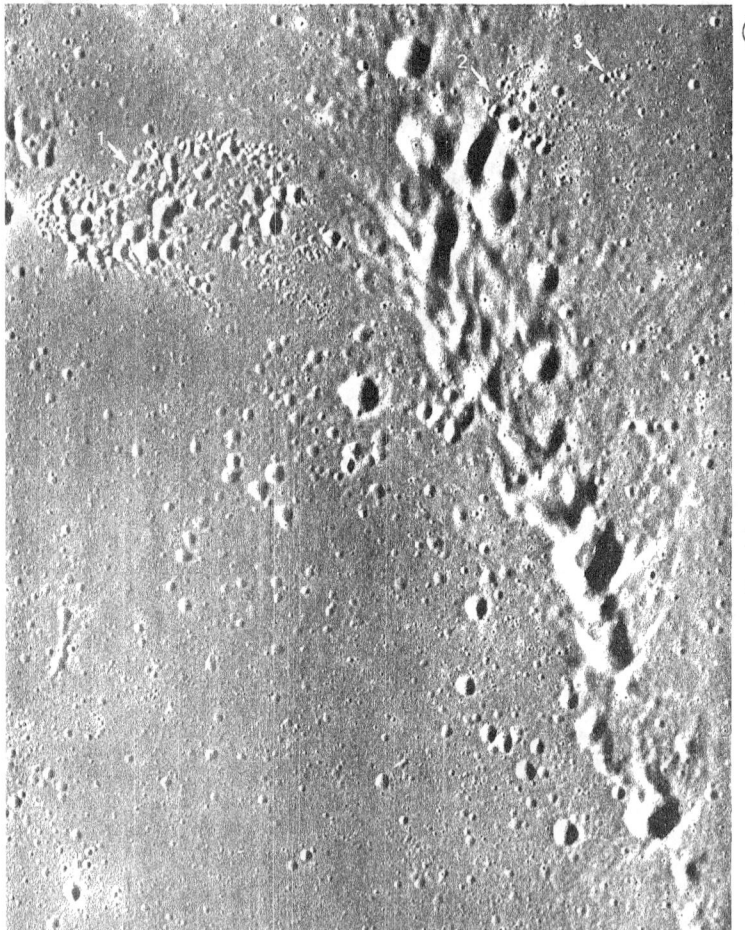

FIGURE 127.—A secondary Copernicus crater chain is enlarged in this view. (See fig. 126.) Its elongate shape, the irregular form of the individual craters, and the splashed appearance of their ejecta are clearly discernible. Also present in this view are two or three groups of craters (arrows) that are also of secondary impact origin, but are different from the Copernicus chain. They occur in clusters, not chains. The craters within the clusters are smaller and more regular in shape and do not have the splashed appearance of the Copernicus secondaries. At arrow 2, some of them are superposed on—and hence are younger than—the Copernicus chain. A few have faint herringbone ridges that veer toward the west, indicating that the primary crater lies in that direction. Except for the familiar crater Aristarchus (fig. 165), which is 580 km west of this area, there are no other large craters that are also young enough to have been the cause of these secondaries. Aristarchus is, furthermore, younger than Copernicus. For these reasons Aristarchus crater is the most likely source of the material that landed here to form the clusters of craters.—G.W.C.

AS17-3093 (P)

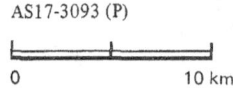

0   10 km

FIGURE 128.—These three clusters of secondary craters (see arrows) are on the east flank of the larger crater Ptolemaeus near the center of the Moon's near side. Each cluster has a ridged and hummocky appearance. The primary crater has not been identified in this case, but the configuration of the clusters tells us that it must be to the south of Ptolemaeus. Note that the south-facing side of each cluster is more sharply defined than the north-facing side. This is a consequence of the oblique trajectory of impacting fragments that causes the ejecta of the secondary craters to be propelled away (down range) from the primary crater. Observations of manmade impact craters have shown that the individual fragments within a cluster of secondary debris strike the surface nearly simultaneously. In the process, ejecta from one secondary collides and interferes with ejecta from adjacent craters, producing a ridged and hummocky surface.—H.J.M.

AS16-4653 (P)

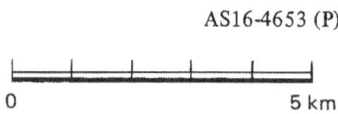

0   5 km

131

FIGURE 129.—The secondary crater cluster near the lower edge of figure 128 is enlarged to show more detail. The largest crater within the cluster is about 600 m wide. The small, sharp, circular craters are primary craters that post-date the cluster. Some poorly defined V-shaped ridges at the south edge of the cluster point southward toward the primary crater. The effects of the secondary cratering are much more extensive than was apparent in the smaller scale picture. The finely lineated terrain west and north of the cluster is caused by a great many parallel and subparallel ridges, gouges, grooves, and barely visible small irregular craters. They are apparently caused by the impact and deposition of material ejected from the secondary craters. Craters formed during this stage are tertiary craters but are rarely distinct enough to be recognized. Several possible examples are indicated by arrows.—H.J.M.

AS16-4653 (P)

0   3 km

FIGURE 130.—The Davy crater chain (arrow) is one of the most spectacular chains of craters on the Moon. It extends for about 50 km across the floor of the large, very old crater Davy Y (Y) and onto its eastern rim. The chain may be related in origin to the pair of irregular craters Davy G (G) and Davy GA (GA), 75 km from the furthest end of the chain. Two origins have been proposed. Some lunar geologists believe it is a chain of secondary impact craters, and others believe it is a line of volcanic craters. The simple geometry of the Davy chain, the symmetry and uniform spacing of its individual craters, and its alinement with Davy G strongly support, in this writer's opinion, a volcanic origin. Also arguing against a secondary impact origin is the fact that the Davy chain is a lone feature. There are no other similar chains with this trend in the area. As was shown earlier in this chapter (figs. 124 and 125), secondary crater chains tend to occur in large numbers within the belt of secondary craters surrounding a large primary crater.

On Earth some rocks from deep within the crust have been brought to the surface through volcanic orifices, thus providing a means of studying material that would otherwise be inaccessible. For this reason the Davy area was once seriously considered as a landing site. However, when the originally planned number of Apollo missions was reduced, the Davy area was one of those eliminated.—H.M.

AS16-1973 (M)

0   50 km

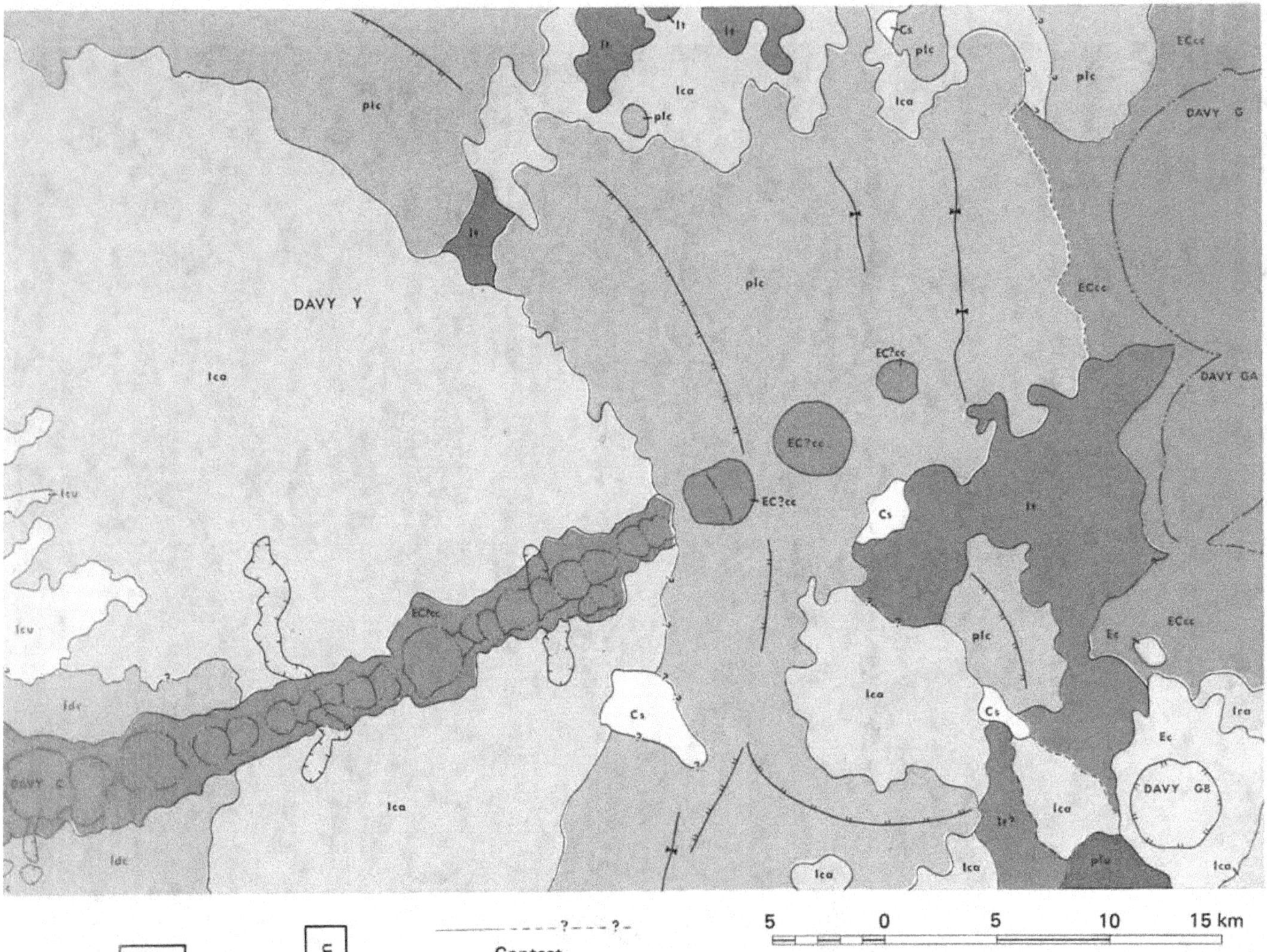

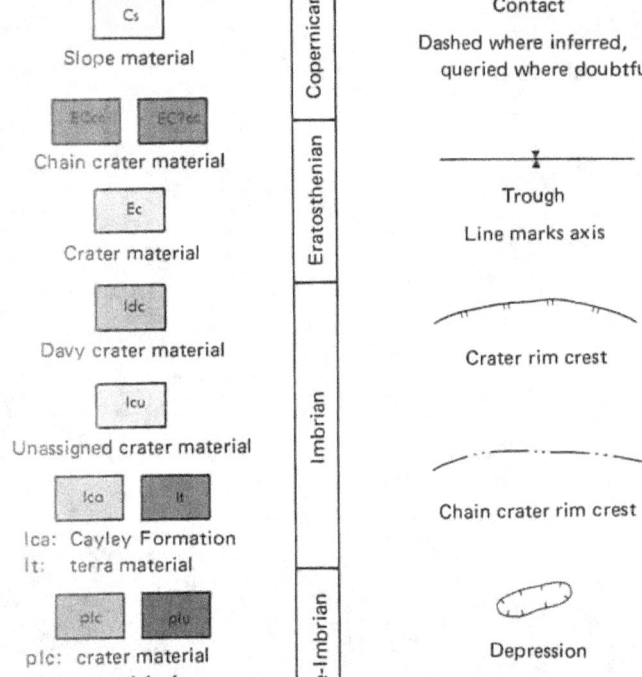

Map labels (Davy area):

DAVY Y · DAVY C · DAVY G · DAVY GA · DAVY G8

Units shown on map: plc, Ica, It, Cs, Icu, Idc, EC?cc, ECcc, Ec, Ira

Explanation (legend):

| Cs |
Slope material

| ECcc | EC?cc |
Chain crater material

| Ec |
Crater material

| Idc |
Davy crater material

| Icu |
Unassigned crater material

| Ica | It |
Ica: Cayley Formation
It: terra material

| plc | plu |
plc: crater material
plu: material of uncertain origin

Time-stratigraphic column (top to bottom): Copernican · Eratosthenian · Imbrian · Pre-Imbrian

------ ?---?-
Contact
Dashed where inferred,
queried where doubtful

———I———
Trough
Line marks axis

⌒⌒⌒
Crater rim crest

⌢⌢⌢
Chain crater rim crest

Depression

5   0   5   10   15 km

FIGURE 131.—This is a reduced version of part of a geologic map of the Davy area compiled by R. Hereford of the U.S. Geological Survey. Photogeologic mapping did not provide an unambiguous answer to the question of the origin of the Davy chain, but did provide much information on the succession of events that shaped this part of the Moon's surface. Following the map explanation, the succession is briefly as follows: (1) An early period of intense cratering during which the large crater Davy Y was excavated and its rim uplifted. (2) A presumably brief period during which the Cayley Formation and one other plains unit were deposited. (The origin of the Cayley as a semi-fluid cloud of ejecta from the Imbrium basin was discussed in figs. 46, 54, and 55.) (3) A long period when scattered medium-sized craters and their deposits were formed. (4) The formation of the Davy crater chain and the larger, presumably volcanic, craters Davy G and Davy GA. (5) Finally, mass movement of material on steep slopes to form relatively fresh slope deposits and sporadic impact by very small bodies that made craters too small to be shown at the scale of the map.—G.W.C.

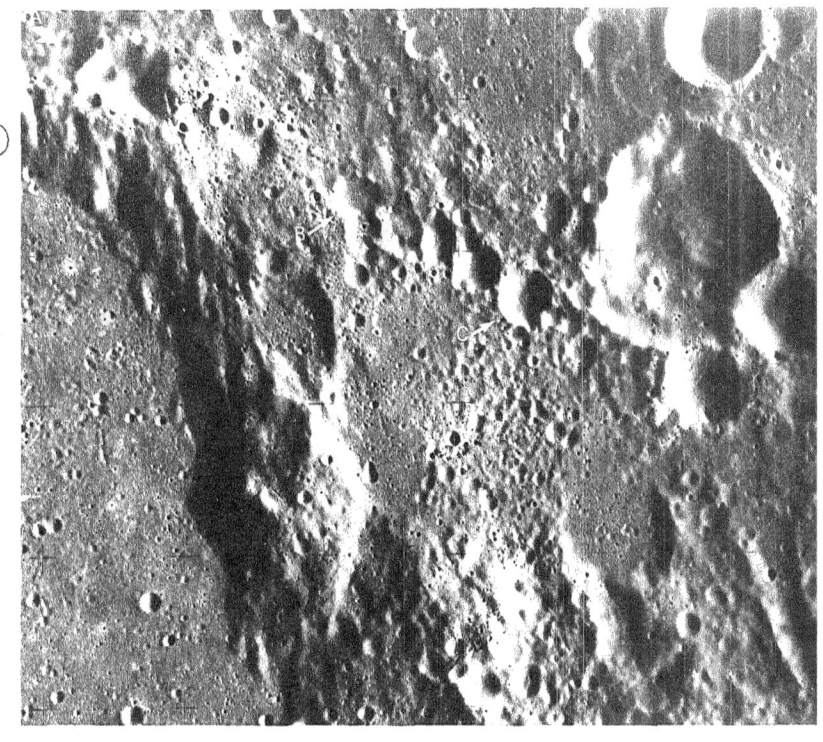

FIGURE 132.—The crater chain extending from the large arrow at *A* is part of a discontinuous lineament that grazes the northeast rim of Ptolemaeus (left), and that may extend as far as 720 km beyond the edge of the picture. The origin of the chain is perplexing. In some respects it resembles Imbrium sculpture (figs. 36, 47, 48, and 53), except that the individual craters in the chain are more circular and more distinct. An apparent difference in freshness of craters within the chain (as at *B* and *C*) suggests that the chain may have formed by volcanism localized along a tectonic fracture. On the other hand, the lack of visible faulting in this area and the strong resemblance of this chain to some secondary crater chains suggest the more likely alternative that it originated by secondary impact, perhaps by ejecta from Schrödinger, a relatively small double-ringed basin in the south polar region.—C.A.H.

AS16-1671 (M)

```
|   |   |   |   |   |
0                 25 km
```

FIGURE 133.—This photograph shows the crater chain of figure 132 in its regional context (large arrows). Trending east-southeast, the chain sharply transects the Imbrium sculpture, which trends south-southeast (*smaller* arrows near margins). The difference in trends argues strongly against a common origin and somewhat less strongly against a related origin. The laser altimeter on board the Apollo 16 CSM showed that the area directly east of Ptolemaeus was one of the highest areas on the near side of the Moon along the ground track. The other anomalously high area was in the vicinity of the Apollo 16 landing site near Descartes. Also in the Ptolemaeus area the Apollo 16 mass spectrometer recorded an exceptionally high ratio of aluminum to silicon. The relationship between high elevation and high aluminum content and the presence of a prominent crater chain are probably more than coincidental. One interpretation is that volcanism in this part of the highlands extended into a later period of time than in most other highland areas on the near side and that more differentiated materials were extruded. The crater chain may be related in origin to this late-stage eruption.—H.M.

AS16-1276 (M)

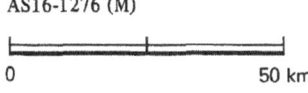

```
|                 |
0                 50 km
```

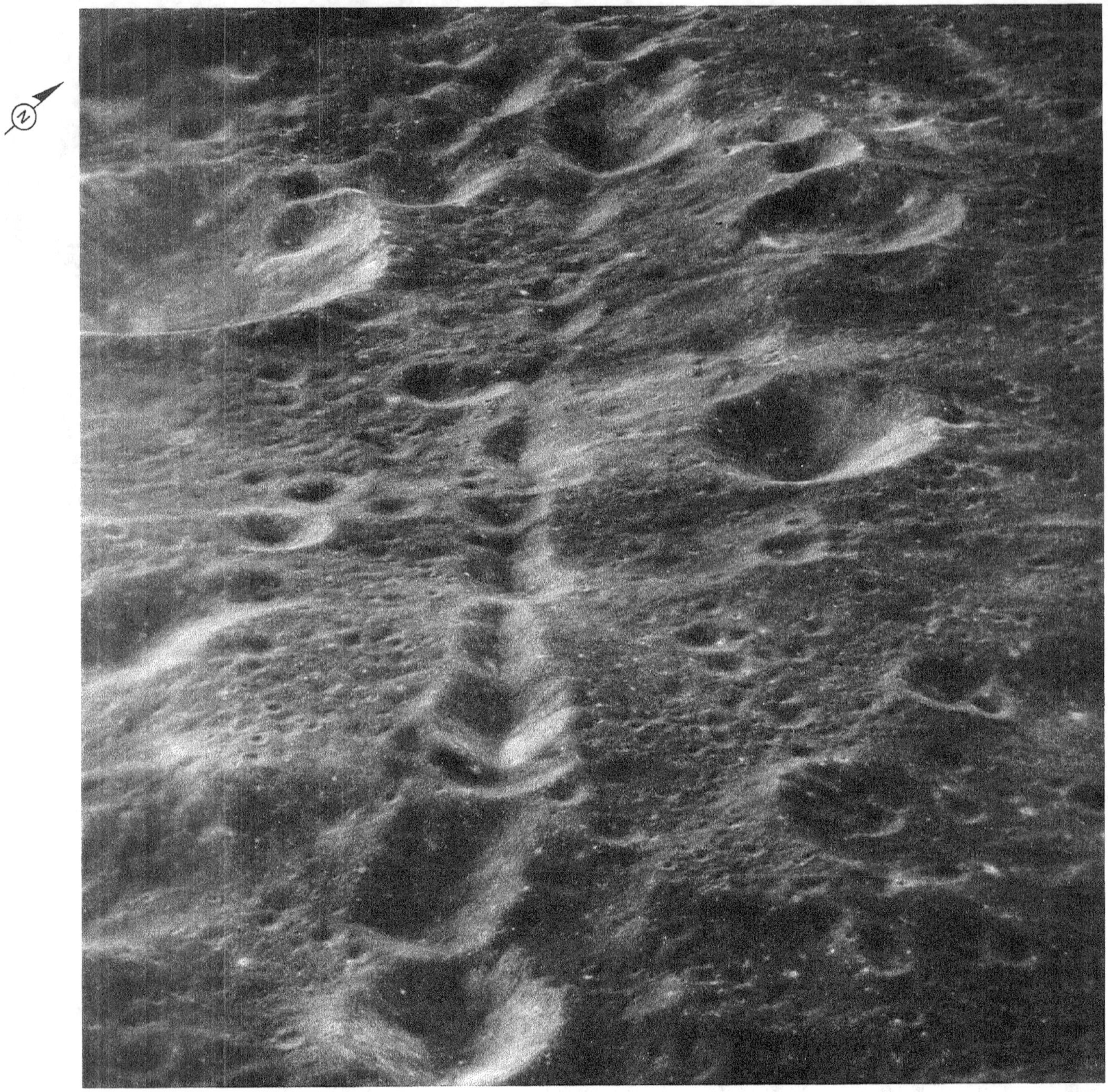

AS17-149-22838 (H)

FIGURE 134.—This oblique view taken with the Hasselblad camera shows a crater chain on the far side, about 500 km north of Tsiolkovsky. For an idea of the scale, the large crater near the upper left corner is about 26 km wide. The origin of this chain is controversial. To some geologists, the irregular shape of many of the craters suggests that the chain was formed by the impact of a stream of ejecta from a large primary crater. The presence of herringbone ridges would have strengthened this interpretation, but none are visible; perhaps the high Sun angle and the oblique viewing angle of this scene have obscured them. To others the simple geometry of the chain suggests a volcanic origin. However, there is an apparent lack of faulting to control the alinement of the craters and an apparent absence of a blanket of volcanic ejecta. The origin of this chain may not be decipherable until, and unless, additional photography becomes available.—G.W.C.

FIGURE 135.—This crater chain in southern Mare Serenitatis is clearly of internal origin because it is lined up parallel to several fault valleys or grabens. The craters in the chain do not appear to have any rims; consequently, they may have formed by collapse and not by the explosive ejection of volcanic material. The large crater in the right side of this scene, however, has a rim and so cannot be the result of collapse alone. The finely lineated texture across the left side of the photograph is caused by ejecta from the crater Dawes to the south.—K.A.H.

AS17-2321 (P)

0   5 km

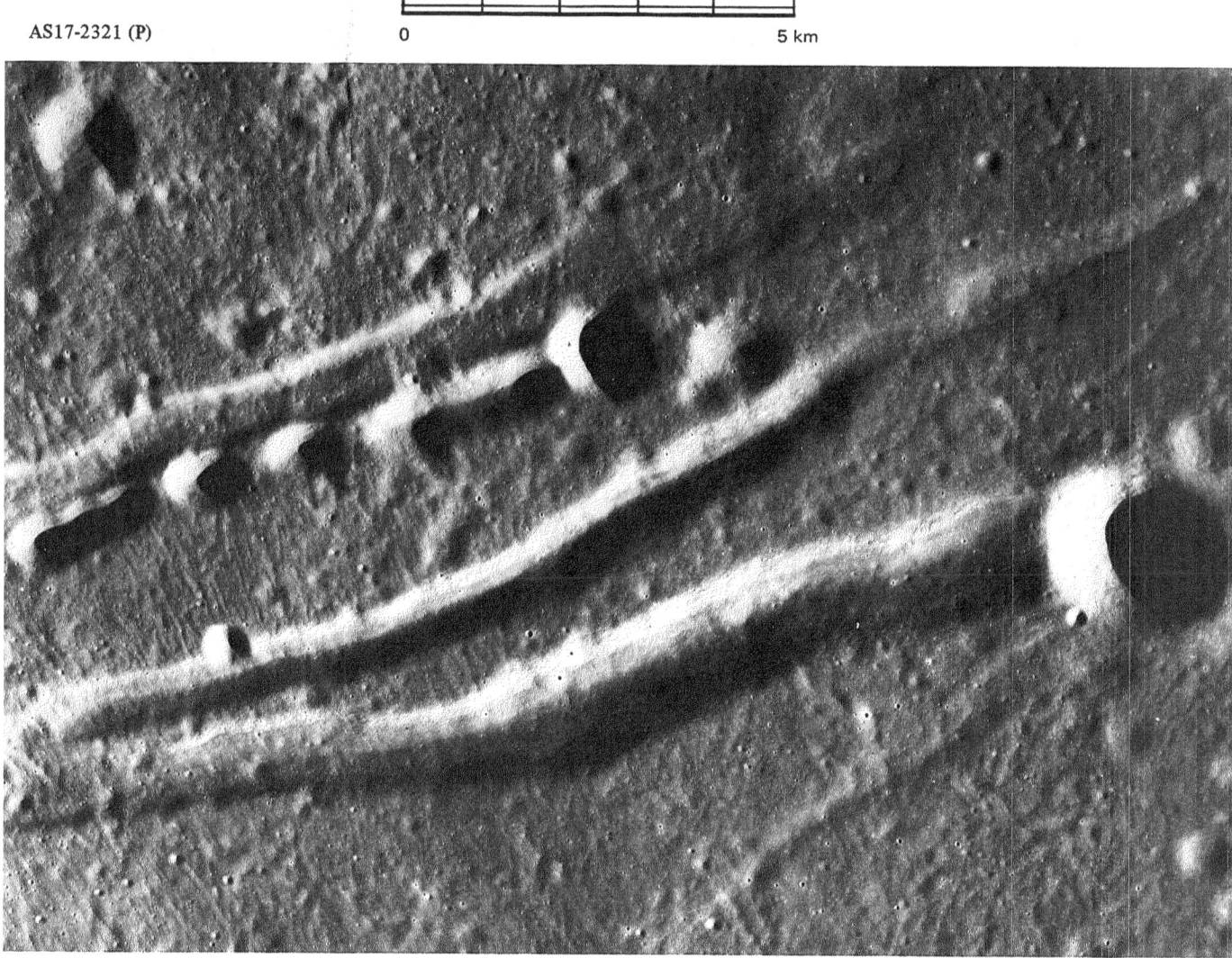

137

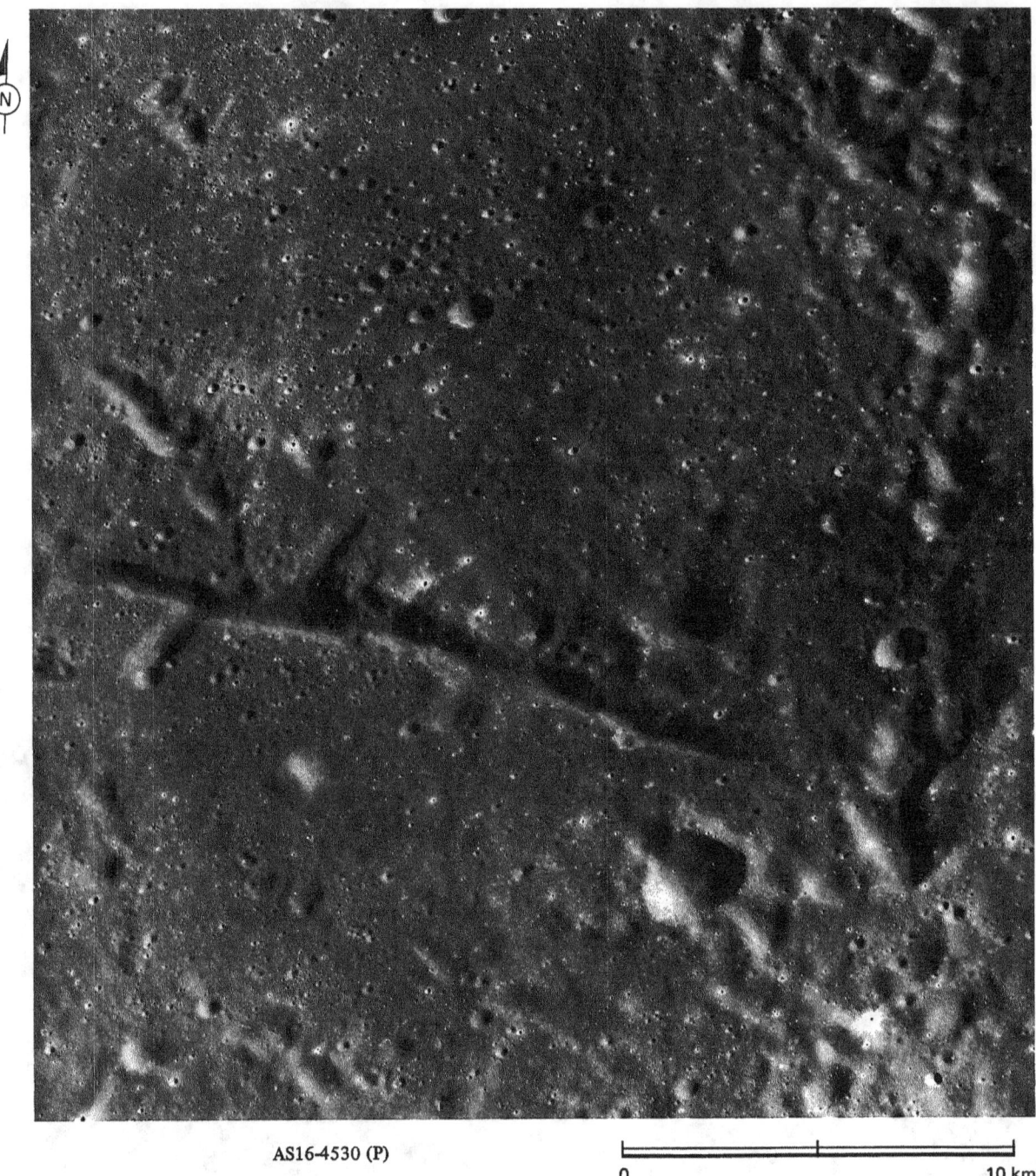

AS16-4530 (P)

0   10 km

FIGURE 136.—Linear features of external and internal origin are contrasted in this area of southernmost Mare Tranquillitatis. The northtrending line of overlapping, very irregular craters along the east edge of the picture is clearly a chain of secondary impact craters. Its trend is radial to Theophilus, a large crater of early Copernican age that lies about 105 km south of this area. The flaring shapes of some of the craters and their state of preservation also suggest that Theophilus is the primary crater. The narrow, straight rille or graben that extends westward across the picture is clearly of internal origin. It formed when tensional forces ruptured the crust, causing the floor of the rille to subside along faults. As is discussed in chapter 6, straight rilles are commonly the sites of volcanic cones or of blankets of volcanic ejecta; however, there are no signs of volcanism here that can be related to this rille.—M.J.G.

FIGURE 137.—Parts of three frames from the Apollo 17 panoramic camera were mosaicked to form this high-resolution view. The crater Euler in southwestern Mare Imbrium is an exceptionally fine example of a young medium-sized crater. Twenty-seven km in diameter, Euler has most of the features that typify young craters in this size range. Its sharp rim shows little evidence of rounding. A solid blanket of ejecta is visible for approximately one-half crater diameter outside the rim, and the radial pattern of secondary craters, crater clusters, ridges, and grooves is visible outward to a full crater diameter. Terraces formed by slumping of the steep crater walls, probably contemporaneously with the formation of the crater, are clearly evident. The steepness of the walls and the fact that the crater floor is below the level of the surrounding mare surface indicate that relatively little erosion and infilling have occurred. Other features typical of medium-sized craters are the central peak and the level floor surrounding the central peak. The pattern of ejecta around Euler is notably asymmetric because the area was later flooded by mare lavas that inundated parts of the ejecta blanket and other ejecta features.—M.C.M.

0   25 km

AS17-3107 (P)   AS17-3105 (P)   AS17-3103 (P)

140

FIGURE 138.—This similar view of Euler, taken with the Apollo 17 mapping camera, is included because it shows even more clearly the relationship between the ejecta from Euler and the surrounding mare lavas. The youngest lava flows in the Imbrium basin (Schaber, 1973) have overlapped and embayed the Euler ejecta from the north, west, and south—especially at places marked by arrows. The long sinuous rille south of Euler is associated with the late stage lava flooding. The clusters of large secondary craters (S) are from Copernicus, 400 km to the southwest. In the southeast part of the picture some of these secondary craters and their associated ray deposits overlie the mare lavas. The following sequence of events therefore took place: (1) the Euler impact, (2) flooding by lava, and (3) the impact of material ejected from Copernicus.—G.G.S.

AS17-2922 (M)

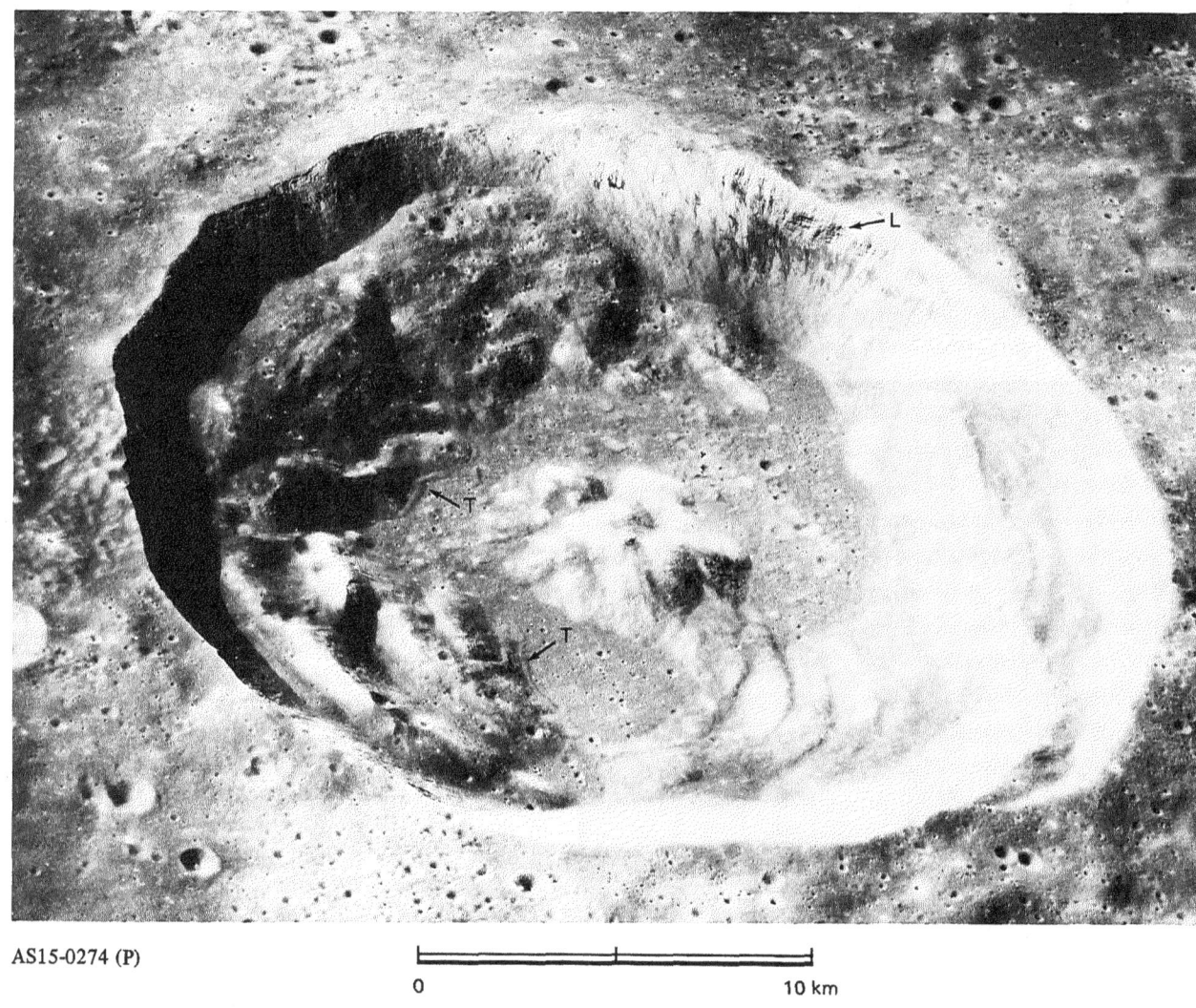

AS15-0274 (P)

0   10 km

FIGURE 139.—In this oblique view of Euler some details are shown that are not visible in figures 137 and 138. Note, for example, the ledges (L) of bedrock cropping out along the south wall and the low terraces (T) at the points of contact between the slump masses and the floor. They may be aprons of debris or "bathtub rings" of lava like those shown earlier in figure 69. This oblique viewing angle also enhances the polygonal outline of Euler's rim crest and the size and ruggedness of the huge masses that have slumped from the walls.—G.G.S.

FIGURE 140.—This crater on the lunar far side is similar in age and size to the near-side crater Euler (figs. 137 to 139). It is located midway between the craters Bečvář and Langemak. Thirty-six km in diameter, it was informally called the "Bright One" by the Apollo 14 astronauts because of its bright ejecta and ray pattern. The bright halo that surrounds the crater is about 150 km in diameter. Its brightness is not evident in this view because the picture was taken when the Sun angle was low. The radial pattern of dunelike ejecta around the crater is most apparent where the Sun's rays are perpendicular to the direction of ejecta flow, as in the lower part of the picture. The hummocky or bumpy floor of the crater is caused largely by material that has slumped from the walls. Stuart A. Roosa, the Apollo 14 CMP, used a handheld camera with an 80-mm lens for this photograph. Later, using a 500-mm lens, he photographed in much more detail that part of the floor of the crater outlined in this photograph and shown in figure 141.—M.C.M.

AS14-70-9671 (H)

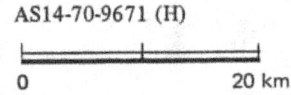

0   20 km

FIGURE 141.—When photographed with the 500-mm lens, the abundance of blocks (bright spots with shadows extending to the right) attests to the freshness of the materials on the floor of the "Bright One." Material that has flowed and in some instances formed smooth-surfaced "pools" is evident in much of the area. Arrows mark the edge of a major flow distinguished by its surface texture, color (in the original negative), location in a topographic low, and clearly defined border. Note that the abundance of boulders in the flow is much less than in nearby areas, presumably because the flow has buried most of the boulders in its path. Scientists generally agree that material has flowed here, and on the floors and flanks of many other craters, but the nature of the material that has flowed is a matter of debate.—M.C.M.

AS14-72-9975 (H)

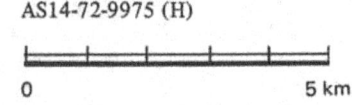

0   5 km

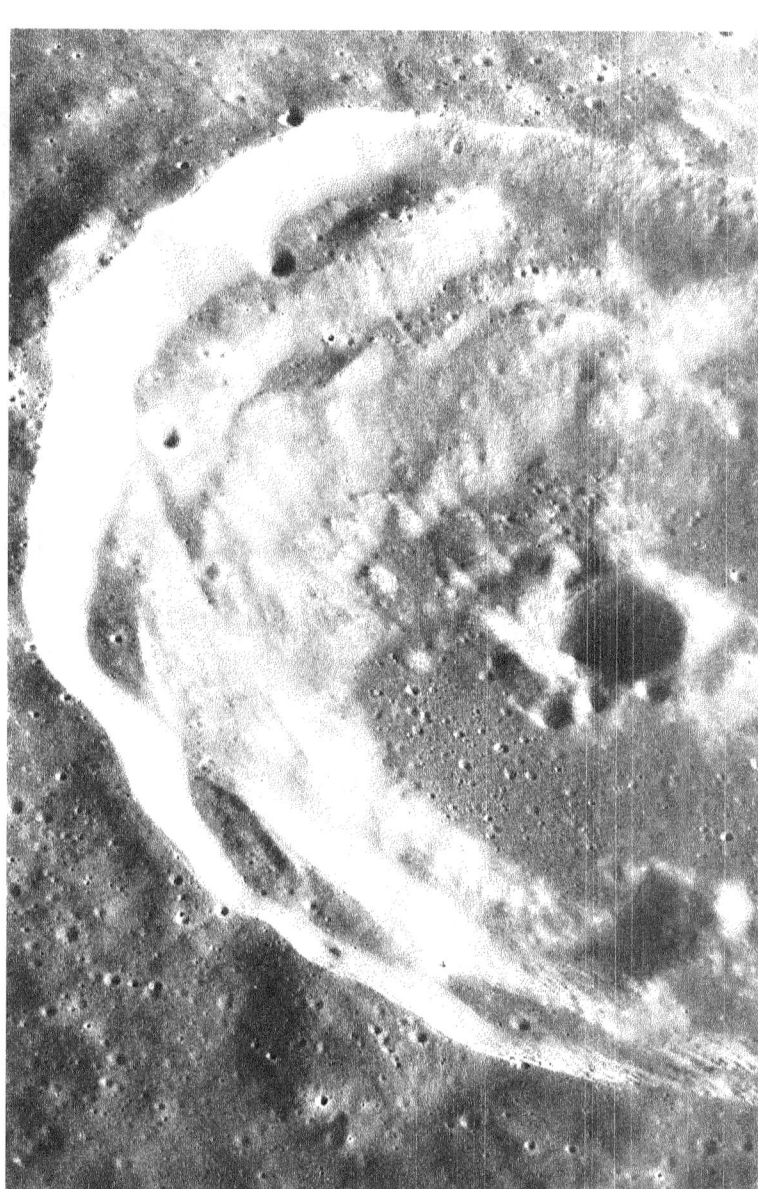

FIGURE 142.—This oblique view of the crater Timocharis in southeastern Mare Imbrium illustrates how the original diameter of a crater is enlarged by slumping of its walls. Its present diameter is about 35 km. The sparsity of small superposed craters on the walls of Timocharis—in contrast to their density on its floor and rim—is caused by the erosive effect of downslope movement of material on the steep walls. Timocharis, like many other young impact craters of similar size, possesses a well-defined central peak complex. Such structures are believed to result from elastic rebound of the bedrock immediately after the impacting event. However, the central peak of Timocharis apparently has been substantially modified by a large superimposed crater.—G.G.S.

AS17-3062 (P)

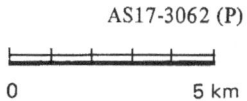

FIGURE 143.—High Sun views such as this often show fascinating dark and bright patterns that would be overwhelmed by highlights or shadows if the Sun were lower in the sky. This view of the 17-km-wide crater Jansen B shows numerous bright avalanche deposits on the steep crater walls, apparently originating at outcrop ledges near the top of the wall. Most avalanches stop in a moat at the base of the wall, but a few in the foreground extend out onto the irregular, inward-sloping floor. The floor is a jumble of slump blocks. Avalanching appears to be a major means of erosion on steep lunar slopes.—K.A.H.

AS15-9866 (P)

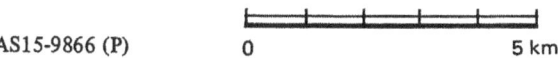

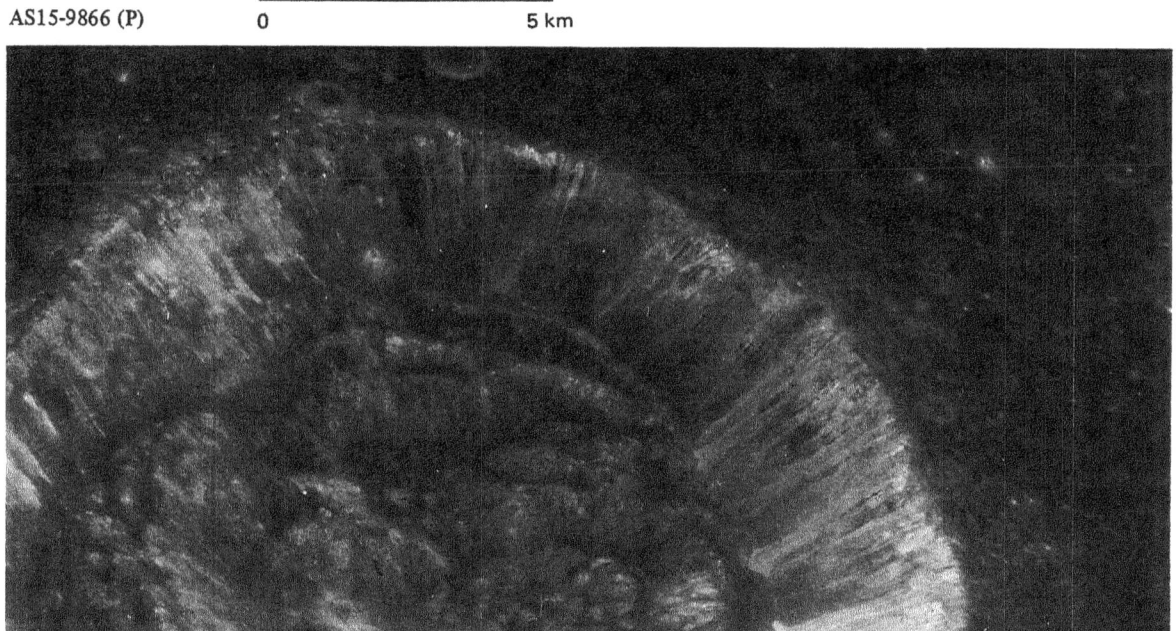

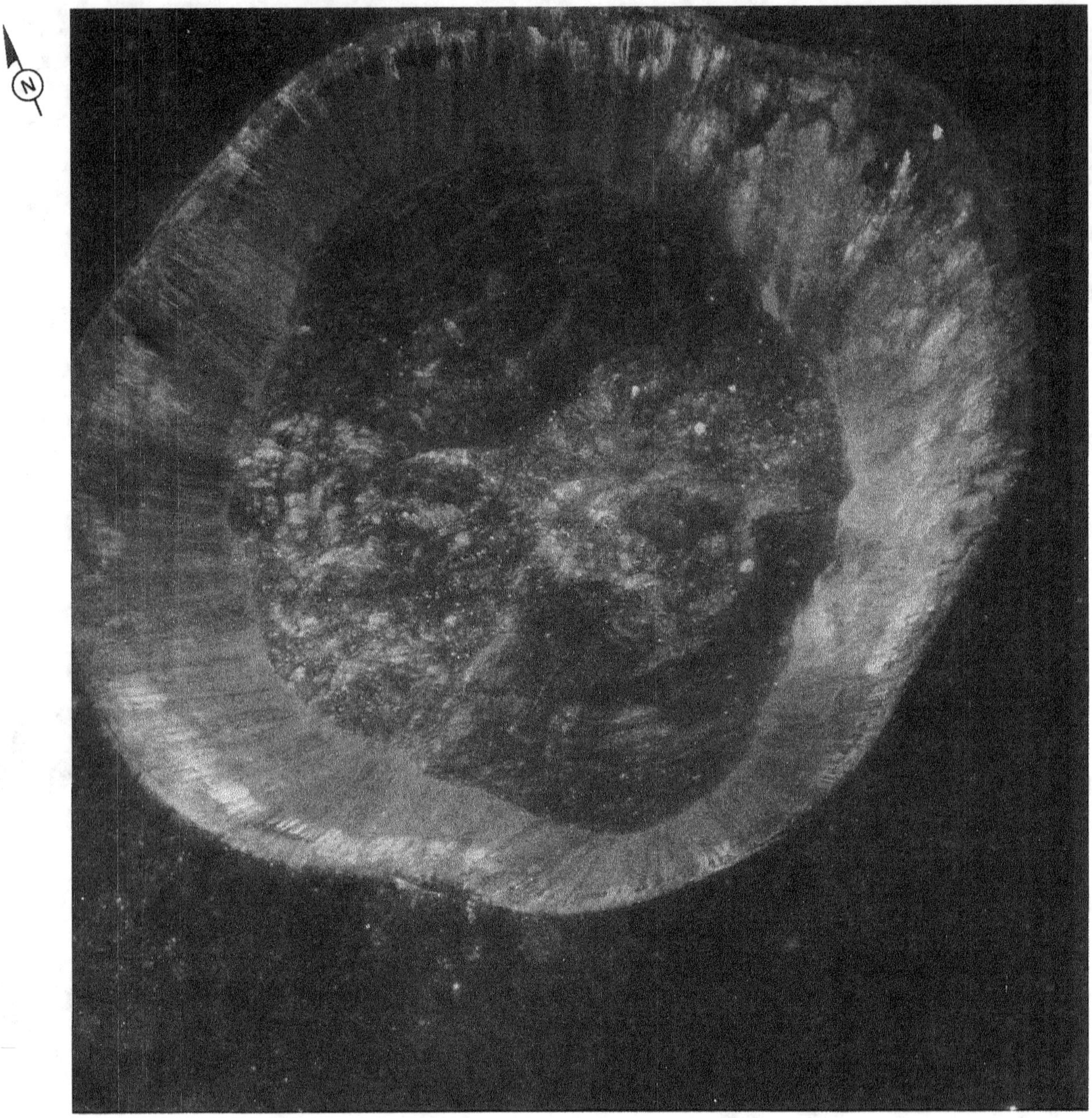

AS15-9874 (P)

0   5 km

FIGURE 144.—This is a near vertical view of the crater Dawes, 18 km in diameter. Morphologically it is typical of many lunar craters in the 15- to 20-km size range. It lacks terraced walls and distinct central peaks but has an extremely rough floor. Small terracelike structures on the crater floor (upper left, lower right) occur where the wall is bowed outward and probably represent slump deposits where portions of the crater wall have collapsed into the crater. Local stratigraphy is revealed in the walls of the crater, and material of different albedo is seen streaming down into the crater from various levels. The dark layer clearly visible in the upper part of the crater wall represents the thin mare deposits in this part of northern Mare Tranquillitatis. The lighter gray material below it is a combination of underlying submare material and talus from units higher on the crater wall. The highest unit (white and gray) probably represents the ejecta blanket and may consist primarily of lighter lunar crustal material excavated from beneath the mare.—J.W.H.

144

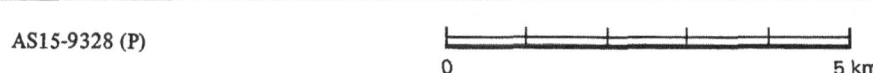

N

0   5 km

FIGURE 145.—Outcrops of layered rock are strikingly evident in the upper part of
the far wall of the crater Bessel (17-km diameter) in south-central Mare Serenitatis.
The outcrop is most evident where it forms shadows; however, the dark debris that
streams downslope from the layered rock is visible even on parts of the crater wall
where the Sun has washed out all details of relief. The outcrop is at a uniform
distance below the crater rim, indicating that the strata are horizontal. Thus, Bessel
furnishes convincing evidence that mare surfaces are underlain by dark layered
rock. The dark rock is now known to be basalt that accumulated as successive flows
or layers of lava. Bessel is youthful enough that boulders are abundant on its rim
and floor. An anomalously high number of boulders is visible in and around the
750-m diameter crater (arrow) on the floor.—M.C.M.

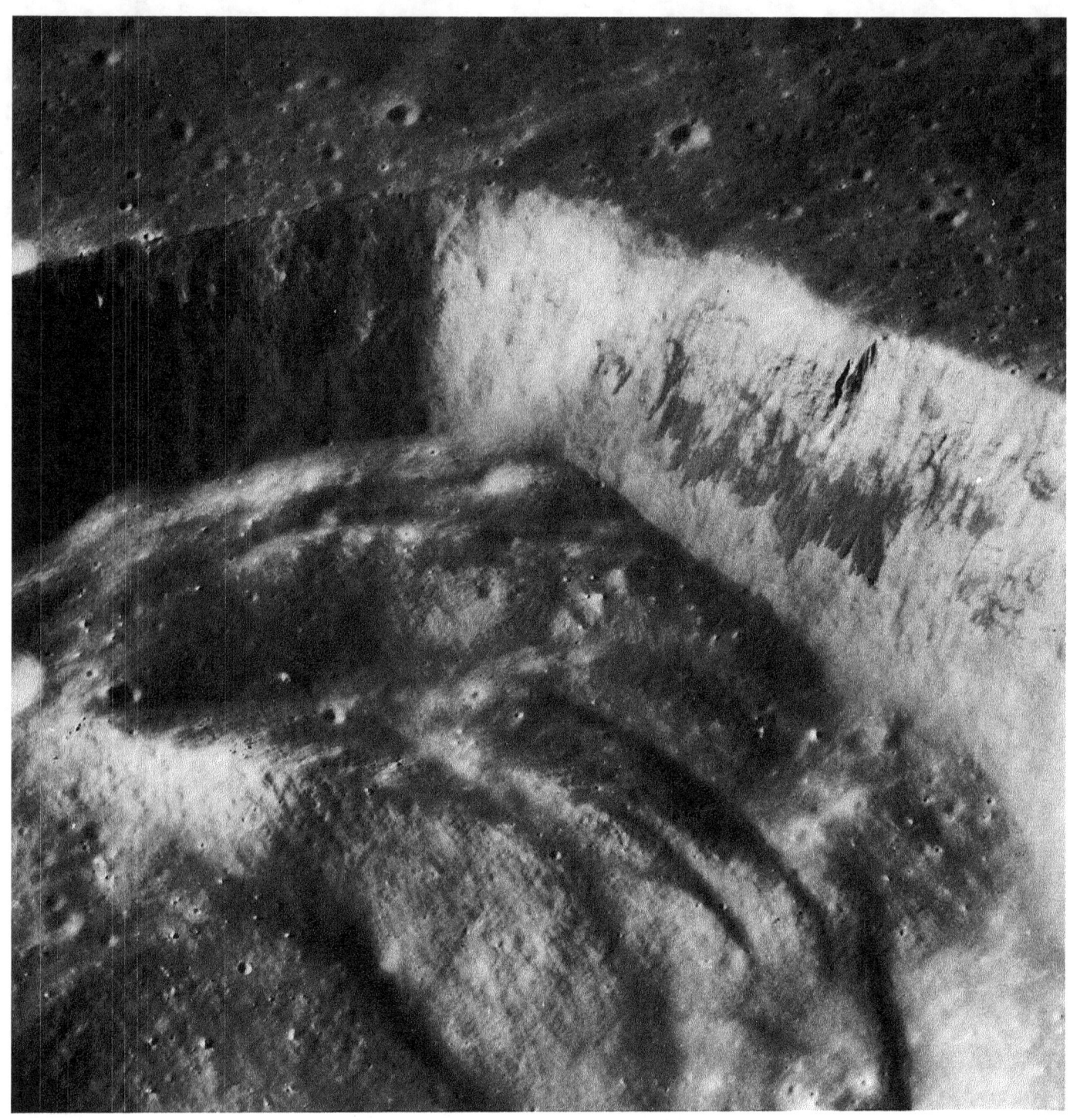

AS17-3081 (P)

FIGURE 146.—The Apollo 17 panoramic camera provided this high-resolution, enlarged view of the south wall of Pytheas. Pytheas is about the same size as Bessel (fig. 145), but is located in south-central Mare Imbrium almost 1100 km west of the latter. The outcrops in the walls of the two craters are remarkably similar. These and the many other craters in mare areas that contain outcrops of dark horizontally layered rock demonstrate the moonwide uniformity of conditions in the upper part of the mare basins.—M.C.M.

FIGURE 147.—This oblique view looks south over the 26-km-diameter crater Proclus in the highlands at the western edge of Mare Crisium. Proclus is a young rayed crater that is distinctive because of the marked asymmetry of its ray system—a characteristic visible even in Earth-based telescopic views. The excluded zone is along the southwest edge (top of photograph) but is visible in this moderate Sun photo only as a slight albedo change. Laboratory experiments suggest that a low trajectory angle might account for the asymmetry. A number of large blocks can be seen at the edge of the crater rim. The exceptionally large block (arrow) is about 200 m wide and, judging from the length of the shadow it casts, nearly as high. As in several other craters shown in this chapter, a darker layer is present in the upper part of the crater wall.—J.W.H.

AS17-2265 (P)

0   10 km

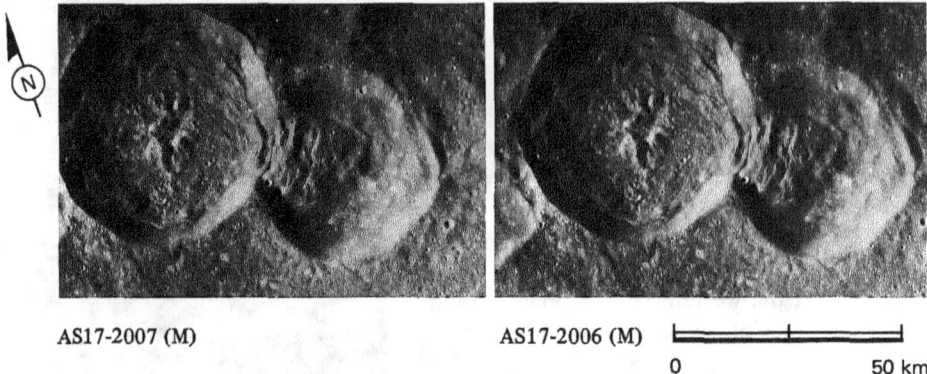

AS17-2007 (M)   AS17-2006 (M)

0   50 km

FIGURE 148.–This pair of overlapping medium-sized craters illustrates some of the criteria used to determine relative ages. Material ejected from the larger polygonal crater on the left partially fills the smaller crater on the right; thus, the crater on the left is younger. Furthermore, the wall of the large crater is complete, whereas the west wall of the smaller crater is absent, obviously having been destroyed by the larger crater. Even if the two craters did not overlap, the sharp rim, terraced walls, and prominent central peak of the larger crater clearly identify it as the younger of the two. The frames used in the stereogram were selected to show exaggerated relief, a technique very helpful to photointerpreters in determining shapes and relative elevations of surface features. The two craters are located in the rugged terrain of the far-side highlands approximately 250 km north of the prominent crater Tsiolkovsky. The photograph below is an oblique view of the same pair of craters. It is included, not because it is a more dramatic view, but because it shows from another perspective that one crater clearly is superposed on the other. –M.C.M.

AS17-2773 (P)

Approximate

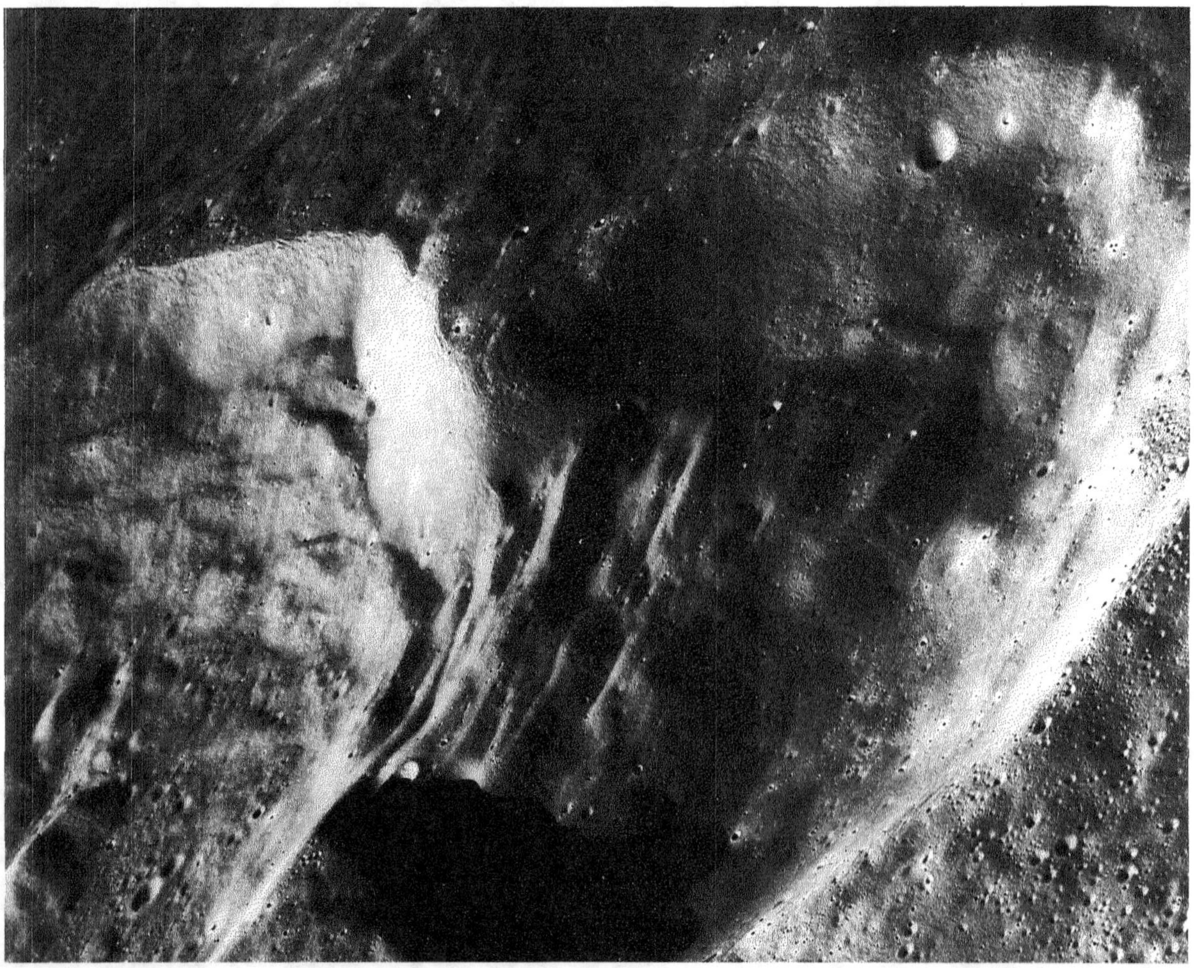

AS16-122-19580 (H)

0   50 km

FIGURE 149.—This vertical view of the crater King on the lunar far side was taken with the Apollo 16 Hasselblad camera. King, approximately 75 km in diameter and 4 km deep, is one of the most interesting features on the far side. It is a superb example of a youthful, large crater. It attracted much attention and was the object of numerous scientific studies (e.g., El-Baz, 1972*b*; Young, Brennan, and Wolfe, 1972). King is the freshest crater on the far side in its size range. Among its many interesting features are (1) a unique lobster-claw-like central peak, (2) a flat pool-like area of dark material on the north rim believed to have once been molten, (3) a very-well-developed field of fine ejecta extending outward for approximately two crater diameters, and (4) a massive landslide on the southeast rim (see arrow). In this view the southern part of the central peak has a distinctly ropey appearance and is segmented parallel to the terraces of the adjacent crater wall. The low Sun illumination enhances the fine texture of King's ejecta. Northeast of King the ejecta mantles an old large crater and in the southwest corner of the picture it mantles a relatively smooth terra unit. The slightly raised plateau on which the crater is situated may be part of the ring of an old basin.—F.E.-B.

149

AS16-120-19268 (H)

FIGURE 150.—The similarity in appearance of the southern part of the central peak and the slump terraces on the southern wall of the crater is emphasized in this oblique view of the crater King. The parallelism of the two arms of the central peak and the southern segment of the peak suggests that the unique shape of the structure is caused by a preexisting tabular body that was excavated during the formation of the crater (El-Baz, 1972b). Numerous domical structures with summit pits are present on the crater floor in the lower right part of the photograph.—F.E.-B.

150

FIGURE 151.—The northern part of the central peak complex of the crater King is shown in this enlarged view. The massifs forming the arms of the central peak trend approximately north-south. They are crossed by linear valleys that are the surface expressions of faults. The blocks that litter the peaks attest to the freshness of the structure and, therefore, of the crater King itself. The floor material surrounding the massifs is very hummocky and shows many flow patterns. Arcuate flow boundaries, flow channels, and cooling cracks indicate that this material was once molten, perhaps because of shock melting at the time of the crater formation.—F.E.-B.

AS16-4998 (P)

0   5 km   N

AS16-5000 (P)

0   5 km

FIGURE 152.—A close look at the western part of the floor of the crater King reveals many similarities to the floors of other fresh craters such as Copernicus or Tycho. The crater wall rises steeply along the left. Cracked and corrugated lavalike material occupies low areas between hills and knobs on the floor. Some cracks extend across hills, suggesting that those hills are completely covered by the lavalike material. Rocky outcrops on other hills (some examples are identified by arrows) show that they are made of hard rock.—K.A.H.

FIGURE 153.—This greatly enlarged part of a panoramic camera photograph shows a small area in the eastern part of the floor of the crater King. The large mass in the left part is one example of the many domed structures that occur in the floor of King. It is trisected by depressions in the shape of an inverted "Y" representing three faults; one trends north-south (*1*), another northeast-southwest (*2*), and the third northwest-southeast (*3*). In the area of the fractures and the resulting grabens and depressions, there are numerous blocks indicating that tectonic movement has occurred relatively recently. The northwest-southeast trending fault continues across the level floor to the lower-right corner of the picture. An intricate pattern of finer fractures trending in many directions is also present in the level part of the floor. The partly arcuate nature of the fractures suggests that some may represent cooling cracks.—F.E.-B.

AS16-4996 (P)

0   2 km

FIGURE 154.—This oblique view looking north across the northern part of the crater King was taken by the Apollo 10 crew. The very dark patch on the northern wall of King near the center of the photograph was first observed on that mission and has since been the subject of detailed visual study and orbital photography. The rocks of this dark patch are the darkest material yet recognized in the far-side highlands and are surrounded by brighter material. Two bands of this brighter material extend northward for approximately 30 km beyond the crater. In the near field, at the lower left, is the northern half of the Y-shaped central peak of King. A dark pool of relatively smooth material is visible just beyond the alternating bands of dark and light rocks on the north rim of the crater. The alinement of the arms of the central peak, the bands of dark and white rocks, and the concentrations of blocky material on the central peak complex and on the crater floor all suggest that when this crater was formed, a preexisting linear body was encountered and partly exhumed.—F.E.-B.

AS10-30-4349 (H)

FIGURE 155.—The Apollo 16 astronauts captured this spectacular view of the large dark "pool" on the north flank of the crater King (left) as they approached from the east. The pool (also known as a lake, pond, or playa) is in an old crater swamped by King ejecta. The maximum width of the pool is about 21 km. The peculiar dark material that forms the large pool and also coats adjacent hills was first discovered on Apollo 10, and was later seen again from Apollo 14. The most exciting part of the discovery had to wait until the mapping and panoramic cameras of Apollo 16 showed that this material contains some of the freshest and most spectacular flow structures on the Moon. These structures, some of which are seen in the following figures, show that the material behaved like lava. The material is very similar in appearance to that filling parts of the floor of King. —K.A.H.

AS16-120-19266 (H)

FIGURE 156.—Here the Apollo 16 panoramic camera views the same large pool of dark material. For orientation, the hill identified by an arrow is the "island" near the center of the pool in Figure 155. In the highlands surrounding the main pool are many small pools. Their surfaces merge with a veneer of similar material on adjacent slopes, as if most of the rim of the old crater had once been coated by a fluid rock that partly drained off to accumulate in depressions. Cracks (C), flow channels (FC) near the middle of the picture, and numerous wrinkles show that the material flowed like lava toward the main large pool. Whether the lavalike material is volcanic lava or rock melted by the impact is in vigorous debate. The material is restricted to the north rim of King where ejecta is also most concentrated. If the ejecta is concentrated on the north because north was downrange from an obliquely impacting body, that condition would support an impact origin for the dark material.—K.A.H.

AS16-5000 (P)

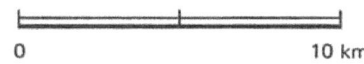

0   10 km

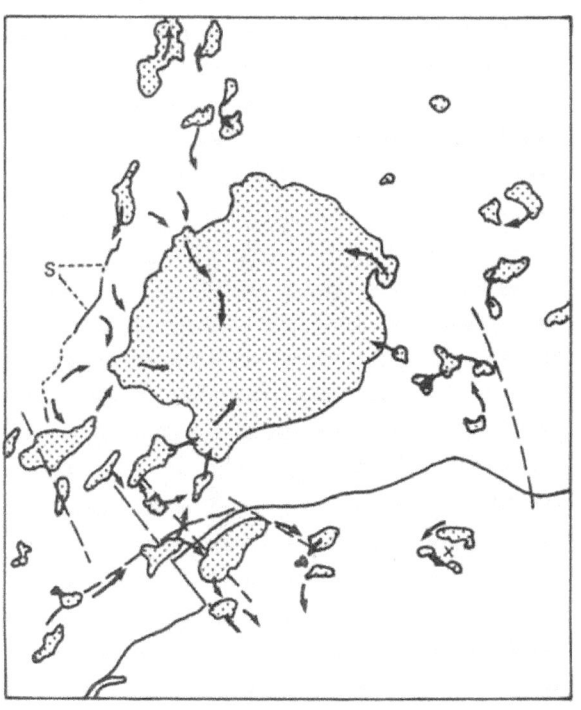

FIGURE 157.—A schematic map based on figure 155 shows some of the geologic relationships. The solid line marks the rim crest of the crater King and the stipple pattern delineates the many pools of smooth dark material. Dashed lines represent lineaments that probably are the surface expressions of faults. Arrows indicate the directions of flow of the molten or partially molten material from higher to lower levels; most finally accumulating in the large pool in the center. A scarp whose origin is not fully understood, but which may possibly be the result of splashing of the molten material from the impact site onto a slope is indicated by S.—F.E.-B.

0   20 km

155

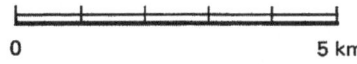

FIGURE 158.—This enlargement of figure 156 shows details in the lavalike materials. The flow features are so young that younger cratering events have not obliterated them. The prominent leveed flow channel in the center of the picture indicates that the material flowed downhill in the direction of the arrows toward the large pool. Like terrestrial volcanic lava, the material is locally compressed into festoonlike corrugations that arch "downstream." In other areas, especially where draped over hills, the material has been stretched and dense patterns of cracks have formed. A linear texture trends northwestward across much of the area shown, but is mostly clearly visible in the lower right corner. It is the expression of King's ejecta blanket underneath the veneer of the lavalike material. If the lava-like material is impact melt, it must have coated the ejecta and drained downhill after the ejecta flowed radially away from King.—K.A.H.

AS16-5000 (P)

FIGURE 159.—The ejecta blanket of King fills this entire scene except for the crater itself and a few small superposed craters. The ejecta coats larger craters, such as the two on the right edge, and completely swamps smaller ones. A large lobe in the ejecta at lower center has been compared (El-Baz, 1972b) to an apparent landslide at Tsiolkovsky (figs. 175 and 176). Close inspection of the surface of the ejecta blanket reveals fine lineations that are concentric near the rim crest and radial further out. Where slopes are encountered, the radial lineations veer from their normal direction to swerve down the slopes.—K.A.H.

AS16-1578 (M)

AS16-1579 (M)

0   20 km

FIGURE 160.—An enlarged view of the area outlined in figure 159 shows the concentric dunelike lineations more clearly. A few of the lineations veer from concentric to radial in orientation. The outward change on a broader scale from concentric to radial patterns also occurs at other fresh craters, and probably records changing conditions of flow as the ejecta was transported outward at high velocity. On the crater wall, numerous fine fractures parallel the terraced slump blocks, much as in landslides on Earth.—K.A.H.

AS16-120-19231 (H)

Approximate 0   25 km

FIGURE 161.—The lobate scarps in the foreground are a lunar feature seen as Apollo 16 flew over the margins of King's ejecta blanket. This view looks away from King and is centered about 50 km from its rim crest. The lobes appear to border thick flows or slide masses of King ejecta. Beyond the lobes in the upper left part of the picture the ejecta is thin or absent.—K.A.H.

158

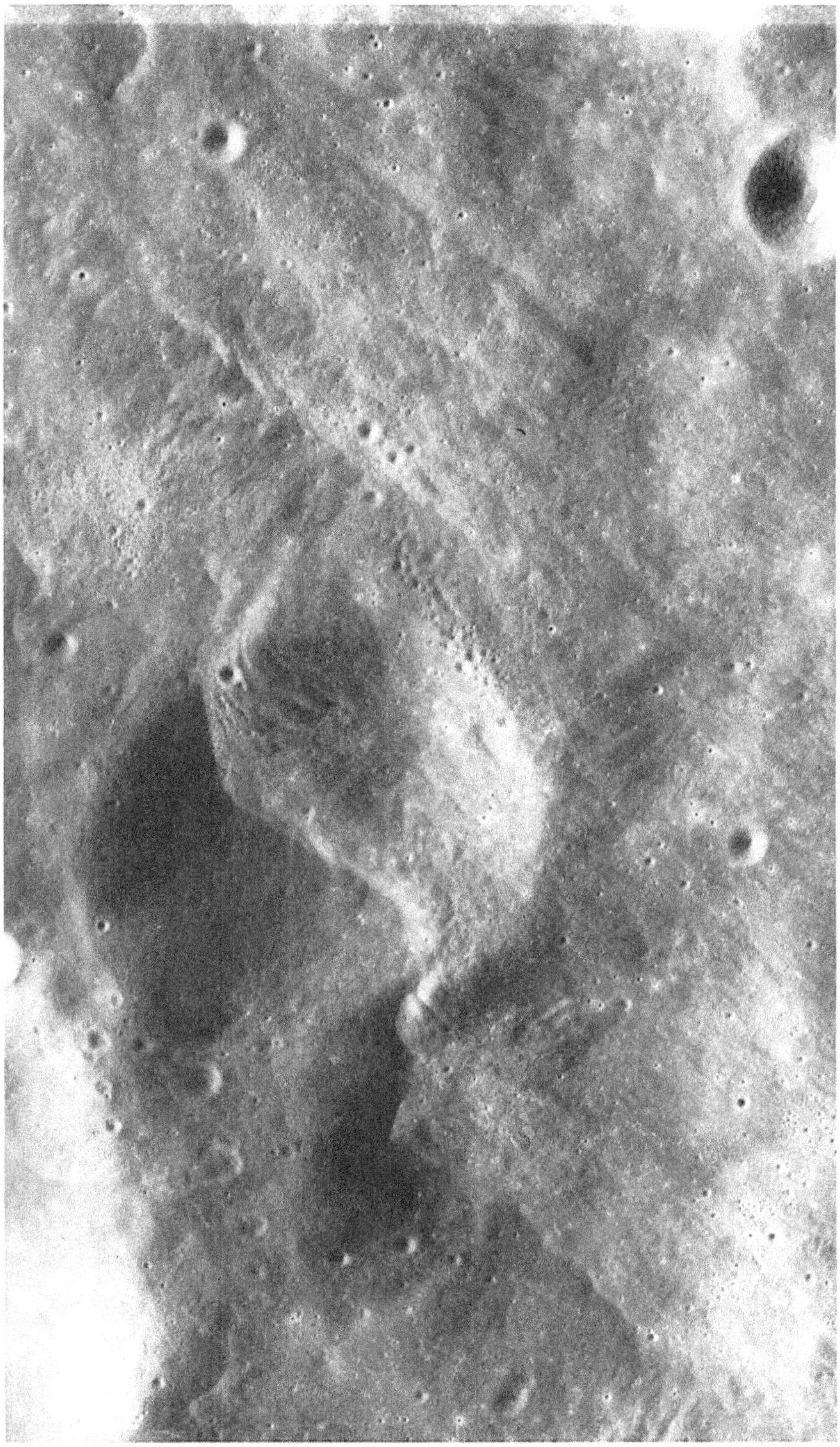

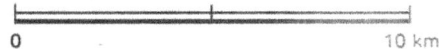

FIGURE 162.—Here is an enlarged vertical view of more flow lobes immediately south of those in figure 161. Fine lineations radial to King are prominent in the ejecta blanket behind (southeast of) the lobate fronts. The term "deceleration lobe" has been applied because the lobes occur only where the ejecta slowed down and came to rest on slopes that face toward King. They resemble terrestrial rock avalanche deposits that came to rest after climbing a small slope. Some lobes overlap each other outward like shingles. The sketch shows what would probably be seen in a cutaway view. The arrow shows the direction of movement of the ejecta over the old landscape.—K.A.H.

AS16-5006 (P)

```
0                                10 km
```

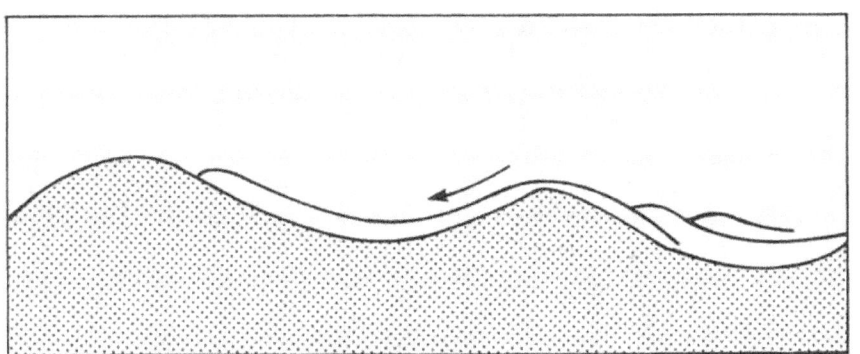

AS16-4815 (P)

0   5 km

FIGURE 163.—Here the panoramic camera sees the edge of King's ejecta blanket 75 km northeast of the crater. The bright slope along the left side of the photo faces both the Sun and King. It apparently acted as a barrier for the ejecta surging outward from King, for there are small deceleration lobes on the slopes. Most of the plateau east of the slope is pocked by small craters that were not covered by the ejecta blanket. However, a belt of linear dunelike features radial to King crosses part of the plateau from lower left to upper right. Careful inspection shows that the lineations are downrange from a cluster of small irregular craters (arrow) at the left edge of the plateau. These are part of a field of secondary impact craters that completely surround King's ejecta blanket. Debris thrown out on low-trajectory paths from King made the secondary craters and apparently splashed debris downrange. Closer to King, the continuous ejecta blanket is lineated in the same way as is the secondary ejecta. Perhaps much of the continuous ejecta blanket is also formed of material splashed out by secondary impacts.—K.A.H.

FIGURE 164.—The large crater Copernicus has served as a type example of lunar impact craters since the classic analysis was made by E. M. Shoemaker (1962). Bright rays of ejecta radiate outward from Copernicus across a large part of the Moon's near side. Material from one of the rays may have been sampled at the Apollo 12 landing site, 370 km south of the center of the crater. This photograph shows how the crater appeared from the Apollo 17 spacecraft looking southward over the Montes Carpatus (Carpathian Mountains). Notice that the rim deposits immediately adjacent to the crater have a very crisp, blocky appearance in contrast to the softer appearance of the rest of the ejecta blanket. This crisp zone is also found on many other craters and suggests the ejecta here was swept clean by some erosion process late in the cratering event. The terraced slumps on the crater wall appear like giant stair steps leading to the floor, 3 to 4 km below the rim. The 1-km-high central peaks were made famous in 1966 by a "picture of the century" view looking into the crater from the south by Lunar Orbiter 2. Now Apollo has given us scores of even more spectacular photographs.—K.A.H.

AS17-151-23260 (H)

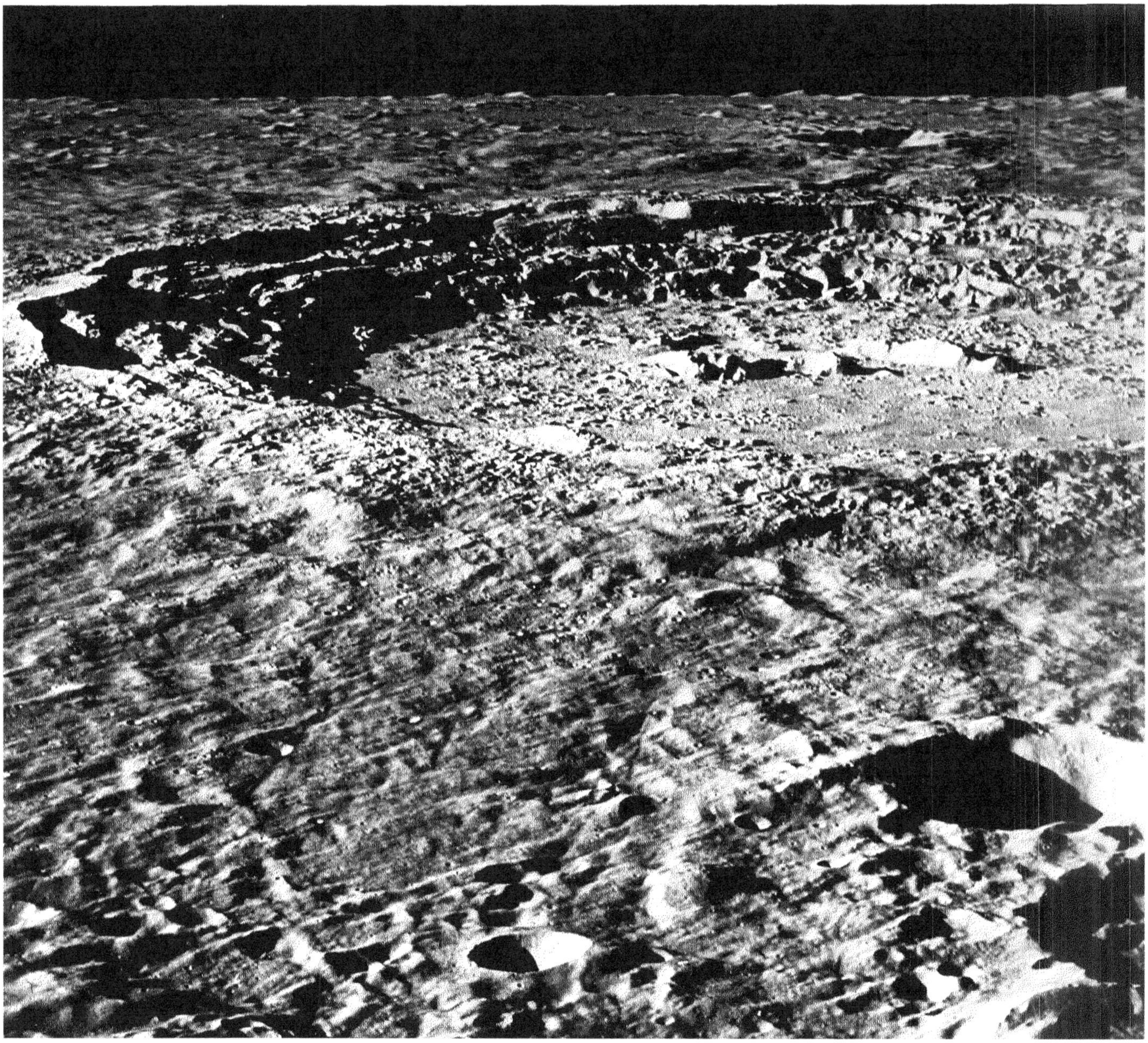

AS15-0326 (P)

Approximate   0   10 km

FIGURE 165.—Aristarchus is a large crater on the edge of a plateau within northern Oceanus Procellarum. In this scene the crater is viewed obliquely from the north. One of the brightest and youngest craters of its size on the near side of the Moon, Aristarchus is believed to be younger even than Copernicus. The general appearance of Aristarchus and of parts of the plateau around it led Alfred Worden, the Apollo 15 CMP, to describe this part of the Moon as "... probably the most volcanic area that I've seen anywhere on the surface." For many years before the Apollo missions, Earth-based viewers had reported telescopic sightings of transient events centered on Aristarchus. These brief, subtle changes in color or in sharpness of appearance have been suggested as evidence for volcanic activity or the venting of gases from the lunar interior. The sightings are controversial, but Aristarchus remains a center of interest. —M.C.M.

About 39 km in diameter, Aristarchus is on the borderline between medium-sized and large-sized craters. We have included it among the large craters because its well-developed concentric terraces are characteristic of most large craters that have not been too severely degraded. Its terraced walls, as well as its arcuate range of central peaks, are particularly well shown in this view. The walls and parts of the crater floor are extremely rough and cracked, a characteristic feature of other young impact craters of this size range, such as Tycho and Copernicus. The rough deposits in the floor are probably made up largely of shock-melted material formed at the time of the impact. The inner, rougher portions of the rim show a series of channels, lobate flows, and smooth puddlelike deposits that may represent shock-melted material deposited on the crater rim. The outer, smoother portions show the rhomboidal pattern characteristic of crater ejecta blankets.—J.W.H.

FIGURE 166.—Theophilus is a relatively young crater similar in size but slightly older than Copernicus (fig. 164). It lies on the eastern edge of the Kant plateau, an elevated area in the Central Highlands along the northwestern margin of Mare Nectaris. Part of Nectaris is visible as the smooth, dark area near the horizon at the left edge. Like Copernicus and Aristarchus, Theophilus has ruggedly terraced walls and a complex central peak protruding through a level floor. Smooth-surfaced material is present in "pools" at various levels on the terraces, on parts of the crater floor, and on the ejecta that blanket the near (north) side of the crater. As one alternative, the pools may have been emplaced as fluid lava. The pooled material and the prominent central peak complex of Theophilus are shown in more detail in figures 167 and 168.—M.C.M.

AS16-0692 (M)

FIGURE 167.—Theophilus is older than King, Copernicus, and Aristarchus, and many of its original features are more subdued than are those of the younger craters. Here on the north rim, the entire width of the Theophilus ejecta blanket can be seen, and irregular-appearing secondary impact craters are visible beyond the blanket at the top of the picture. In the middle of the scene, smooth material occurs in pools on top of the ejecta blanket. Smaller pools of similar material are present on the terraces in the wall of Theophilus. D. J. Milton (1968) originally discovered this pooled material in the mid-1960's from telescopic studies and suggested that it might be volcanic. This Apollo 16 view shows that the pools are very similar to, but more degraded than, those on the rims of King (figs. 149 and 155 to 158) and other fresh craters such as Tycho and Copernicus. This similarity suggests that the pool material may be rock melted by the Theophilus impact.—K.A.H.

AS16-0154 (M)

0   50 km

AS16-4531 (P)

0   10 km

FIGURE 168.—A detailed view of part of the central peak complex of Theophilus. Central peaks are typical of most young, large impact craters on the Moon—and also of many manmade craters on Earth. From experimental data using controlled explosions, central peaks are known to consist of bedrock originally lying below the crater floor that, during the explosion, was uplifted, faulted, and folded by shock wave action. The irregular light-toned mountainous mass projecting above the floor of Theophilus is split into at least three enormous blocks separated by V-shaped structural valleys. Four or five circular craters without a prominent raised rim are located near or at the bases of the steep slopes. If these craters are endogenic vents rather than impact craters, their presence further suggests structural control along major fault planes. The planar walls of the northwest-trending valley contrast with other sloping surfaces of the central peak complex. They are steeper and, except for a few outcrops of protruding bedrock, are marked by linear grooves not unlike slickensides on many fault planes on Earth. Rock chutes do not seem to be a likely explanation for the grooves because there are no talus deposits or blocks at their lower ends. The debris cover is thin enough along the southern valley wall (top of picture) to show that the southern mountain block consists of layered rocks—at least five thick, light-toned layers alternate with thin, dark layers.—M.J.G.

165

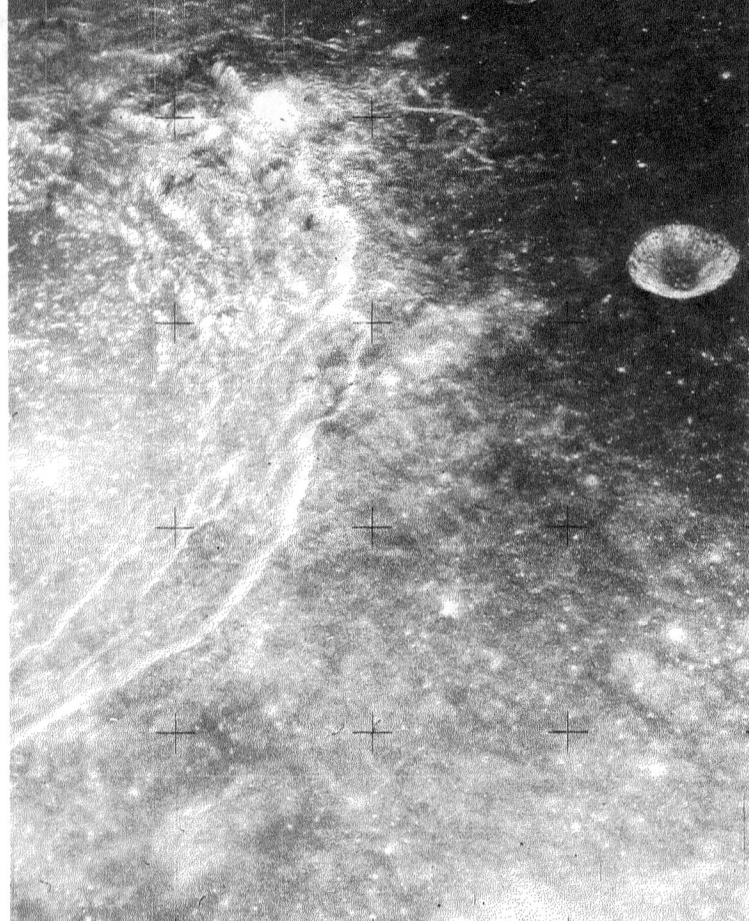

FIGURE 169.—The preceding pictures of large craters were all taken under low Sun conditions. This picture of Langrenus, a large crater near the east limb, was taken when the Sun angle was 71°. The high Sun angle emphasizes the brightness of the steep slopes in the walls and central peak complex. Like the other craters, Langrenus has an irregular rim crest, its walls are thoroughly terraced, and the floor surrounding the central peaks is level but very rough. Langrenus is about 130 km in diameter.—G.G.S.

AS16-0532 (M)

FIGURE 170.—Tsiolkovsky is one of the most prominent features on the far side of the Moon. It is a 190-km-wide impact crater with a large, complex central peak that is offset from the apparent center of the crater. Differences in tone and texture between the central peak, the lava-flooded floor, the terraced walls, and the ejecta blanket are dramatically displayed in this oblique view. The ejecta blanket is dominated by a coarse pattern of ridges radiating outward from the crater; superposed on this pattern are many small level pools of smooth material that are much lighter than the otherwise similar smooth dark mare in the floor of Tsiolkovsky. The pools probably originated differently. They may consist of rock that was melted by the heat and pressure generated during the impact event and that flowed into depressions before it hardened.

Cratering experiments on Earth have shown that central peaks consist of bedrock that has been displaced upward by a distance equal to about one-tenth the diameter of the resulting crater. If samples could be obtained from the central peak at Tsiolkovsky, they might be rocks that were 20 km below the Moon's surface before Tsiolkovsky was created. —H.M.

AS15-0757 (M)

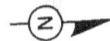

FIGURE 171.—This is clearly not another view of Tsiolkovsky, but there are some remarkable similarities between this picture and figure 170. Some of the similarities, such as viewing angle and Sun direction, resulted from screening many pictures of the same crater to duplicate viewing conditions. Other similarities are inherent in the crater-forming process. The crater above is about 100 m in diameter and 6.5 m deep. It has been intensively studied by D. J. Roddy (1968), U.S. Geological Survey, who provided the picture. It was produced in the plains of western Canada by the Canadian Defence Research Establishment as a cratering experiment. The explosive charge was a 450-Mg hemisphere of TNT resting on the ground. The site is underlain by alluvial and fluvial deposits of interlayered fine-grained silt, sand, and gravel, much of which were below the local water table. The smooth, dark "floor" of the crater is water that flooded the crater immediately after it was excavated.

Similarities between the two craters include (1) complex central peaks of uplifted rock material, (2) complexly terraced walls caused by slumping along concentric faults, (3) irregular but approximately circular raised rim crests, and (4) continuous blankets of hummocky ejecta, both of which are dominated by radial depositional patterns. An obvious difference is the thin, dark, arcuate line in the near field of this picture. It is part of a zone of circumferential fractures extending around the crater. Similar fracture zones may be present around lunar craters, but, if so, they are much less obvious.—G.W.C.

FIGURE 172.—This vertical view shows the central part of Tsiolkovsky in more detail. From the nature of the boundary between the dark mare lavas and the lighter materials at the base of the walls and in the central peak, we know that the lavas must have lapped upon and embayed the lighter materials. The relatively level areas of lighter material in the southwest and northwest parts of the floor have a distinctly different texture than the coarse blocky materials of slumped wall that surround the floor elsewhere. Finely cracked, furrowed, and hummocky, they closely resemble parts of the floor of the crater King (figs. 151 to 154). They probably consist of impact melt that solidified to form the original floor of Tsiolkovsky before it was flooded by mare lavas. —G.W.C.

AS15-1030 (M)

0   50 km

FIGURE 173.—Moderate enlargement of part of a panoramic camera frame provides greater detail of the central peak complex of Tsiolkovsky. A relatively large population of superposed craters has been preserved on level areas of the peaks (near the left-center of the photograph). In contrast, very few craters are present on steep slopes—most have been destroyed by the downslope movement of erosional debris. An intermediate population of craters on the dark mare shows that the mare surface is younger than the level areas of the peak complex but older than the freshly exposed steep slopes of the peaks. The youngest part of the mare surface is the dark, smooth area adjacent to the small angular rille in the upper left corner. Here small craters have been almost completely filled by the flow and are barely discernible. The rille may have served as the vent for the young lavas.—M.W.

AS15-9591 (P)

0   10 km

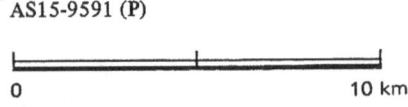

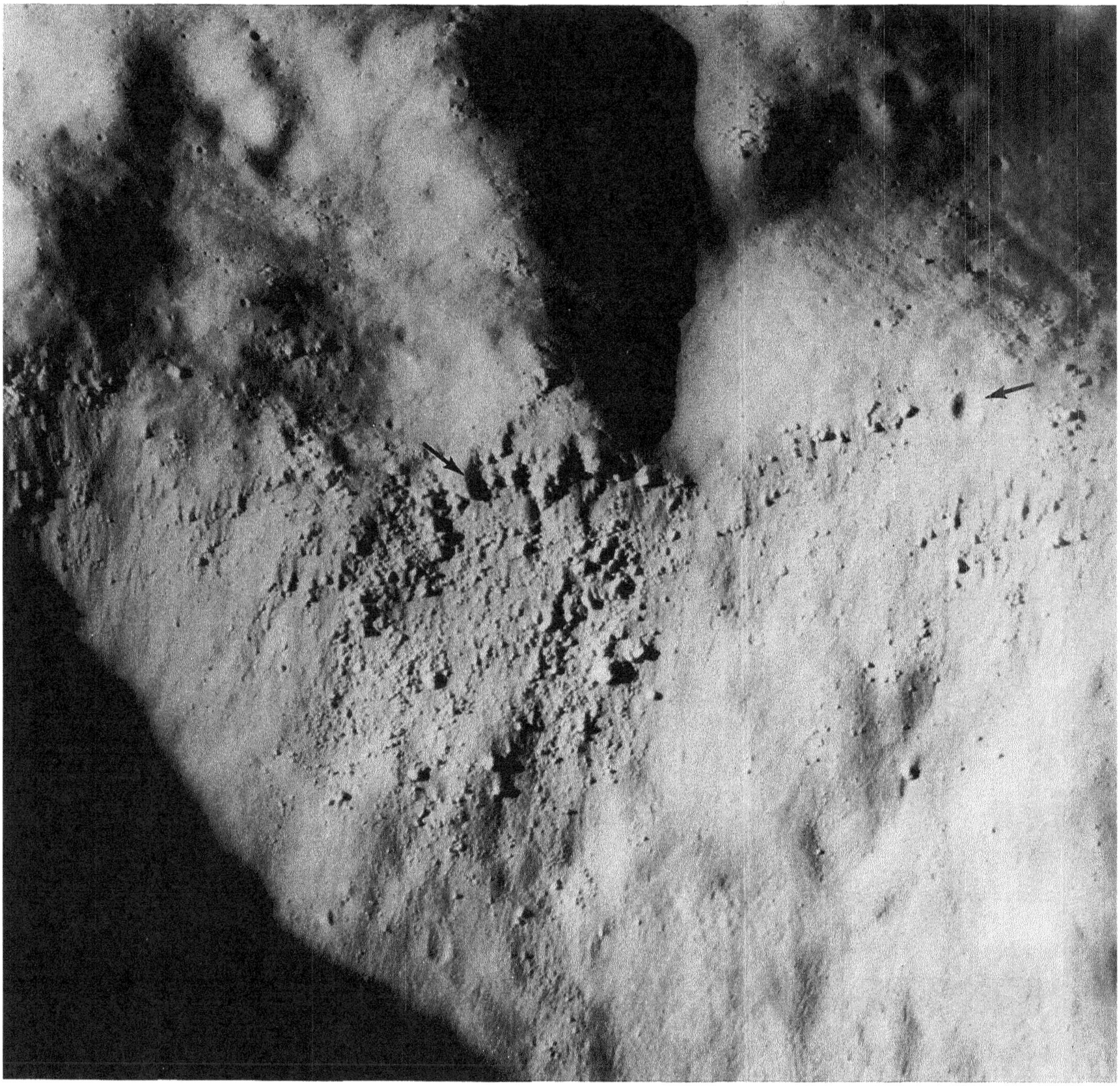

AS15-9596 (P)

0   2 km

FIGURE 174.—Drastic enlargement of a panoramic camera frame provides a wealth of detail within the small area outlined in figure 172. Note the many large blocks on the slope. The largest block is about 125 m wide. Most blocks apparently originated at the discontinuous ledge near the top of the slope. Note also the fillets on the upslope side of many of the blocks. They probably consist of fine-grained debris that was trapped behind the blocks as it moved downslope. The arrows identify what appear to be two craters in the process of being destroyed by erosion. Otherwise, craters are absent on the steeply dipping slope, although numerous craters are present on the gentler slopes above.—M.W.

AS17-2608 (M)

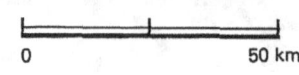

0   50 km

FIGURE 175.–This apronlike expanse of striated material on the northwest flank of Tsiolkovsky is interpreted as a giant landslide. In some respects it resembles terrestrial landslide deposits. A good example is the Sherman landslide in Alaska, which was triggered by the "Good Friday" earthquake in 1964 and which covers part of the Sherman glacier. The striations apparently outline individual filaments or jets of debris like those that have been observed in experiments involving the flow of particulate matter traveling at high velocity. The flow originated at the west-facing scarp marked by an arrow. From the scarp on Tsiolkovsky's rim to the floor of the ancient crater Fermi, where the landslide came to rest, the difference in elevation is about 3000 m. The maximum distance of travel was about 50 km.–H.M.

0   50 km

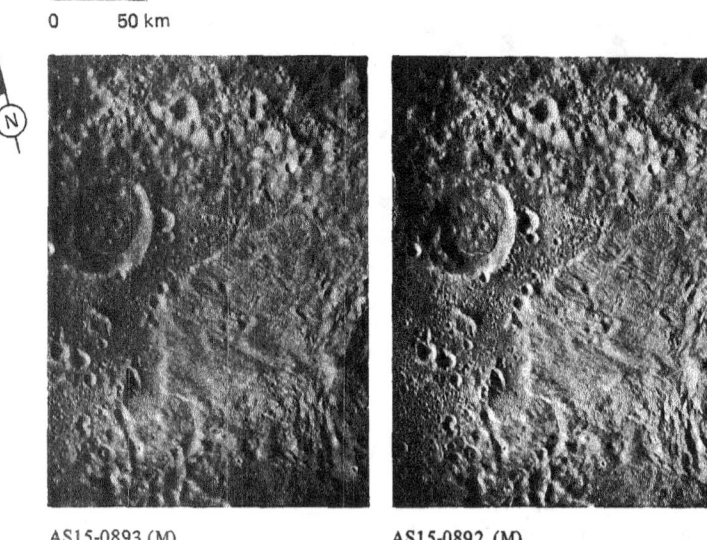

AS15-0893 (M)   AS15-0892 (M)

FIGURE 176.–The setting of the landslide is much more evident in this stereogram provided by L. J. Kosofsky. Here it is evident that the northwest rim of Tsiolkovsky is appreciably higher than the area where the landslide came to rest. The northern wall of Fermi is clearly recognizable as the very rugged south-facing slope near and parallel to the top edge of the stereogram. To those without prior knowledge of the area, Fermi's wall probably was not evident in monoscopic figure 175.–G.W.C.

FIGURE 177.—This topographic contour map (Wu et al., 1972) was compiled from two mapping camera pictures like those in figure 176. For easier reference part of the rim crest of Tsiolkovsky has been marked with a heavy dashed line. From this map the area, the slope, and (after making some assumptions) even the volume of the landslide can be measured. The heavy straight lines across the slide locate the three topographic profiles shown below the map. Profiles are another method of portraying topographic information. On two of the profiles the lower edge or toe of the landslide has been marked with an arrow. (On the third, the lower edge has been destroyed by an impact crater.) Because the landslide is topographically a rather subtle feature, the vertical scale of the profiles has been exaggerated by a factor of 5 to show the configuration of the slide more clearly. However, for comparison, one of the profiles has also been drawn without vertical exaggeration.—G.W.C.

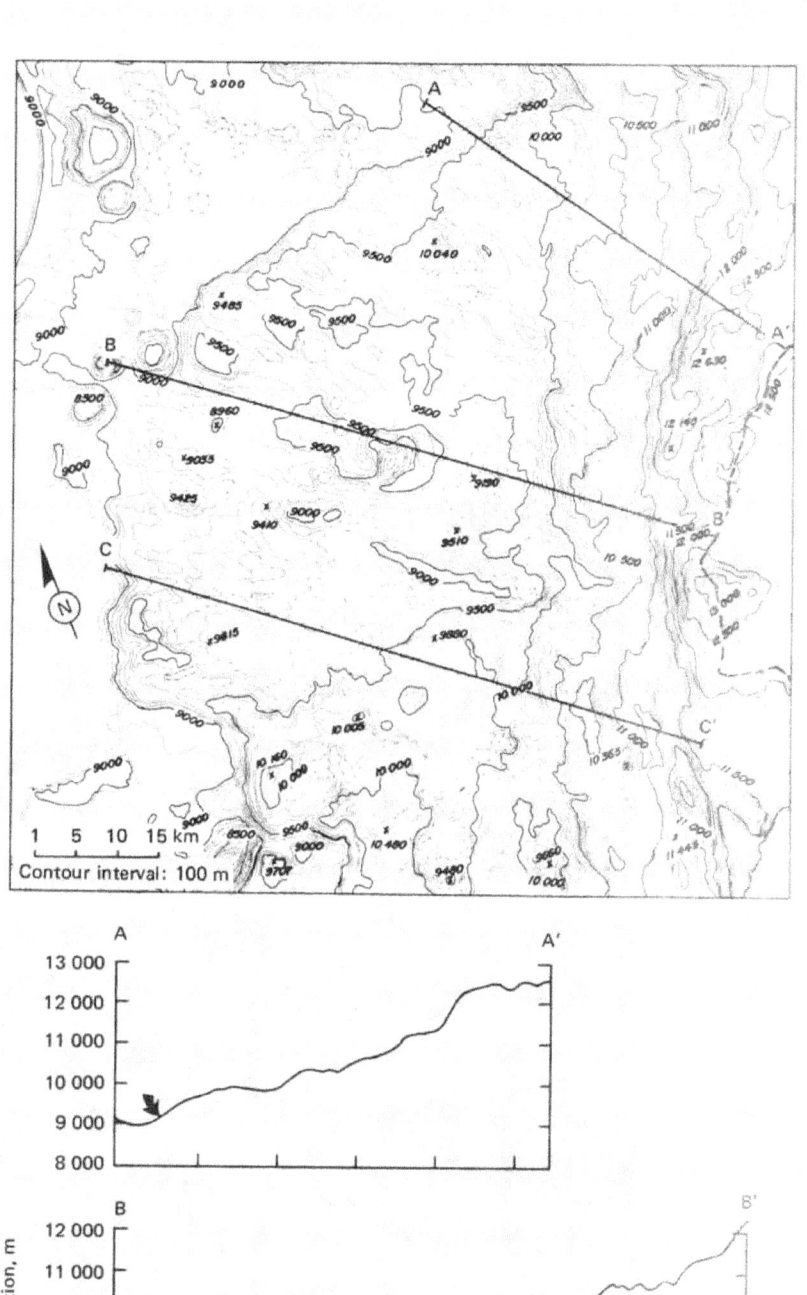

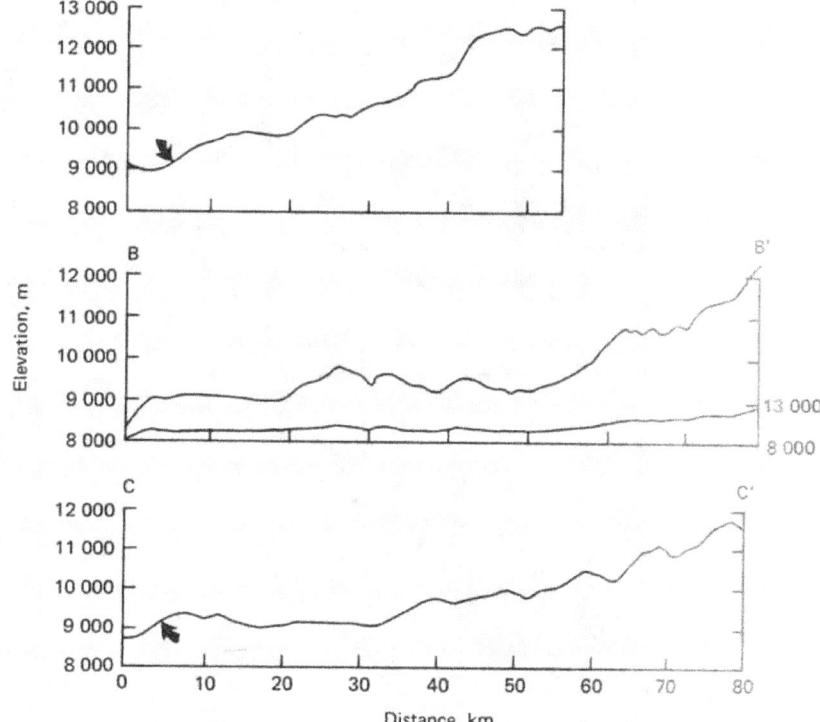

FIGURE 178.—Aitken is a large crater (145 km in diameter) on the far side. Like many other large, young craters it has an outer field of secondary crater clusters and chains, an extensive inner blanket of bright ejecta, bright and rugged walls, a flat floor partly occupied by dark mare material, and a central peak complex. However, in several respects Aitken differs from other craters of comparable age and size. Its central peak has an unusual subcircular pattern, its rim crest is complexly crenulated, and its walls are best described as jumbled (El-Baz, 1973a) rather than terraced. The last two characteristics may have been caused by the shock associated with the formation of the large crater (30-km-diameter) on Aitken's north wall. The shock may well have destroyed the preexisting terraces and caused the additional slumping that gives the rim crest its crenulated appearance. Many interesting features are visible on Aitken's floor. Among them are some unusual ridges (arrow), which were described earlier (fig. 84), and several very unusual craters or depressions, some of which may be the sources of the mare fill (El-Baz, 1973a).—G.W.C.

AS17-0341 (M)

AS15-1541 (M)

FIGURE 179.—The ejecta blanket and secondary impact craters of the mare-filled crater Archimedes (80 km in diameter) are visible on the terrain toward the viewer (south) but not on the mare surface to the crater's left and right. Yet at one time ejecta like that to the south must have completely surrounded Archimedes because similar ejecta surrounds craters such as Aristillus (upper right). Thus, the mare lavas, in addition to filling the interior of Archimedes, obviously have covered the eastern and western parts of the ejecta. In turn, ejecta from Archimedes has covered materials of the Imbrium basin like the rugged hills in the lower left of the picture. These stratigraphic relations prove that time elapsed between formation of the Imbrium basin and its filling by mare—time enough for impacts to create Archimedes, the deeply flooded crater to its right (arrow), and similar "Imbrian-age" craters elsewhere, as was pointed out by Eugene Shoemaker in 1962.

Archimedes is the first large crater described in this chapter that has no visible central peak complex. Presumably the complex exists but has been completely inundated by the mare.—D.E.W.

173

AS15-2510 (M)

FIGURE 180.—The crater Humboldt, on the east limb of the Moon, as seen from
Earth, is 200 km across, a little larger than Tsiolkovsky. This view by the Apollo 15
mapping camera looks southward across Humboldt's ejecta blanket and into the
crater. Irregular secondary craters partly covered by the ejecta are in the foreground, and a long chain of secondaries extends from Humboldt's rim to the
foreground. Humboldt is one of the largest craters known to have a prominent
central peak. If the crater is like terrestrial impact structures, the peak may expose
rock uplifted about 10 percent of the crater's width, on the order of 20 km from
beneath the crater floor. This would be an exciting find for future astronauts. A
spider web of cracks on the crater floor suggested to R. B. Baldwin (1968) that the
floor was bowed up in the middle. Later, dark mare lavas flooded low areas in the
outer part of the floor and covered the cracks. A peculiar "bull's eye" double crater
on the crater floor has several counterparts elsewhere on the Moon. The origin of
these double craters is a continuing puzzle.—K.A.H.

174

FIGURE 181.—This view into the shallow crater Gassendi shows another strongly
fractured crater floor. Gassendi is about 110 km wide. Dark mare lavas in the
distance embay the rim and a little of the interior of Gassendi. They may have
entered the crater through the narrow gap partly in shadow below the arrow. Most
craters that have fractured floors are near areas of mare flooding. This suggests that
the fracturing is a consequence of volcanic activity. An area next to the central
peaks of Gassendi was the runnerup choice for a landing site for Apollo 17.—K.A.H.

AS16-120-19295 (H)

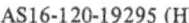

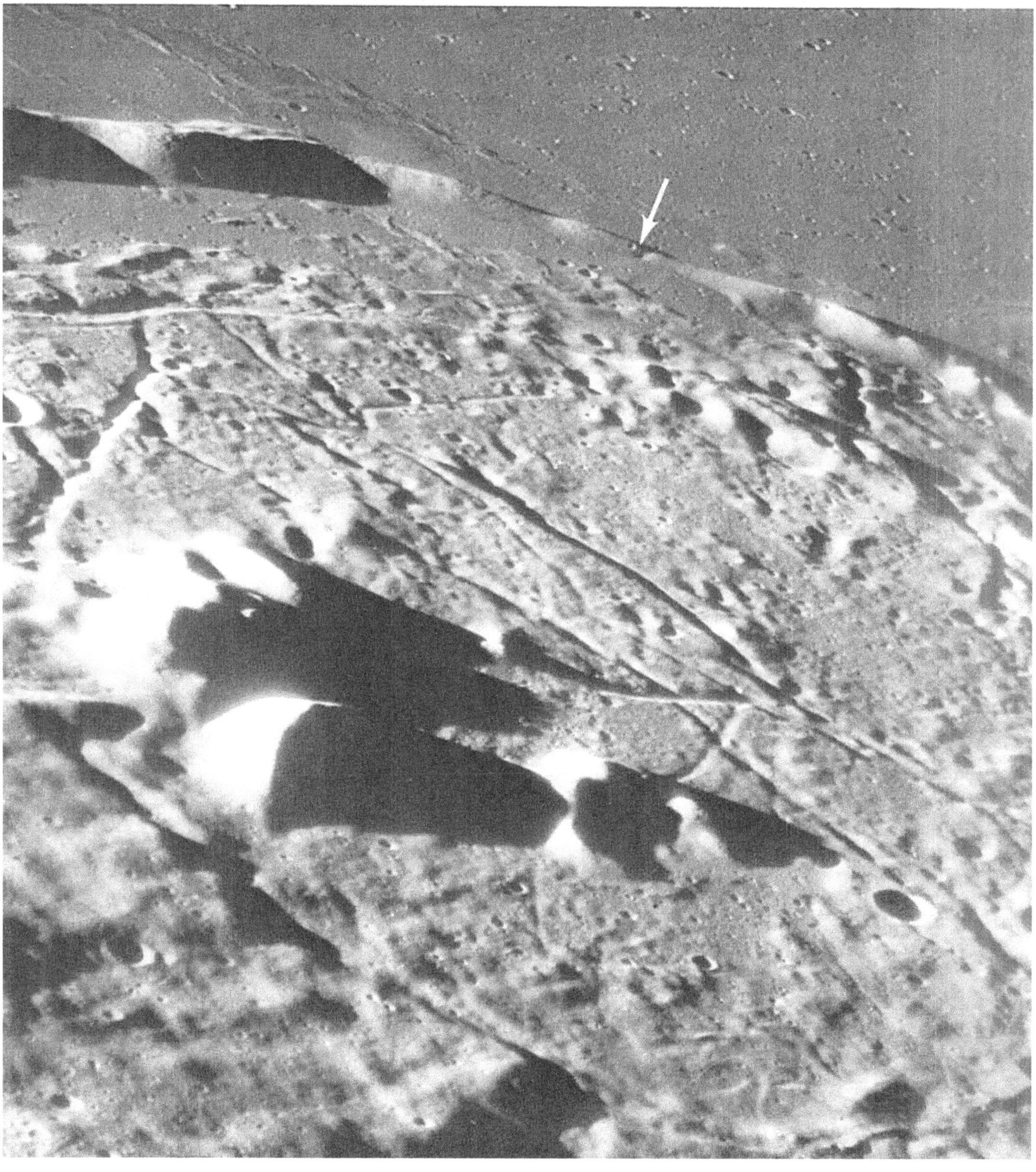

FIGURE 182.—This arcuate structure is Letronne. It straddles the boundary between southern Oceanus Procellarum and the southern highlands. From "horn to horn" it is about 115 km wide. Astronomers have long recognized Letronne as a crater and geologists also interpret it as a crater because those parts preserved have much in common with better-preserved craters. The preserved crater elements include a large segment of a raised rim, a partly preserved blanket of ejecta occupying depressions along the lower edge of the picture, and the tips of three centrally located peaks that presumably represent the top of a buried central peak complex. The largest and steepest slopes along the rim face inward and probably define the wall of Letronne. The northern one-third of the rim and wall has been almost completely buried by the mare lavas of Oceanus Procellarum. An isolated small hill (arrow) and the crudely arcuate band of mare ridges east of the hill mark the approximate position of the buried rim. The abrupt disappearance of the rim beneath Oceanus Procellarum suggests faulting, but vertical movement without faulting is also possible; this part of Procellarum may have been tilted downward or the adjacent highlands upward.—G.W.C.

AS16-2995 (M)

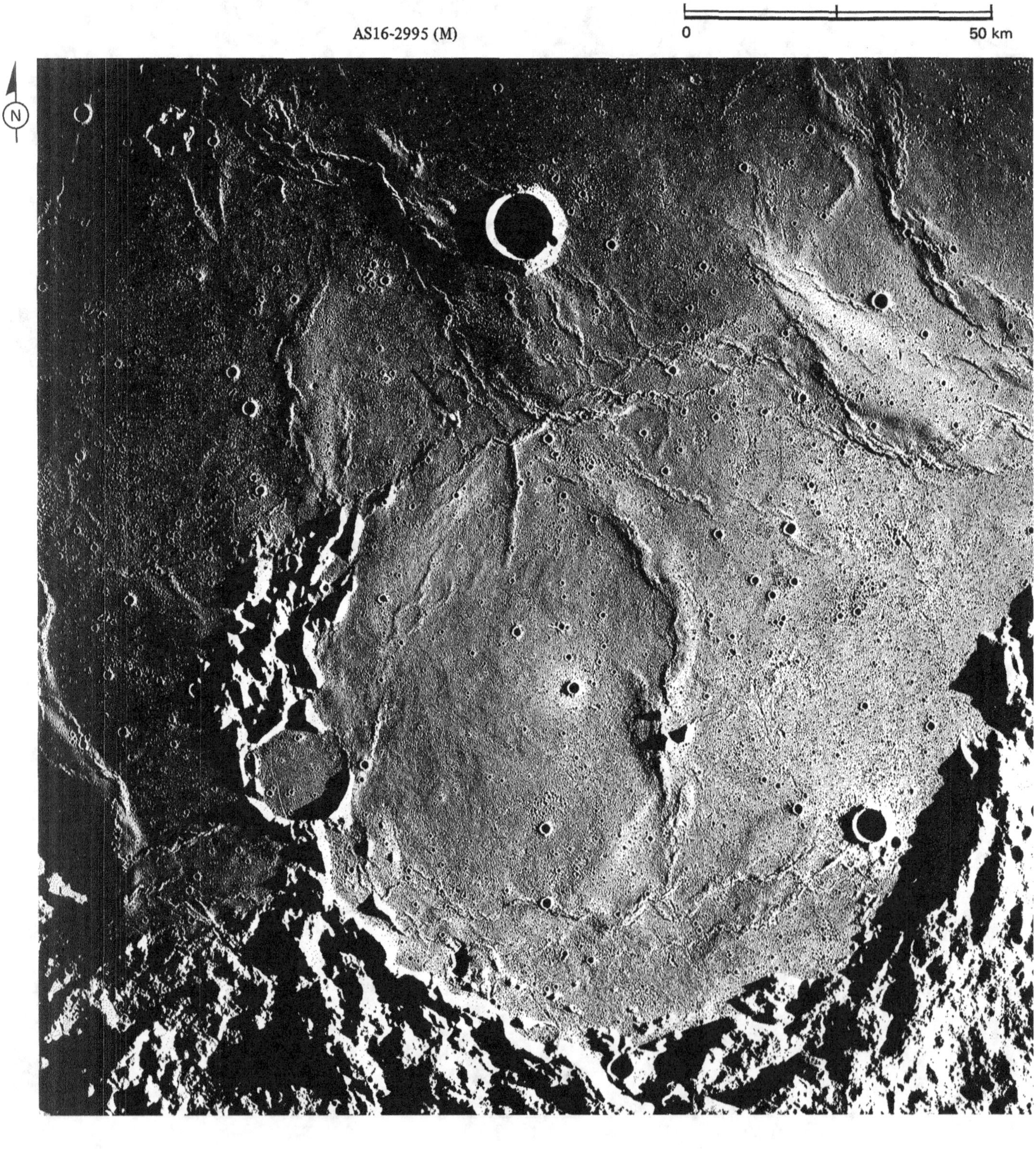

0   50 km

AS16-0839 (M)

FIGURE 183.—The very old crater Hipparchus has been nearly obliterated during eons of lunar change. Except for the circular pattern of subdued mountains (dashed line) that surround most of it, Hipparchus would not be recognizable. It is about 150 km wide. Beyond it near the horizon at the left is Sinus Medii, the smooth dark-surfaced area that lies at the center of the Moon when seen from Earth. Part of the rim of Hipparchus is modified by "Imbrium sculpture," the pattern of ridges and grooves radial to Mare Imbrium, which affects the lunar surface for more than 1000 km from Imbrium. The four arrows show how the pattern radiates outward across this part of the Moon. (The Imbrium basin lies beyond the left horizon.) Craters and hills have been formed in the floor of Hipparchus and subsequently have been partially inundated by marelike plains deposits. The ejecta blanket of the much younger crater Horrocks (identified by *H*) is, in turn, spread over the mare-like filling.—M.C.M.

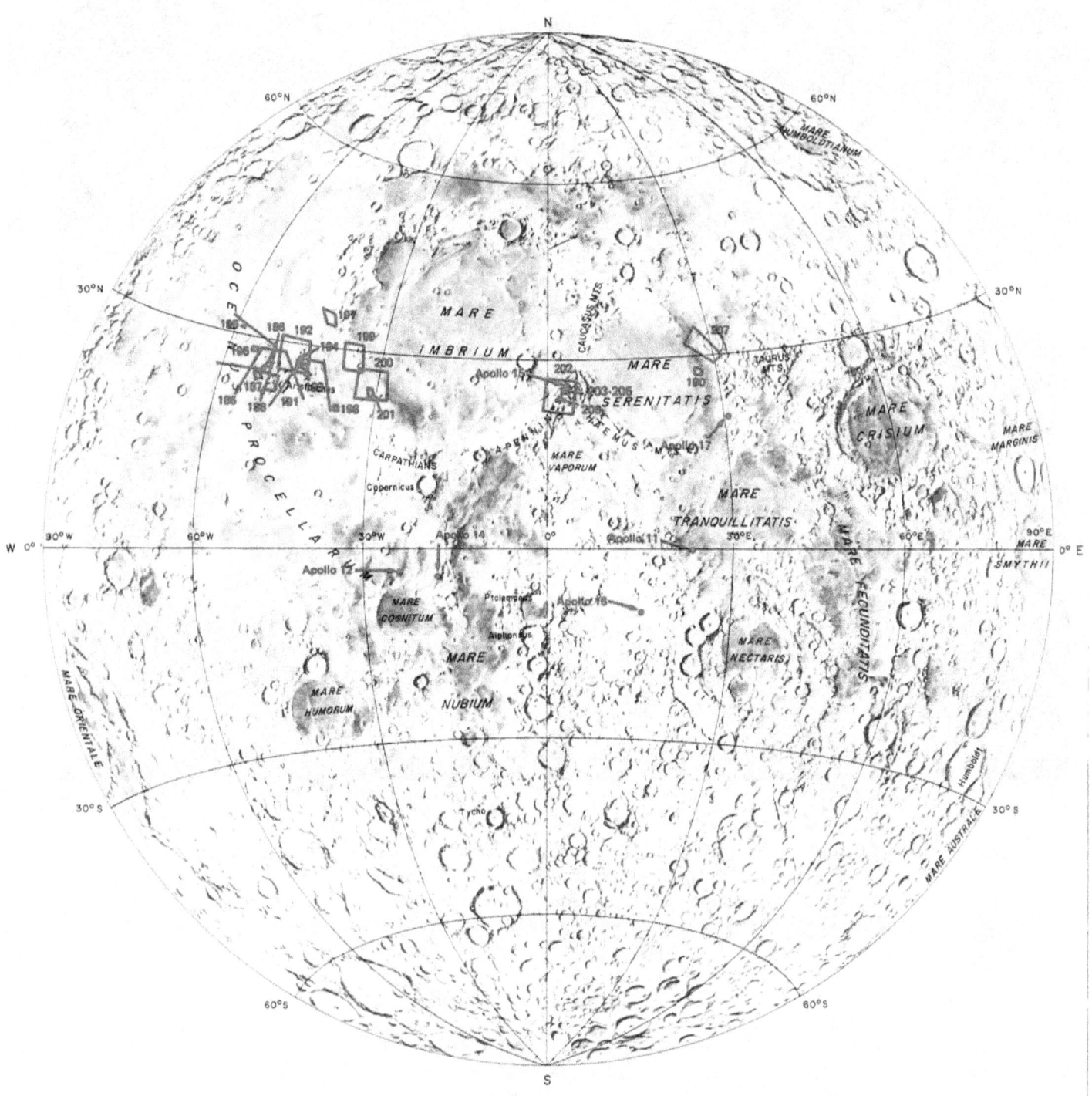

Near side

FIGURE 184.—Location of photographs of sinuous rilles; numbers correspond to figure numbers. [Base map courtesy of the National Geographic Society]

6
Rimae

Sinuous Rimae

The sinuous valleys on the Moon (called sinuous rimae or rilles) have been the focus of much debate. The most extreme view is that they were cut by flowing water. However, study of the details of their shape and the nearly complete lack of water in returned lunar samples make this hypothesis very unlikely. Two other possible hypotheses are that they were formed by faulting and subsidence of the lunar crust and that they are lava channels or collapsed lava tubes. The wide variety in shapes suggests that both of the latter processes have been involved. Some sinuous rilles seem to represent almost pure fault troughs while others represent valleys that were formed by flowing lava and modified by filling of slump materials from the walls.

Data obtained by the Apollo 15 astronauts when they landed beside Hadley Rille indicate that layered basaltic lava flows are exposed in the walls of the valley. These rocks may be the only ones sampled during Apollo that had not been moved one or more times by impact; that is, they were "bedrock" samples. Hadley Rille starts in a volcanic crater on the flank of the Apennine Mountains and flows into the Imbrium basin. The floor of the channel is very irregular; possible eruptions of lava along the channel course may have added to the material flowing from the source crater. Also, the meander pattern of the rille, with points projecting into the channel on one side and rounded on the opposite shores, indicates a great deal of modification by flowing lava of what may originally have been a fault trough.

In some sinuous channels a small valley has formed within a larger valley, indicating at least two episodes of valley formation. Other meandering channels do not start or stop in a crater—they are formed on lava plains that are so flat it is difficult to tell the direction of flow. Still other channels have formed on mountainsides covered by hummocky ejecta thrown outward from major impact basins; these rilles may have been formed by some process other than lava flows.

In summary, there appear to be several different types of sinuous valleys on the Moon. One current theory is that most of them are channels formed by basaltic lavas. The valleys may have originated as fault troughs that were later modified and obscured by lava flows, impact ejecta, or landslide material from the walls. A few of the sinuous valleys (those that have formed on the hummocky mountainous terrain) still are difficult to explain by the processes suggested so far.—H.M.

FIGURE 185.—This south-looking oblique view, centered near 25.5° N, 50.5° W, depicts a prominent "cobra-head" rille, Vallis Schröteri (Schröter's Valley). In the foreground is the Aristarchus plateau, and in the background the smooth surface of Oceanus Procellarum. The two large craters in the middleground are Aristarchus (38-km diameter) on the left, and Herodotus (30-km diameter) on the right. At first look the sinuous, flat-floored Schröter's Valley and the tightly meandering channel within bear a striking resemblance to river valleys on Earth, and some viewers of the Moon have thought that sinuous rilles were formed by flowing water. In detail, however, sinuous rilles differ from river valleys in many respects; for instance, Schröter's Valley becomes smaller toward its downstream end. They are much like an entirely different terrestrial feature—lava channels. (See fig. 189.) Most geologists think that Schröter's Valley is a channel through which lava flowed from the circular crater at the "head" of the rille to the lower elevation of Oceanus Procellarum, a distance of approximately 175 km. Figures 186 to 188 are enlargements of the four-sided areas outlined in this photograph.—M.C.M.

AS15-2611 (M)

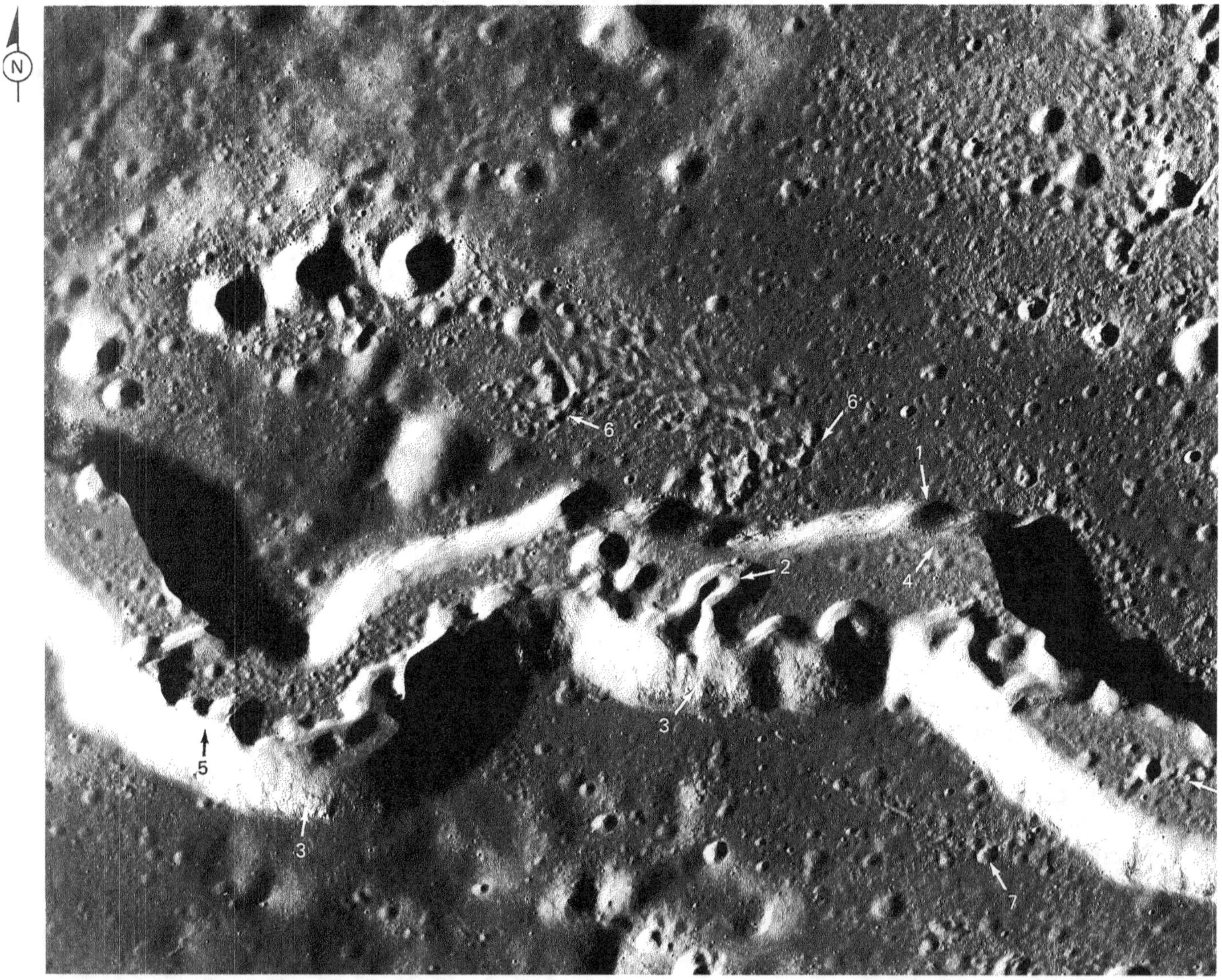

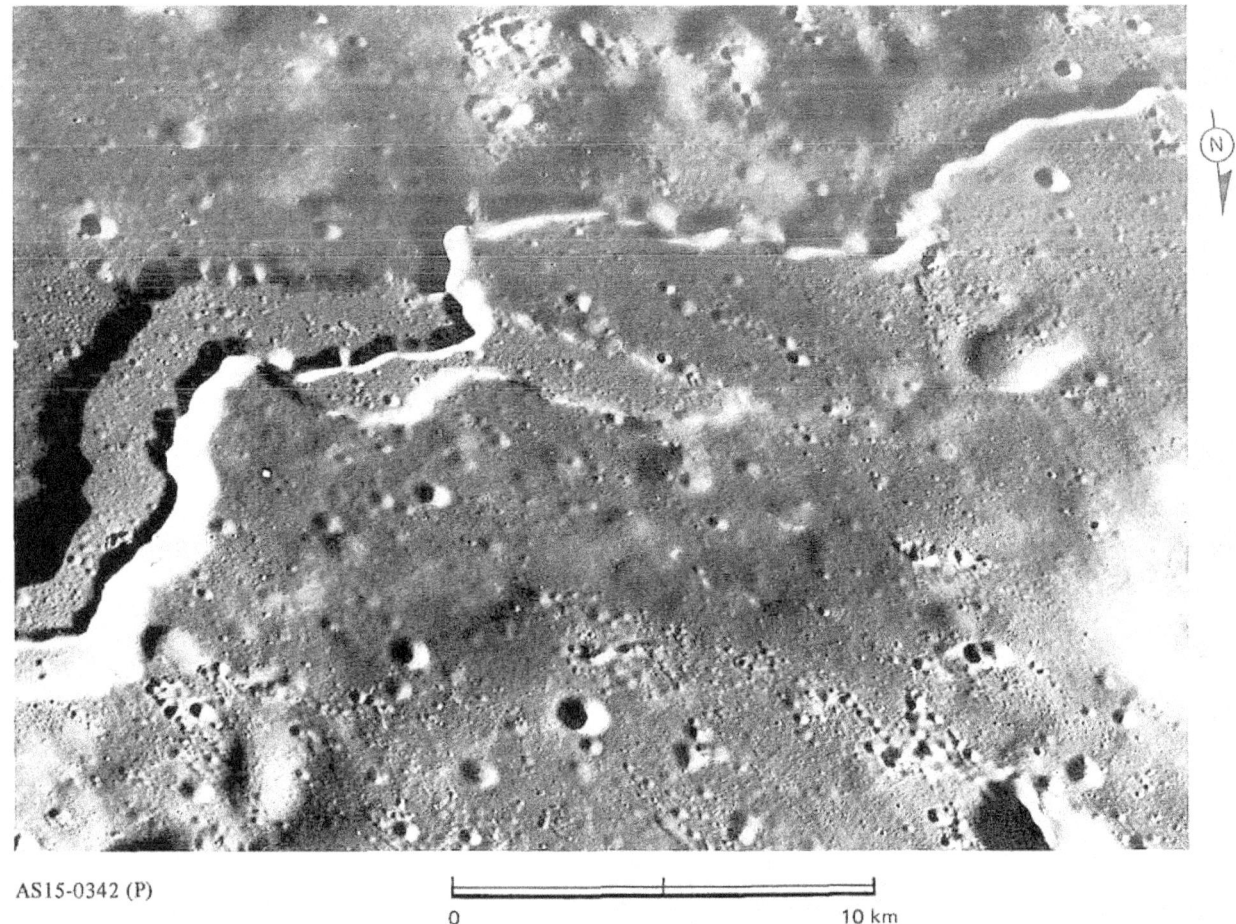

AS15-0342 (P)

0   10 km

FIGURE 187.—A view near the mouth of Schröter's Valley. Clearly shown are the old broad valley stage of formation of the valley—largely controlled by faulting—and the fresher, younger, meandering inner valley that probably represents a basaltic lava channel. Rocks returned by Apollo 15 from the banks of a similar channel are layered flows of vesicular (full of gas bubble holes) basalts. The valley drops 1600 m along its length—a slope similar to terrestrial lava channels. Secondary craters from the bright fresh impact crater to the east show that the impact crater is younger than the valley. The valley—like all terrestrial and lunar lava channels—gets narrower and shallower downstream, possibly reflecting the cooling of the lava and its loss of mobility as it gets farther from its source.—H.M.

FIGURE 186.—Schröter's Valley in the Aristarchus plateau is one of the largest lunar sinuous rilles (width in picture is about 5 km). The valley consists of an arcuate rille (1) that contains a meandering sinuous rille (2). The valley here traverses what appears to be a lava plain embaying low hills in the southern part of the picture. The blocky outcrop ledges, probably lava layers, near the rim (3) and the blocks at the inside base of the slopes (4) are of interest as is the downslope movement of material in the walls, which results in partial burial of the inside rille (5) and shows that the valley is laterally enlarged by mass-wasting processes.

The sinuous shape, uniform width, presence of low levees, irregular depressions at the head of this and other rilles (outside the picture), and uniform cratering of floor and surrounding terrain suggest that the feature originated as lava-flow channels, or collapsed lava tubes (Greeley, 1971). The rilles in the picture probably formed by the draining of a large lava flow channel and a smaller channel in a somewhat later flow that was confined within the boundaries of the larger channel. Incision by thermal erosion of lava streams with turbulent flow is an alternative explanation for the formation of sinuous rilles such as Schröter's Valley (Hulme, 1973).

Other conspicuous features in the pictures are secondary crater clusters from the young crater Aristarchus (6). The secondary clusters cross the rille at (7), and show that the crater Aristarchus is younger than the rille.—B.K.L.

AS15-0341 (P)

0   5 km

181

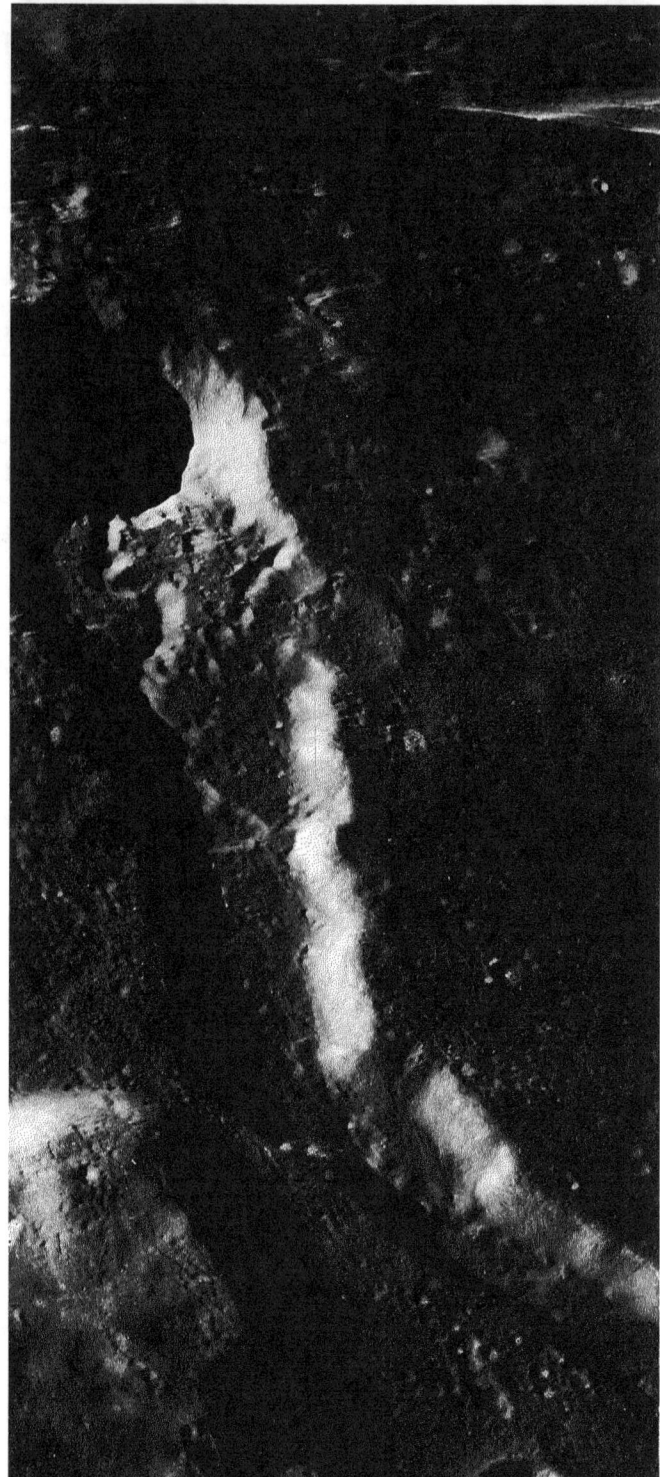

FIGURE 188.—The pile of volcanic rocks that surround the crater at the head of Schröter's Valley has been called the Cobra Head. The crater is 40 km in diameter, the pile of volcanic rocks is 100 km in diameter. The abundant secondary craters on the volcano and the channel show that the crater Aristarchus, from which they were thrown, is younger than Schröter's Valley. The straight line segments of the valley that turn at sharp angles show that the basic form of the valley has been made by faulting, or breaks, in the lunar crust. The sinuous, meandering valley indicates that the basic straight valley shape has been modified heavily by later lava flows coursing down the valley. Samples returned by Apollo 15 from the side of a similar valley, Rima Hadley, strongly indicate that the rocks in the valley walls are basaltic lava flows. The rocks are layered because of the flow and have many holes (vesicles) formed by gas escaping from the rocks when they were still molten.—H.M.

AS15-0332 (P)

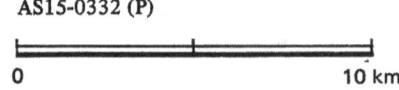

0   10 km

FIGURE 190.—Pictured here is a lunar sinuous rille located in the eastern part of Mare Serenitatis. It is interpreted to be a lava channel because it resembles terrestrial lava channels like the one in figure 189. The rille starts in a crater (presumably the volcanic source) and follows a sinuous course downhill. The channel edges are marked by low ridges resembling those that form on actively flowing lava channels on Earth; the terrestrial ridges are formed when the outer edges of the molten lava chill and make natural levees. The lower reaches of the channel may be a lava tube that formed when the upper surface of the flowing lava chilled and froze while the hotter lava in the interior continued to flow. On the other hand, this ridge at the end of the channel may be a fracture along which lava has been extruded. Similar features are called "squeeze-ups" in terrestrial lava flows. A possible small lateral lava flow is visible near the end of the feature.—H.M.

AS15-9309 (P)

0   5 km

FIGURE 189.—This volcanic crater and lava channel are near Craters of the Moon National Monument in southeastern Idaho. The resemblance to lunar "cobra head" rilles (like the one in fig. 188) is immediately apparent. The crater lies in the gently sloping Snake River Plain, a broad expanse of volcanic rock with craters and linear and sinuous features thought to have formed in the same manner as their lunar counterparts. Apollo astronauts studied this area in preparation for lunar missions. The crater pictured is approximately 100 m in diameter, and the associated channel is over 5 km long.—M.C.M.

S-69-42867

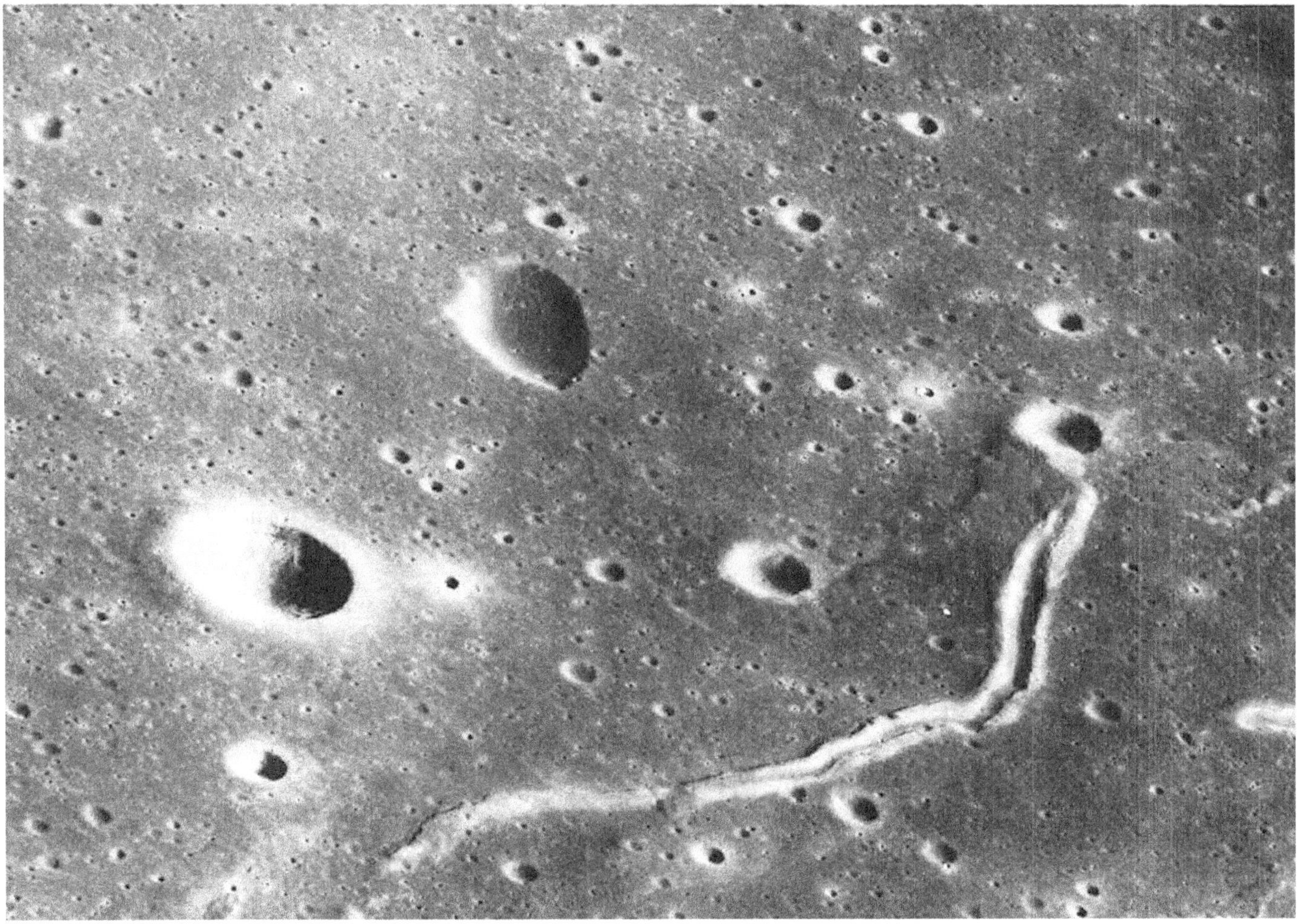

AS15-2606 (M)

FIGURE 191.—The sinuous rilles in this photograph, east of the Aristarchus Plateau, are a particularly interesting complex of these unusual features for which no entirely convincing terrestrial analogs have yet been recognized. The rilles are at least partly controlled by fracture and most originate in craters. Rille *A*, beginning on the flank of the crater Prinz, appears to have had a distinct two-cycle history, producing crater-in-crater and rille-in-rille structures. Rille *B* crosses a ridge of highland material without deviation or deformation, suggesting that the feature was superposed—that is, let down—from an earlier higher mare surface; the ridge appears to have been eroded. A shallow narrow rille (not visible in this photograph) occurs within the broader valley of rille *C* and is traceable across the elongate collapsed depression that bisects the main rille. Craters at the heads of rilles probably represent source vents for fluids that either eroded the rilles or formed lava tubes that drained, contributing to the volume of mare lava in Oceanus Procellarum.

Ejecta from the young crater Aristarchus forms light-colored streaks or "rays" across the dark mare surface; the high albedo of the rays may be due in large part to disruption of the surface by secondary craters.—C.A.H.

FIGURE 192.—A low Sun angle, larger scale view of part of the area shown in figure 191. Rima Prinz I (*1*) graphically displays many of the features considered to be indicative of lunar basaltic lava channels. The rille starts in a crater on the side of the ancient crater Prinz (just off the photograph) and descends about 300 m, becoming narrower and shallower downslope. It is a "two story" channel with a broader older channel and crater inside of which is a younger, more sinuous, channel with its source vent. Samples returned by Apollo 15 from the very similar looking Rima Hadley were from a vesicular (full of holes formed by gas bubbles) flow of layered basalt. The next channel to the west (*2*) also gets narrower and shallower downslope. It is the best example of distributaries—that is, a branching network of smaller channels at the downstream end of a larger channel.

Krieger (*3*) is a "Gambart type" crater inferred to be volcanic in origin. Its flat floor, irregular shape, and highly irregular external deposits resemble the crater Gambart south of Copernicus, which was studied in 1967 by Apollo 17 astronaut H. H. Schmitt. The deposits from Krieger lie on the surface of the mare basalts, indicating that the crater is quite young. Its youthfulness is confirmed by the freshness of the crater floor deposits and the characteristic shape of these deposits. A nice example of a sinuous rille, interpreted as a lava channel (*4*), runs out of the crater onto the mare surface. This lava surface is marked by wrinkle ridges (*5*)—complex mare ridges, generally asymmetric, with a braided ridge along one edge. These ridges are interpreted to be faults or breaks in the mare lava flows along which a later generation of molten lava has been both intruded, raising the already cooled mare lava flows, and extruded onto the mare surface.

A small (8-km-diameter), young impact crater (*6*) is excavated into the mare material. The continuous ejecta blanket formed by the base surge—turbulently flowing ejecta riding on the surface—is particularly well shown. It forms a typical dune pattern (*7*); the crest-to-crest distance (200 m) is an index of the velocity of flow of the base surge. Similar dune lava features have been seen forming around terrestrial experimental craters.—H.M.

AS15-2083 (M)

0   50 km

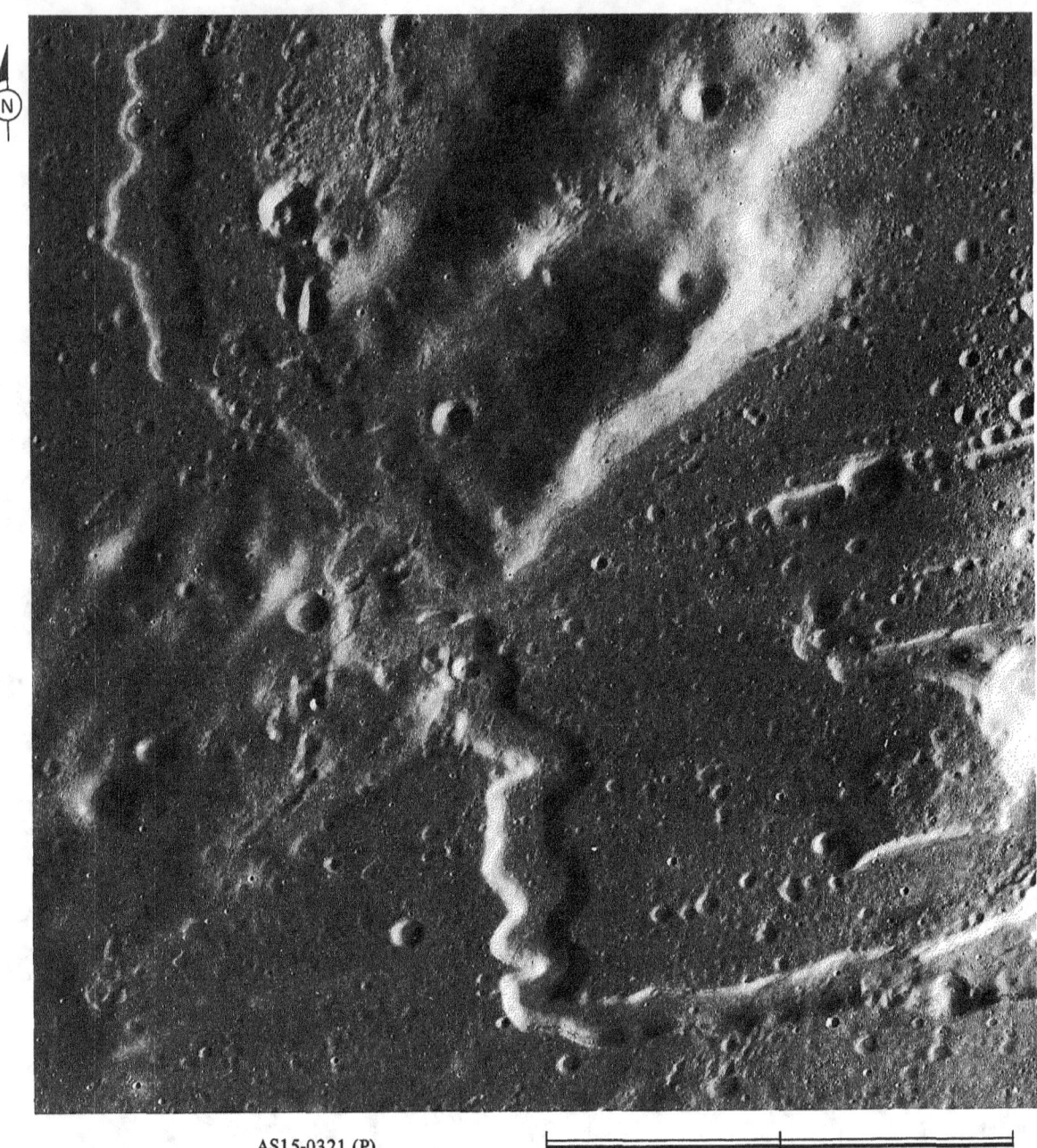

AS15-0321 (P)

0   10 km

FIGURE 193.—A larger scale view of part of the area of figure 192 shows the bend in Rima Prinz II. It and Rima Prinz I form subparallel linear depressions that originate in small craters on the north flank of the crater Prinz. Both make right angle bends about 1/4 to 1/3 of the way from their source. The maximum width of the part of Rima Prinz II shown in this photograph is about 1.5 km. It is at least 100 m deep in the dark mare materials, and shallower in the rugged and older circumbasin (terra) materials. The rille is normal to the terra ridge where it cuts across the ridge, and is younger than the youngest materials it incises.

The processes by which lunar rilles form are open to controversy. Their sinuosities lack the characteristics of meanders in most terrestrial streams. The rims along the north portion of Rima Prinz II have both a rectilinear zigzag pattern and subdued arcuate sinuosity. In general, a concave reentrant is opposite a protruding wall. The rille floor displays the same crater density and the same crater size distribution as the mare materials adjacent to the rim. The morphological evidence, therefore, suggests that tensional stresses in the lunar crust probably caused the rille floor to subside between the steep normal faults that form the rille walls. The zigzag pattern of the walls probably is caused by irregular faulting along conjugate joints and fractures in the lunar grid.—M.J.G.

FIGURE 194.—The sinuous rilles here are part of a network controlled to some degree by fractures. In an area view (fig. 192) it may be seen that the east-west segment at *A* parallels prominent linear trends of several rilles northeast of the Aristarchus Plateau. Fracturing alone, however, cannot explain the origin of the rilles. As shown by the larger rille (*B*), material has clearly been removed from the walls so that, despite their parallelism, they cannot be fitted together. In addition, a terrace occurs at *C*, suggesting two cycles of rille formation. The larger rille is nearly obliterated at its juncture with rille *A*, suggesting that formation of *A* may have been slightly later, and that some sort of fluid erosion and overbank flooding may have been involved.—C.A.H.

AS15-0320 (P)

0   10 km

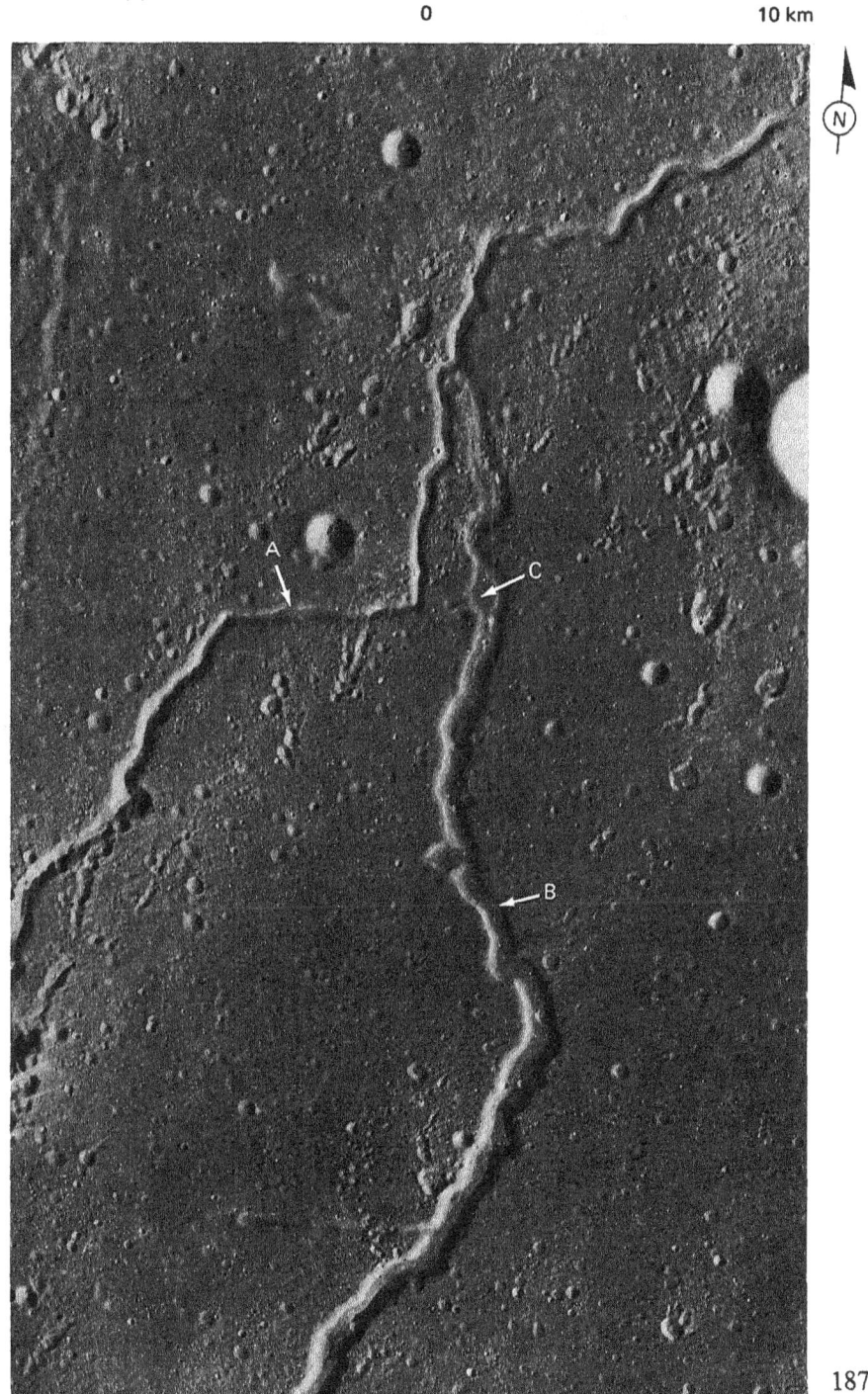

187

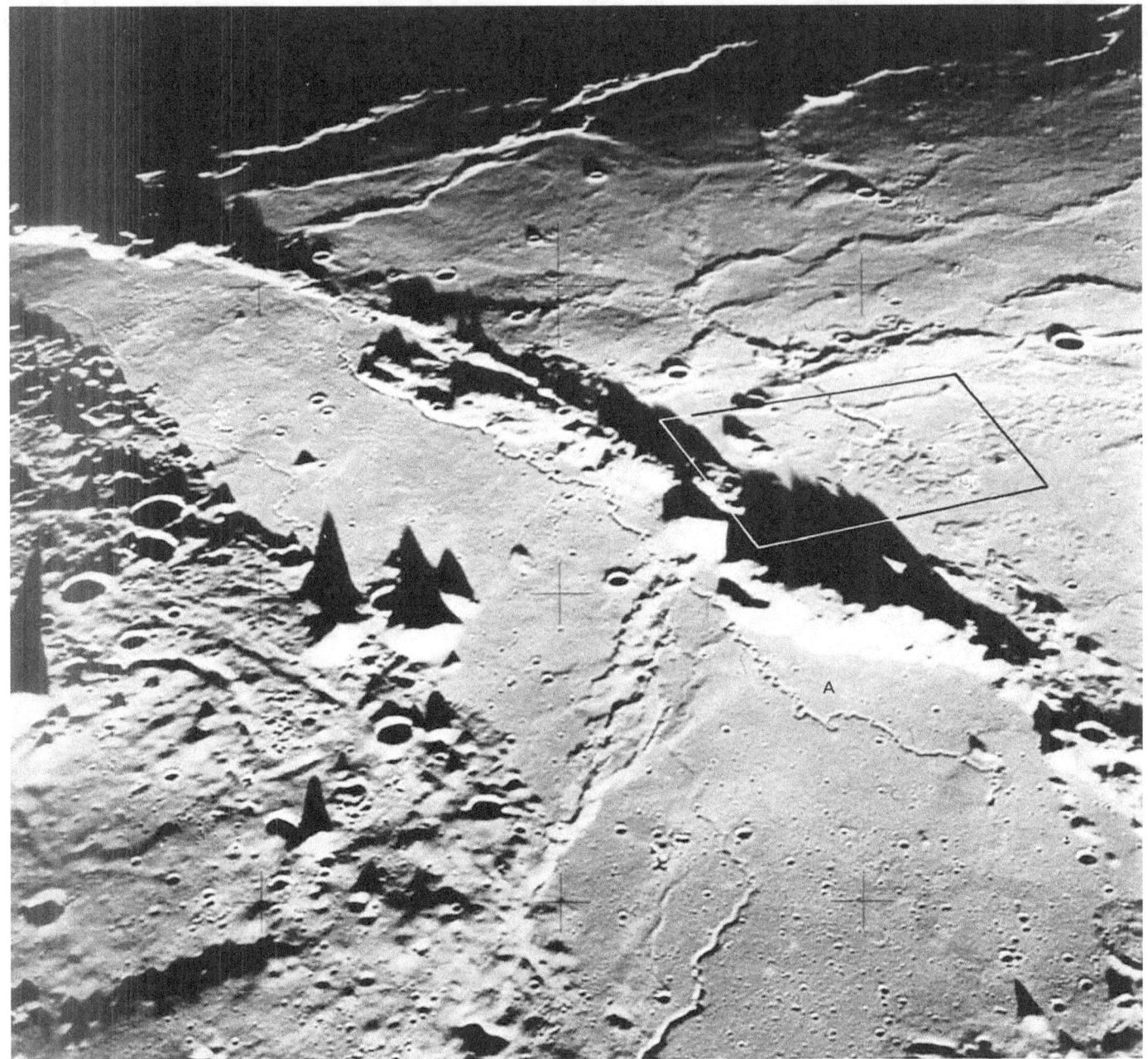

AS15-88-11982 (H)

FIGURE 195.—Relief is exaggerated in this near-terminator (low-Sun) photograph of the northwest margin of the Aristarchus Plateau. A broad graben (approximately 15 km wide) separates the rugged linear mountain chain from the sloping plateau surface. The crenulate profiles of mare wrinkle ridges (some as high as 200 m) are visible on the horizon; one mare ridge bisects the graben. Sinuous rilles are prominent in the graben; several originate on the plateau. The surface of the mare in the graben appears to be higher on the west side of the mare ridge, suggesting faulting associated with ridge formation. The rille A crossing the ridge is nearly 100 km long and has interlocking meanders, which preclude its formation as merely a crack in the mare surface; erosion by downstream transport of a fluid seems necessary to explain such sinuosity. Origin of the diamond-shaped Aristarchus plateau itself is enigmatic, although its relatively straight edges suggest fault control.—C.A.H.

FIGURE 196.—Sinuous rille (*A*) is unusually shallow for its width and has probably been mantled or filled since its formation. The entire area shows evidence of an older mantled and subdued topography on which secondary craters (*B*) from the 40-km primary crater Aristarchus (to the southeast) have been superposed. Among the mantled features are a straight rille (*C*), craters (*D*), and numerous polygonal depressions (*E*), which may be subsidence features, possibly caused by withdrawal of magma at depth, or differential compaction of lava over small highland blocks. The Aristarchus plateau is to the southeast.—C.A.H.

AS15-0349 (P)

0   10 km

FIGURE 197.—A number of mare-related features are present in this oblique view looking north across a mare surface between Mare Imbrium and Oceanus Procellarum. A broad mare arch is visible in the foreground. As it is followed toward the northwest the arch gives way to a series of small ridges, which then converge on an unusual sinuous chain of craters, elongate ridges, and elongate depressions. This sinuous chain leads into a large arcuate elongate depression at the boundary between the highlands and the mare. Similarity of parts of this structure to parts of some sinuous rilles, for example, Rima Hadley, suggests that its strange morphology may be a poorly understood variation of a lava channel, possibly a partially collapsed lava tube.—J.W.H.

AS15-93-12725 (H)

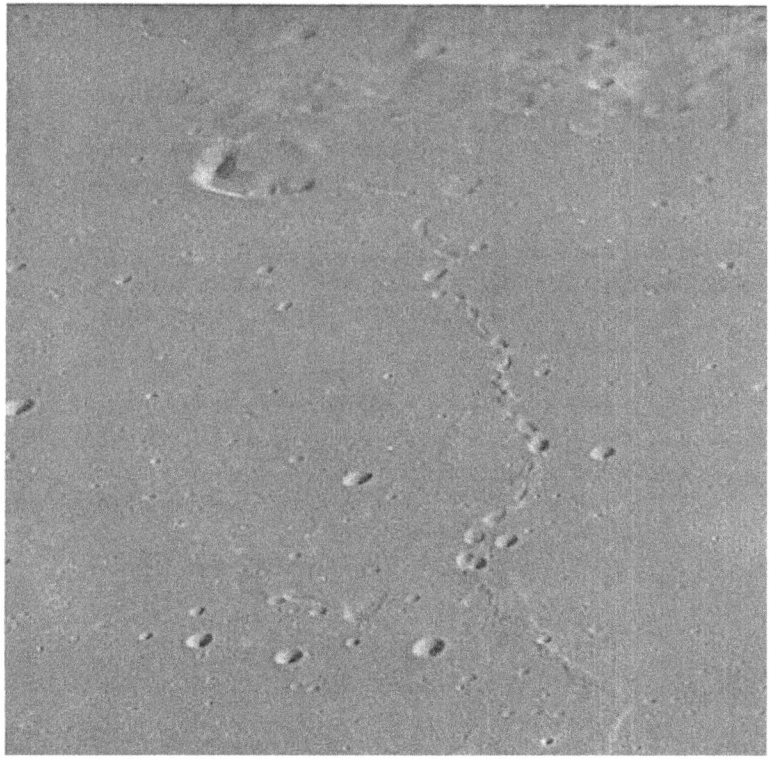

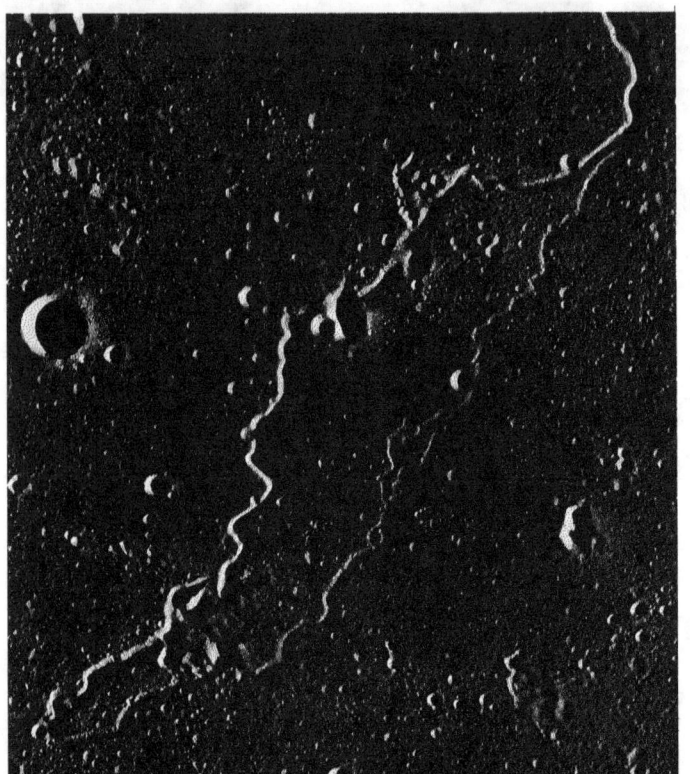

FIGURE 198.—The branching sinuous rille in this photograph is part of a continuous single rille extending more than 300 km across the mare in northeast Oceanus Procellarum. The width of the rille is essentially constant (approximately 250 m) throughout its length. The secondary branches, all of which rejoin the main "stream," are shallower than the larger channel. Slope is probably northeast toward the center of Mare Imbrium. Like some other lunar rilles, this one crosses several mare ridges with no apparent deviation or deformation; unlike many rilles, particularly those near the Aristarchus plateau, there is no associated crater at either end. The origin of lunar sinuous rilles remains controversial. Among the alternatives proposed are lava channels and lava tubes, but fracture control is decidedly apparent in some places. Some sort of fluid erosion, however, seems necessary to account for the configurations of many rilles with exactly parallel walls from which material has been removed; lava may be the most plausible agent for erosion—inasmuch as no evidence of water exists in the lunar samples. The diversity among rilles suggests that several genetic hypotheses may be required to explain all of them.—C.A.H.

0   10 km   AS17-3128 (P)

FIGURE 199.—Delisle and Diophantus are two relatively young craters, 27 km and 19 km in diameter, located north of the area shown in figure 198, in the western reaches of Mare Imbrium. Both are younger than the mare materials in which they are excavated, and also younger than the narrow sinuous rille between them. The rille is sculptured by lineaments radial to the rim crests of Diophantus or Delisle; the lineaments presumably were produced by ejecta from one or both craters.

Both craters exhibit characteristics of young lunar impact craters that have undergone relatively little degradation. The rim crest is sharp, slump terraces inside the walls are distinct, and the density of younger superposed craters is low: only one very young rayed crater, 3 km in diameter, is superposed on Diophantus ejecta.

Five major morphologic facies of crater ejecta can be distinguished in concentric zones away from the crater rim crests: (1) closest to the rim crest, a zone of pitted terrain, dotted with tiny circular craters and large blocks; (2) a zone of rolling troughs and ridges with a smooth surface, grading outward into the third type; (3) elongate or chevronlike secondary impact craters, the rim crests of which form ridges radial to the rim crest of the primary crater; (4) a reticulate network of discontinuous ejecta, forming ridges radial to the rim crest of the primary crater; and (5) isolated secondary impact craters that dot the surface of preexisting mare materials.

The scarp-bounded mountains west and southwest of Delisle are massifs of light-toned circumbasin materials that are older than the mare and crater materials around them.—M.J.G.

AS15-2076 (M)

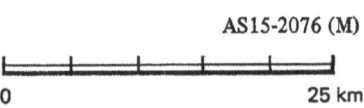

0   25 km

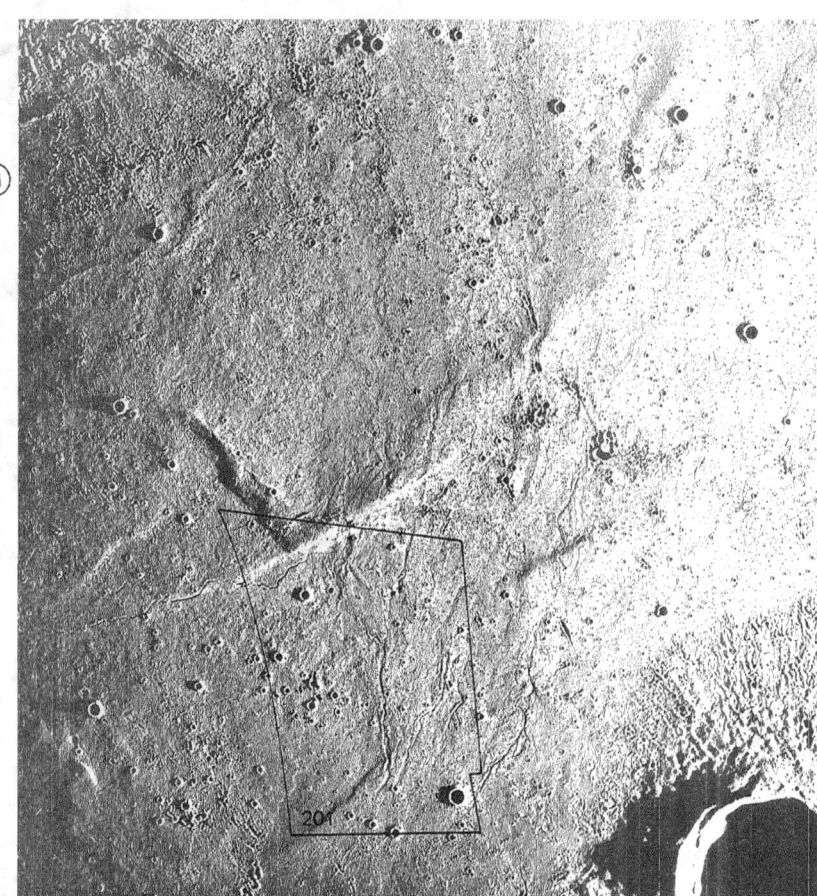

FIGURE 200.—Low Sun angle photograph of mare surface taken by the metric camera. Two large impact craters, Diophantus and Euler, occur in the upper left and lower right corners of the frame; the terminator lies at the left margin of the picture. Many secondary crater chains and smaller primary craters pepper the mare surface, and lava flow fronts, mare ridges, and rilles are common. The rilles exhibit a marked branching or dendritic network pattern. The lava channels become narrower and shallower downstream, to the northeast; terrestrial lava channels also commonly become shallower downstream. The lobes of the lava flows also point northeastward. Apparently the material filling the basin moved from southwest to northeast in this region. The lava flows and rilles obviously antedate the impact craters that pockmark the surface. The outlined area shows the location of figure 201.—H.M.

AS15-1702 (M)

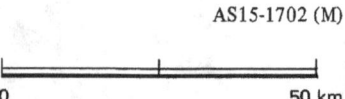

0   50 km

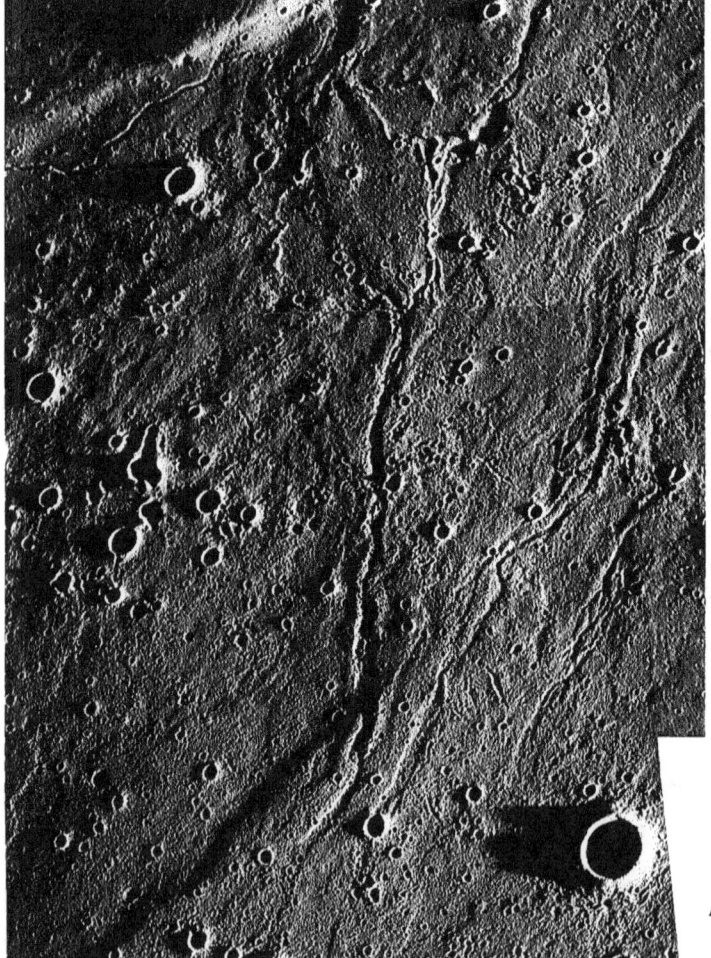

AS17-23717 (H)

FIGURE 201.—This low Sun angle, high-resolution picture shows in greater detail the branching network of lava channels displayed in figure 200. The long shadows along the channel network are formed by its natural levees of chilled lava. The channels branch downslope, as is customary in lava distributary systems on Earth. Small impact craters that postdate the lava flows riddle the channel margins and bottoms and saturate the lava surfaces. Many form lines of secondary craters. Tectonic displacements (faults) offset the lava surfaces and cut across the lava channels; they are, therefore, considered to be younger than the lava channels.—H.M.

AS17-23718 (H)

AS15-1135 (M)

0   50 km

FIGURE 202.—This Apollo 15 photograph depicts the sinuous Hadley Rille, the Apennine Mountains trending from lower left to upper right, and the smooth surface of Palus Putredinus (Marsh of Decay) in the upper left quarter. The Apollo 15 landing site at 26.4° N, 3.7° E (arrow) was selected because of the variety of important lunar surface features concentrated in the small area. The Apennine Mountains, with almost 5 km of relief in the area pictured, are a part of a ring of mountains that surrounds the Imbrium basin and in which very old rocks are thought to be exposed. The smooth lava flows of Palus Putredinus formed later, and Hadley Rille, sharply etched in the mare surface, is thought to be one of the youngest rilles on the Moon. The V-shaped rille originates in a cleft at the base of the mountains in the south and gradually becomes shallower and less distinct to the north and west. Layered rocks crop out in its walls at several places. Near the landing site, the width of the rille is 1.5 km and its depth is more than 300 m. —M.C.M.

193

FIGURE 203.—Astronaut James Irwin and the Apollo 15 rover are perched here above the rim of Hadley Rille. This scene looks northwestward down the rille from the flank of St. George crater (the largest crater in fig. 204). The astronauts discovered that layered basalts crop out in the upper walls of the rille. Talus blocks line most of the walls and the floor of the rille. The rille apparently was once narrower and deeper but has widened by backwasting of the rims.—K.A.H.

AS15-85-11451 (H)

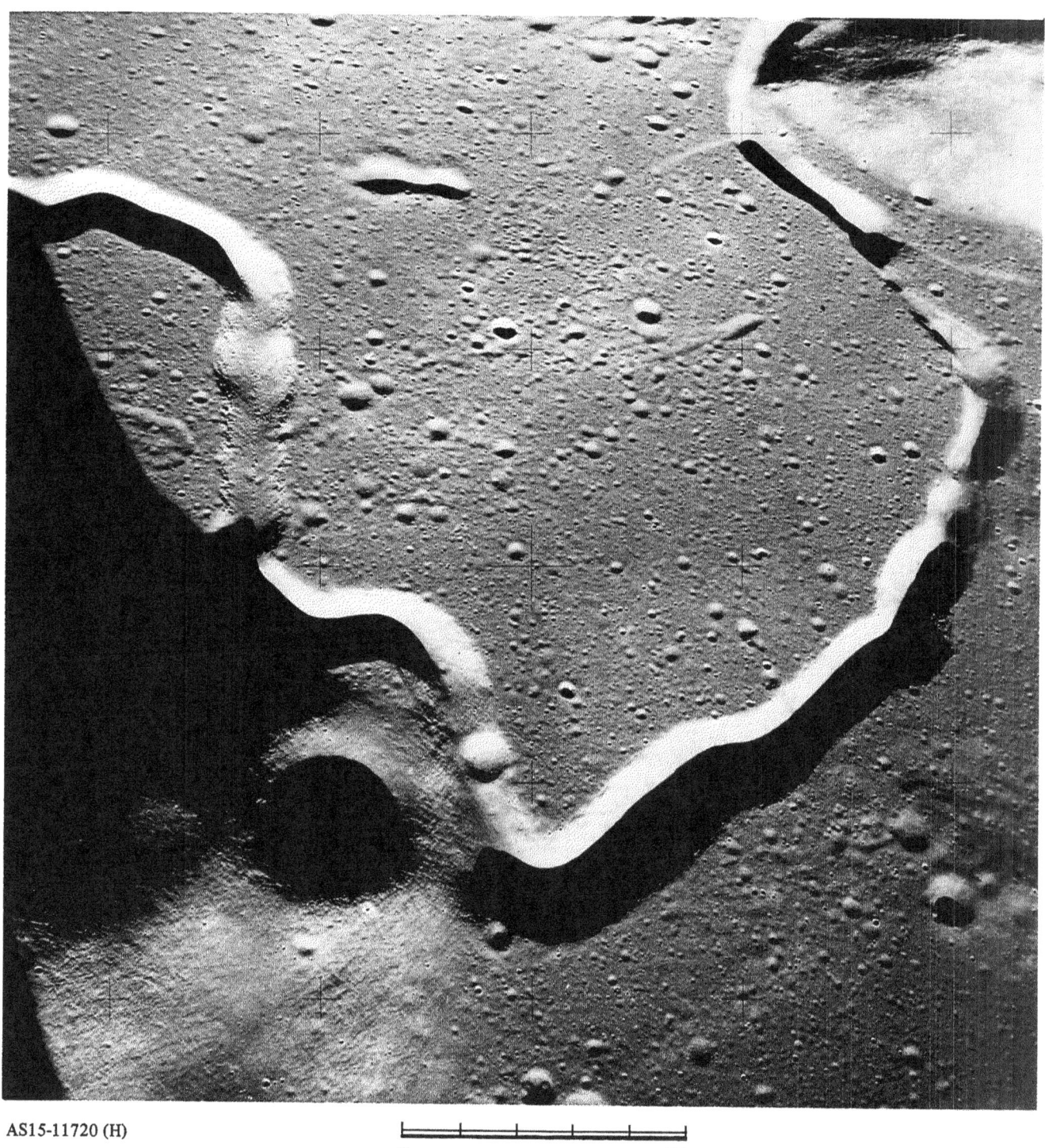

AS15-11720 (H)

0   5 km

FIGURE 204.—Here is how Hadley Rille first appeared to the Apollo 15 astronauts in the LM. They landed in the right foreground. The large subdued crater in the foreground, St. George, is on a mountain to which astronauts drove. Hadley Rille is mainly in the mare basalts of Palus Putredinus. The rille cuts against older mountains in the foreground and at the upper right. Along the top (north) of this photo the rille is discontinuous and is similar in many respects to a partly collapsed lava tube, although it is much bigger than any terrestrial counterpart. The rille is 1.5 km wide and over 300 m deep. It is thought to be a giant conduit that carried lava from an eruptive vent far south of this scene. Topographic information obtained from the Apollo 15 photographs supports this possibility; however, many puzzles about the rille remain.—K.A.H.

AS15-9432 (P)

0   1 km

FIGURE 205.—This closeup of Hadley Rille was taken by the Apollo 15 panoramic camera. This part of the rille appears along the left side (south) of figure 204 under different lighting conditions. Rocky outcrops along the top of the rille walls appear to be cut by fractures, called joints. The blocks that have rolled to the bottom of the rille are huge, up to tens of meters across.—K.A.H.

Straight Rimae

In many places the lunar surface is broken and a portion is down-dropped, forming trenchlike features known as straight rimae or rilles. Some of these rilles are large enough (tens of kilometers across) to be visible on Earth-based photographs; others are so small (a few meters across) that they are barely visible on the highest resolution orbital pictures.

Some of these trenches cut across the surrounding plains, uplands, and craters and may record preferred directions of breakage of the lunar crust caused by internal stresses (the so-called lunar grid). Others ring crater floors and may be related to uplift of the floor caused by crustal readjustment after impact. A few contain low-rimmed dark halo craters that are interpreted to be volcanic vents. Many trenches are curvilinear; some appear to be transitional between straight rilles and sinuous rilles.—H.M.

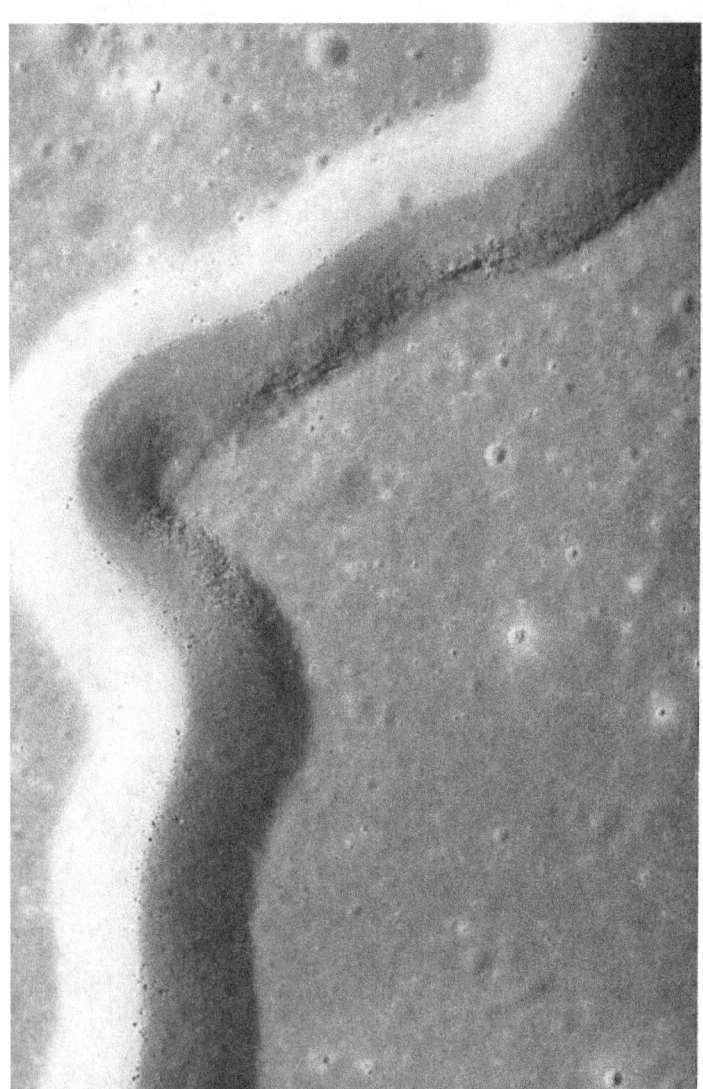

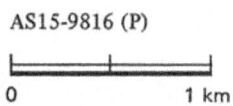

FIGURE 206.—This part of Hadley Rille is far south of the landing site. Lines of outcrops of mare basalt in the upper rille walls suggest thick layering in the basalt. Notice how the edges of the rille stay parallel to each other. One origin suggested for the rille is that it is a graben or fault valley. In its present form, however, the rille could not have formed as a fracture because the sides will not fit back together.—K.A.H.

AS15-9816 (P)

0   1 km

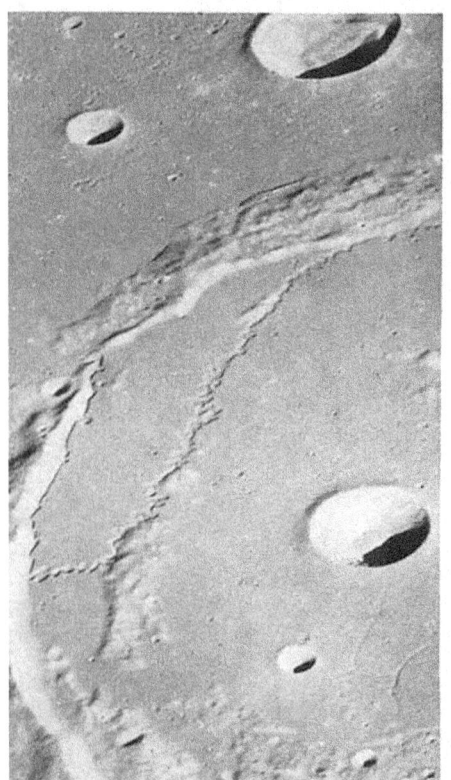

FIGURE 207.—The large (approximately 100-km) crater Posidonius is filled with mare lava to a level higher than the surrounding surface of Mare Serenitatis. The most interesting and perplexing feature of this crater is Rima Posidonius II—the highly sinuous rille that follows a devious course from the north rim of the crater at upper right (outside the photograph) through the breach in the crater rim at center. The rille is topographically controlled in part, hugging the juncture between lava and crater material. Erosion by some sort of fluid may have formed the rille; material appears to have been removed from it. An alternative explanation might be that the feature represents a drained and collapsed lava tube. The fluid involved probably was emitted from the craterlike depression at the head of the rille in the crater's north wall. If the rille is assumed to be contemporaneous with the lava filling, a lava of low viscosity would seem to be required to explain the channel's high sinuosity.—C.A.H.

AS15-91-12366

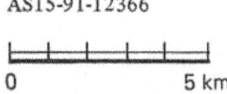

0   5 km

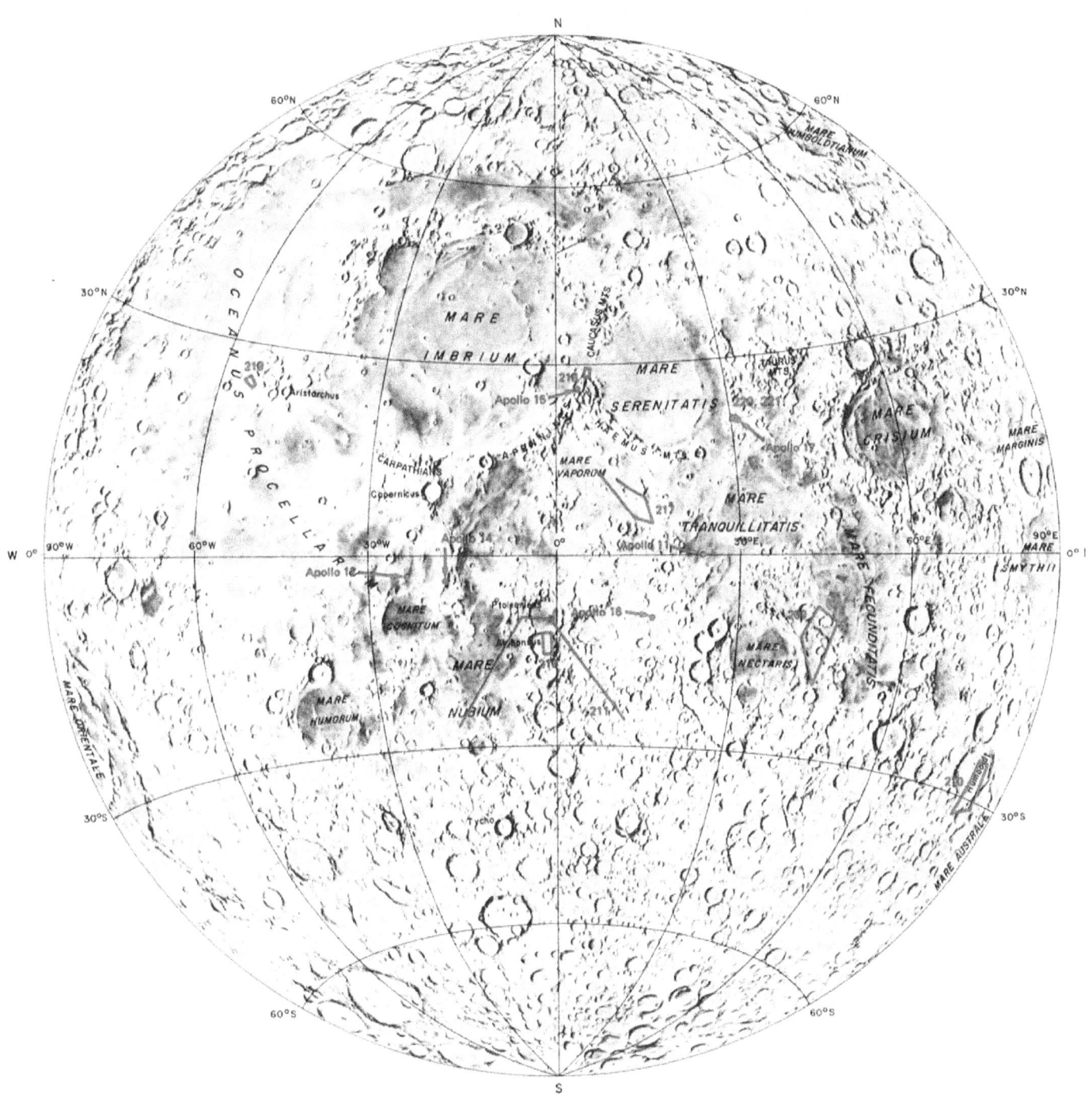

Near side

FIGURE 208.—Location of photographs
of straight rilles; numbers correspond to
figure numbers. [Base map courtesy of
the National Geographic Society]

AS8-13-2225 (H)

FIGURE 209.—This oblique view looks westward over the crater Goclenius (large crater in the foreground) at the western edge of Mare Fecunditatus. The area is typical of the edge of shallow mare basins in that ejecta deposits and interiors of older craters along the margins are partially flooded by mare material. The pitted portions of the intercrater areas in the background are crater deposits that have not been flooded by the smoother mare material. Both the craters and maria are cut by linear rilles (Rima Goclenius I and II); rilles are also seen along other shallow mare basin margins. In this picture, several linear rilles, each about 1 km wide, can be seen crossing from right to left across the mare into the floor of Goclenius. The rilles are thought to be similar to fault-bounded troughs on Earth (grabens) and may have originated when the central portion of the mare basin settled.—J.W.H.

AS15-93-12641 (H)

FIGURE 210.—This picture of the interior of the crater Humboldt (approximately 200 km in diameter) was taken looking southward. The terraced crater wall lies in the background, and the central peaks are visible in the lower right-hand portion of the picture. The crater floor is typical of those that exhibit both radial and concentric cracks, or fissures. The fissures and cracks appear to be related to the uplift of the crater floor subsequent to the formation of the crater. The deformation may be related to upwelling of portions of the crust in an attempt to reach isostatic equilibrium, or it may be coincident with intrusion of lavas below the crater floor. Small patches of dark mare material can be seen along the edge of the crater floor in the lower left and middle right-hand portions of the picture.—J.W.H.

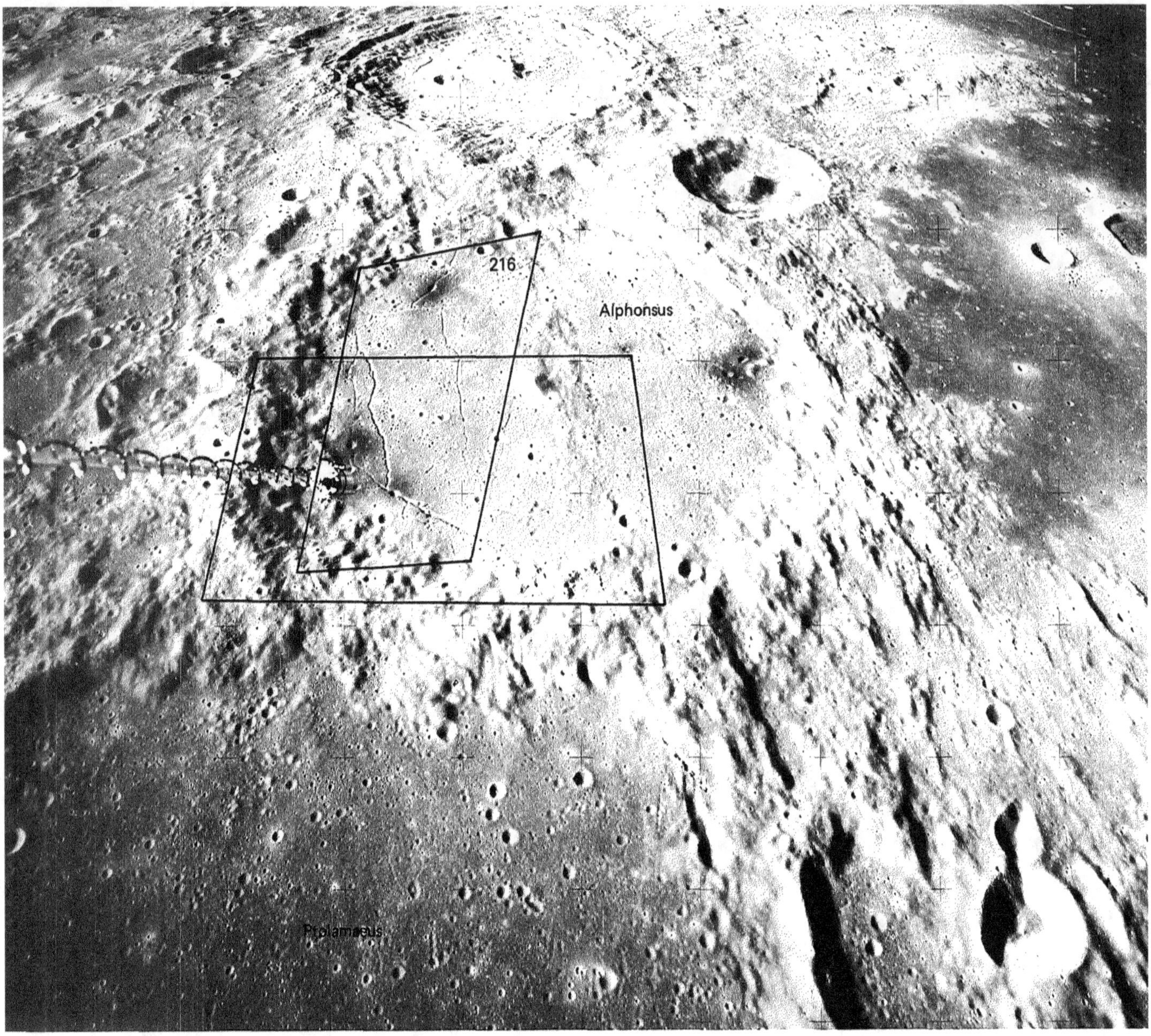

AS16-2478 (M)

FIGURE 211.—This oblique metric photo shows part of the lunar highlands where the ancient crust is saturated with large craters. Portrayed here are the crater Alphonsus (middle ground) and the ancient crater Ptolemaeus (foreground). The floor of Alphonsus is broken by faults that form a polygon roughly parallel to the walls. Dark halo craters lie along these faults. The rims of the dark halo craters fill in the fault troughs. This relationship indicates that the craters must have been formed by material ejected from the central vents rather than by collapse of material into the cracks. However, unlike impact craters with their hummocky ejecta and lines of secondary craters, the smooth rimmed deposits have been interpreted as fine-grained volcanic ejecta.

Lunar transient events have been observed many times in the crater Alphonsus. Red glows have been documented and spectra have been recorded by Kozyrev (1971) that apparently confirm the existence of gaseous emissions. These events are thought to be related to orbital parameters; when gravitational stresses are high, the crust shifts and gas escapes from the interior at regular intervals. If this is true, a low level of activity still continues to affect the lunar crust and interior.

General and detailed contour maps have been made (Wu et al., 1972) of Alphonsus using metric and panoramic photography obtained by Apollo 16. (See figs. 212 to 215. Fig. 212 is outlined in this figure, and fig. 213 is outlined in fig. 212.)—H.M.

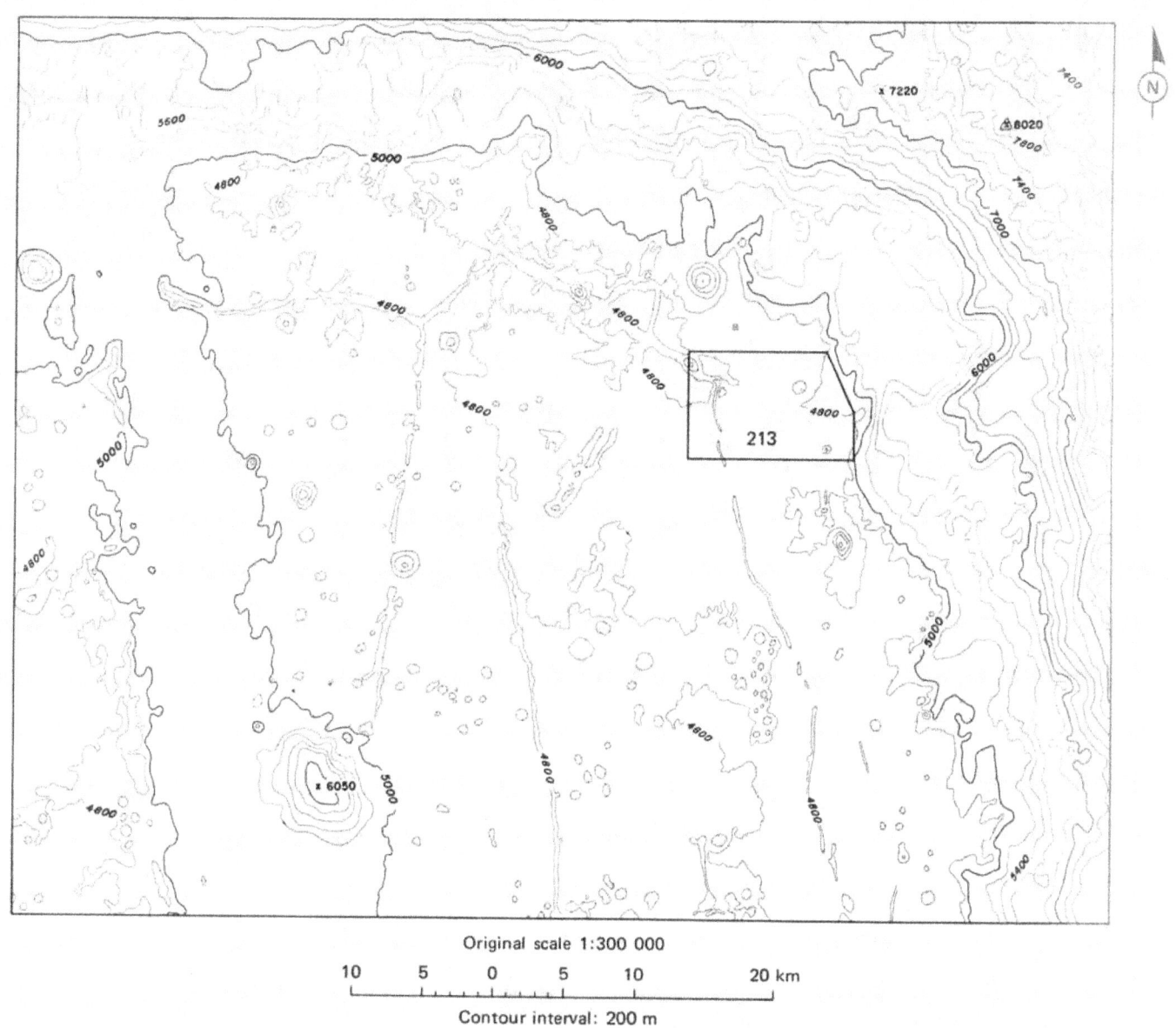

Original scale 1:300 000

Contour interval: 200 m

FIGURE 212.—This map, compiled of the floor of Alphonsus at a scale of 1:300 000 shows an array of faults, volcanic centers, and impact craters. The smooth flanks and youthful appearance of the central peak are apparently the result of the downslope movement of fragmental material that has filled in any impact craters as fast as they formed.—H.M.

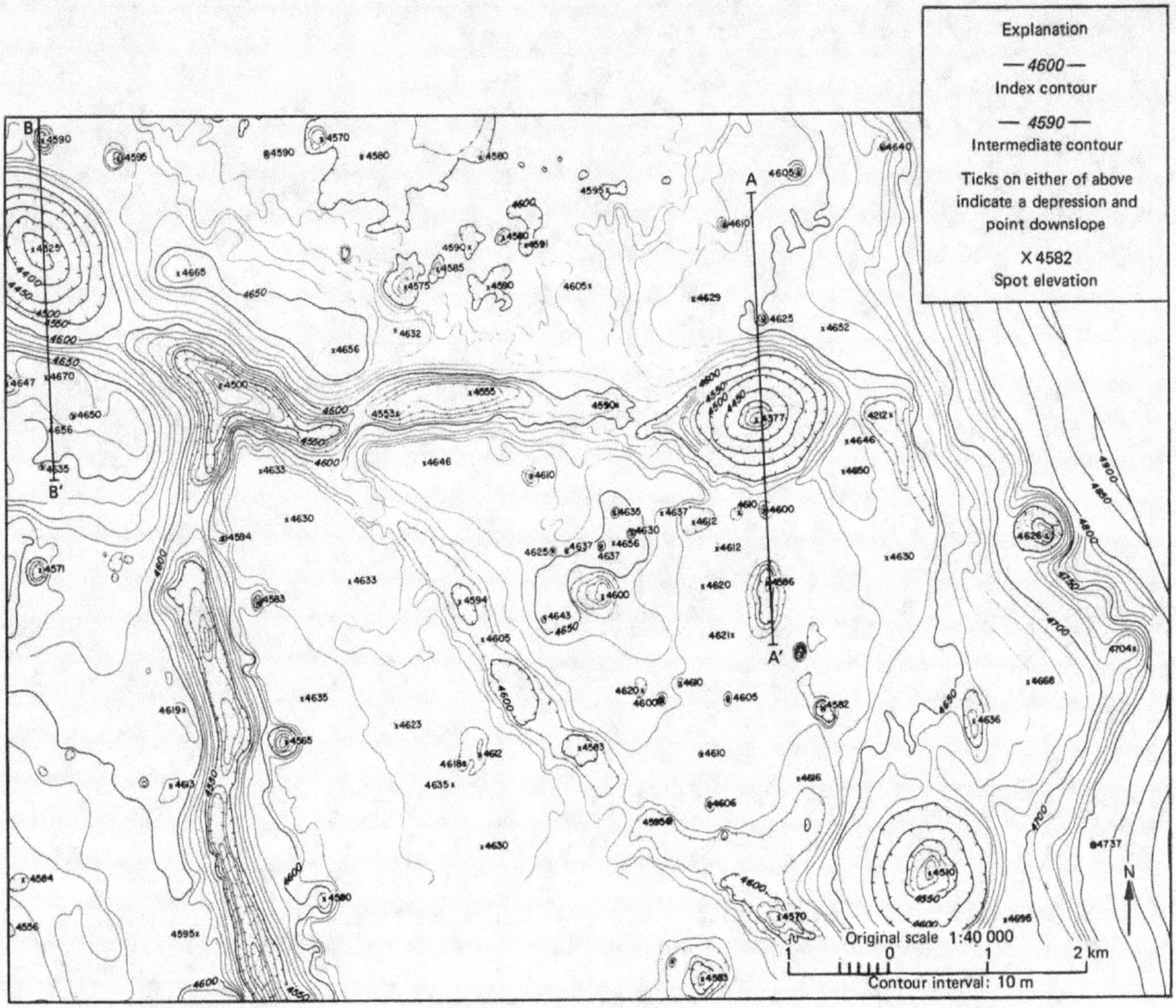

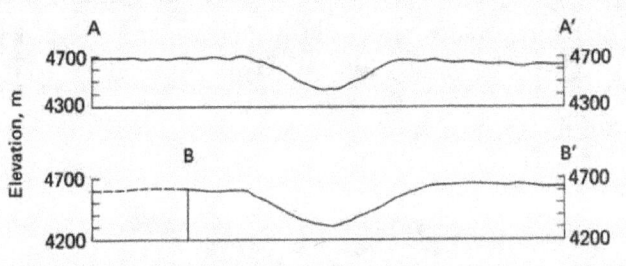

Prepared by the U.S. Geological Survey, Center of Astrogeology,
Flagstaff, Arizona 86001.

Topography compiled from Apollo 16 panoramic photographs,
9383, 9378 and 5380, 5385.

Control established from Apollo 16 metric photographs using
camera orientation data.

Datum arbitrarily set, no specific projection is applied.

FIGURE 213.—This map, compiled at a scale of 1:40 000, shows about 10 times the detail of the floor and crater shapes seen in figure 212 (see outlined area). The locations of profiles of two dark halo volcanic craters (A-A' and B-B') are shown on the map.—H.M.

FIGURE 214.—These sketches of the crater profiles of figure 213 contrast markedly with profiles of young impact craters.—H.M.

204

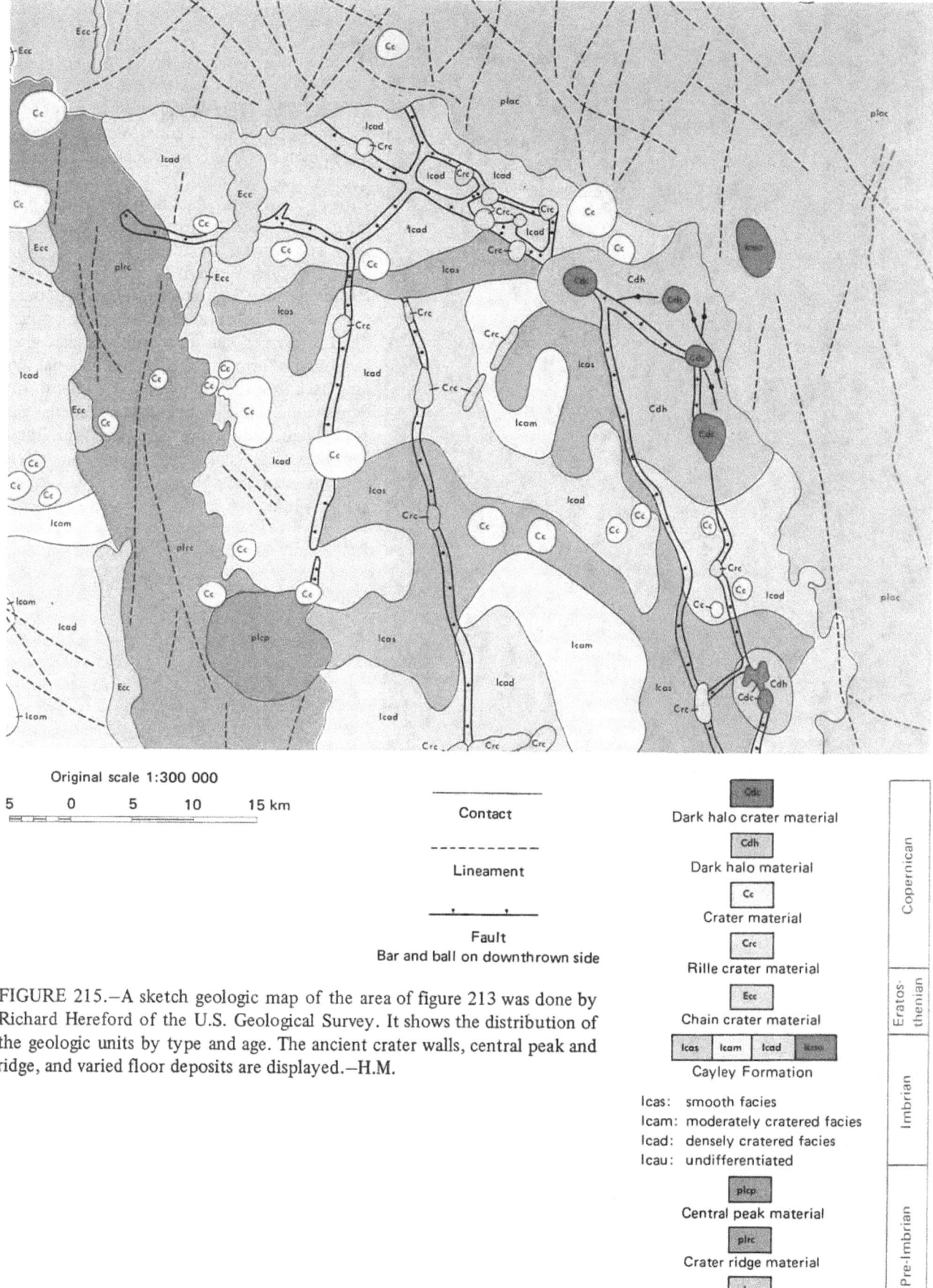

Original scale 1:300 000

5   0   5   10   15 km

——————————
Contact

- - - - - - - - - -
Lineament

—•—•—•—•—
Fault
Bar and ball on downthrown side

FIGURE 215.—A sketch geologic map of the area of figure 213 was done by Richard Hereford of the U.S. Geological Survey. It shows the distribution of the geologic units by type and age. The ancient crater walls, central peak and ridge, and varied floor deposits are displayed.—H.M.

Cdc
Dark halo crater material

Cdh
Dark halo material

Cc
Crater material

Crc
Rille crater material

Ecc
Chain crater material

| Icas | Icam | Icad | Icau |

Cayley Formation

Icas: smooth facies
Icam: moderately cratered facies
Icad: densely cratered facies
Icau: undifferentiated

plcp
Central peak material

plrc
Crater ridge material

plac
Alphonsus crater material

Copernican

Eratos-
thenian

Imbrian

Pre-Imbrian

205

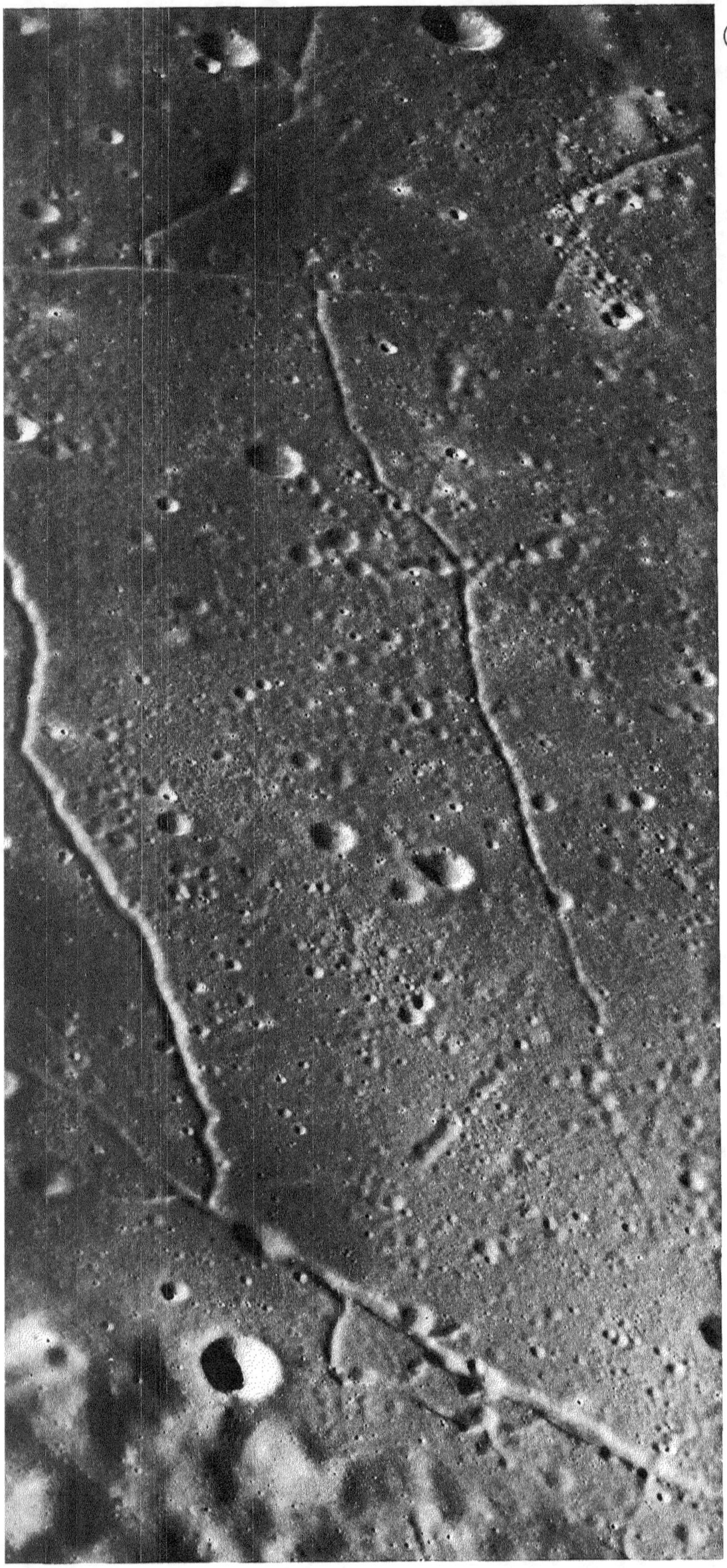

FIGURE 216.—This oblique view of part of the flat floor of the ancient crater Alphonsus shows faults that break the floor along straight line fractures. Spaced at irregular intervals along some of the rilles are dark halo craters that have broad low rims. The first detailed pictures of these craters were taken by Ranger 9 in 1964. This photograph confirms the detail in the Ranger picture and confirms the hypothesis that the dark halo craters are volcanic in origin. Their constructional rims indicate that they cannot have formed by collapse of material running back into the fault zone. The smooth rim deposits indicate that fine-grained material was ejected uniformly from the volcanic vent. These deposits differ markedly from the ejecta patterns around bright, young impact craters.—H.M.

AS16-4656 (P)

FIGURE 217.—Rima Ariadaeus is a fine example of a straight rille. Ariadaeus Rille is over 300 km in length; a portion of the central section of the rille about 120 km in length is pictured here. A linear section of the crust is dropped down along parallel faults or breaks in the crust to form a graben or fault trough. The ridges crossing the trough and the surrounding plains units have been offset by the trough, proving that they are older than the faults. Some craters are cut off by the faults and are, therefore, older. Other craters lie on the wall of the trough and are younger than the faulting. The faulting must be relatively young because so few craters appear to be younger than the faults, and because the edges of the trough appear to be crisp and little affected by slumping and other mass wasting.

There is a gradation between straight rilles, gently curving rilles, and sinuous rilles modified by volcanic flows. This example shows no trace of associated volcanism; it is, therefore, considered to be the end member of the sequence, where only pure faulting is involved.—H.M.

AS10-31-4645 (H)

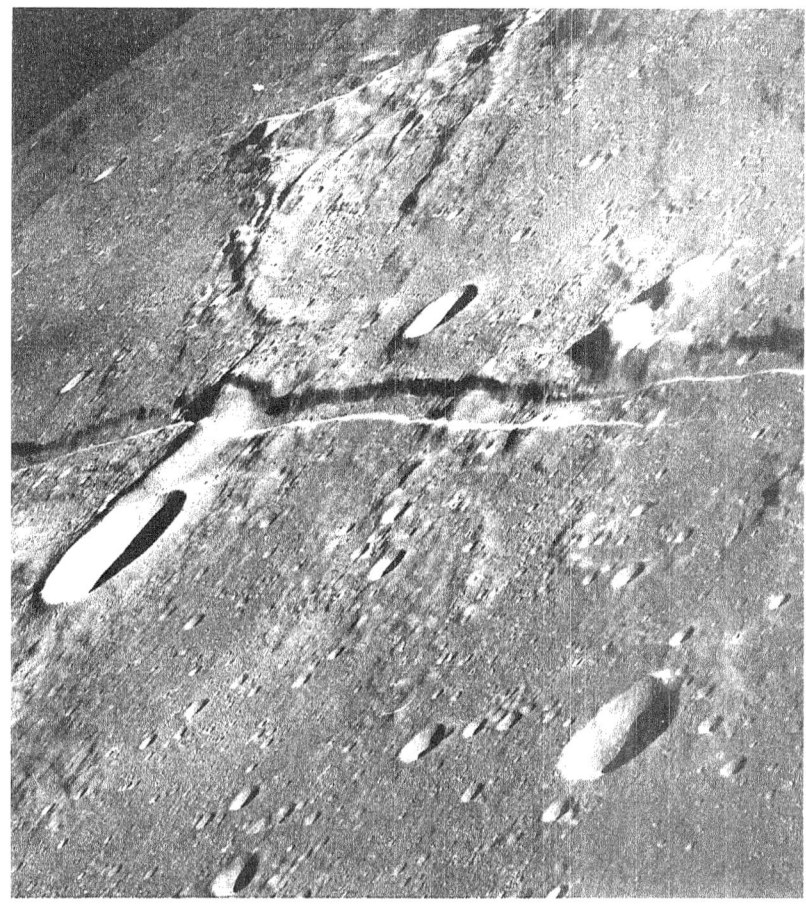

FIGURE 218.—Rima Fresnel I (1) approximately parallels the Apennine Mountain front (2) along the eastern margin of the Imbrium basin. The rille and its several branches at the northern end fall into the class of linear rilles that are usually attributed to a tensional structural regime causing the formation of grabens. Faults in the highlands (3), parallel to some of the rilles, support the contention of structural origin for the rilles. Some of the curvilinear segments of the rilles, however, suggest that they may have served locally as lava channels.

The rille traverses a plain that was called the Apennine Bench (4) and interpreted to be formed of older mare lavas (Hackman, 1966). The inside of the rille is flooded by younger mare lavas (Carr, Howard, and El-Baz, 1971) (5), which also embay its northern truncated terminus (6). Some younger mare lavas may have poured over the bench material and buried a crater (7). Old lava benches on the inside margin of basins are a common feature on the Moon. They are usually crossed by old rilles and embayed and partially buried by younger lavas.—B.K.L.

AS15-9368 (P)

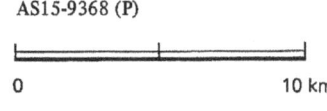

0   10 km

207

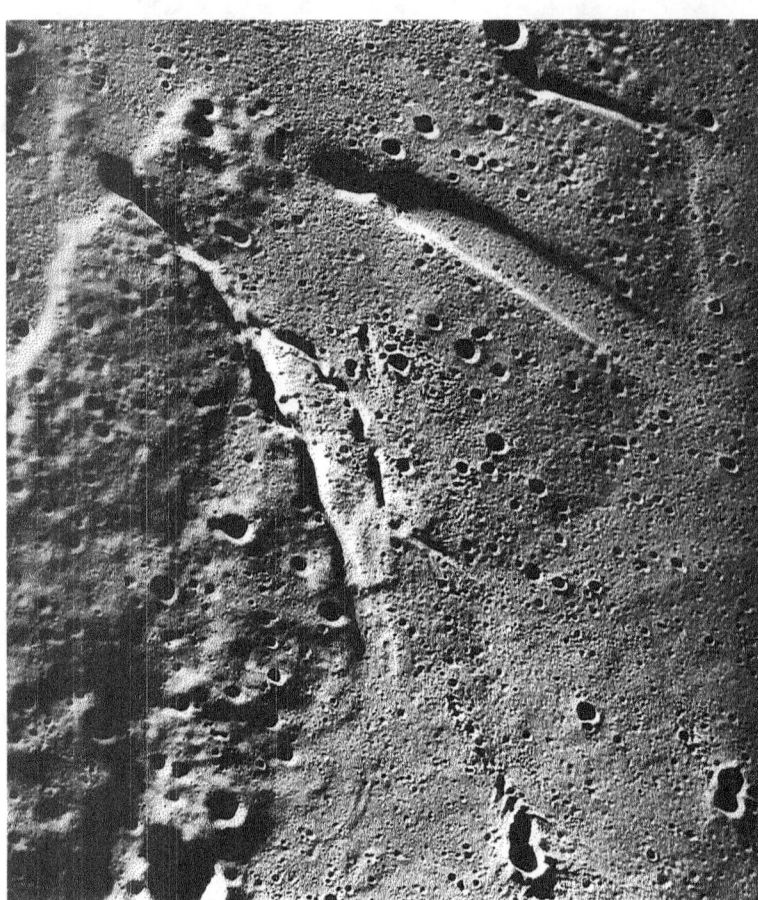

FIGURE 219.—A low Sun angle view of the western part of the Aristarchus Plateau shows three sharp-rimmed linear depressions that occur in the higher unit and terminate at the edge of the plateau. Their floors are filled with mare material that is similar to the surrounding materials of Oceanus Procellarum (lower right corner). The chain of craters in the mare of Oceanus Procellarum is believed to be secondaries from the crater Aristarchus (outside of the view). As described by Alfred M. Worden, the Apollo 15 CMP, the mare materials in this region display a brownish tint, as compared to the metal-gray materials of Mare Imbrium farther east.—F.E.-B.

AS15-13345 (H)

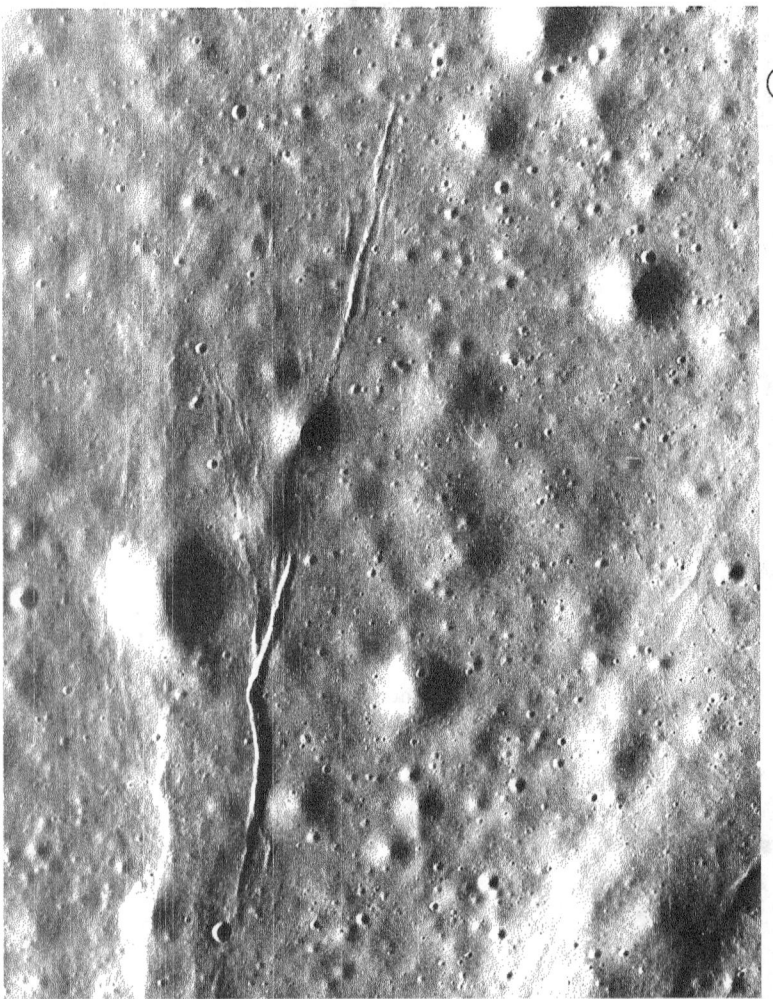

FIGURE 220.—Some of the freshest and youngest fractures yet found are in the Littrow area. The sharpest gash in this view runs along the right side of one of the much older Littrow rilles. The surface material in this area appears to be soft; blocks are very rare and most craters appear subdued. Therefore, it seems likely that the fresh fractures here, which are smaller than those in the preceding figures, will not survive long before they are obliterated by mass wasting.—K.A.H.

AS17-2313 (P)

0   25 km

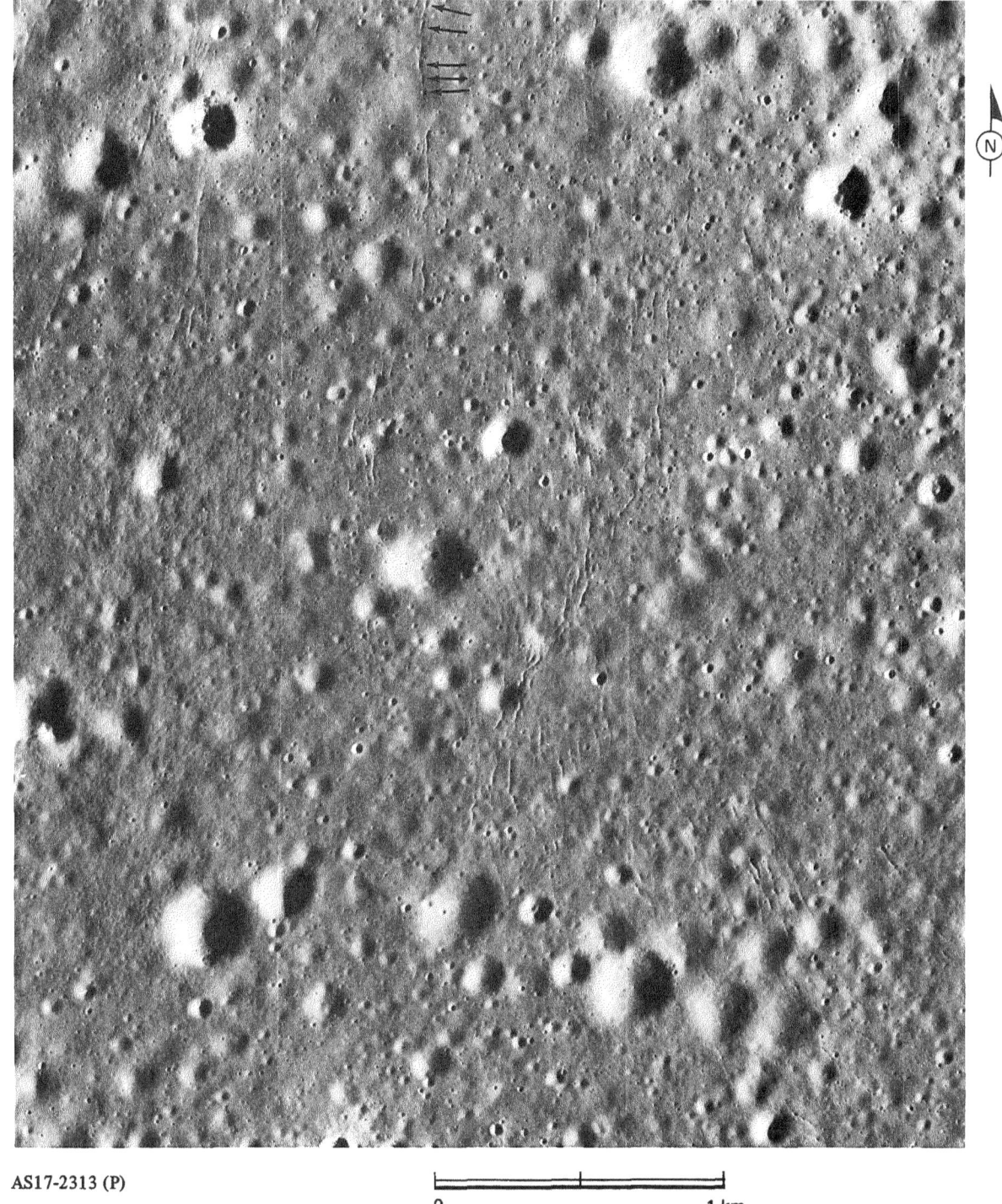

AS17-2313 (P)

0   1 km

FIGURE 221.—This detailed view of a mare surface near the eastern edge of Mare Serenitatis, just west of the Apollo 17 landing site, shows the numerous small grooves. They are unusual features that have not been observed in such large numbers elsewhere. They are here developed in a thick accumulation of regolith overlying an ancient mare basalt. At first glance, resembling chains of secondary impact craters, they are more logically interpreted as structural features. A likely explanation is that they are the result of drainage of unconsolidated regolith into openings caused by fissuring in the consolidated bedrock. At some localities, drainage did not occur at a uniform rate along the fissure but was concentrated at certain points, resulting in a series of unequally spaced, pitlike depressions (arrows) along the groove.—B.K.L.

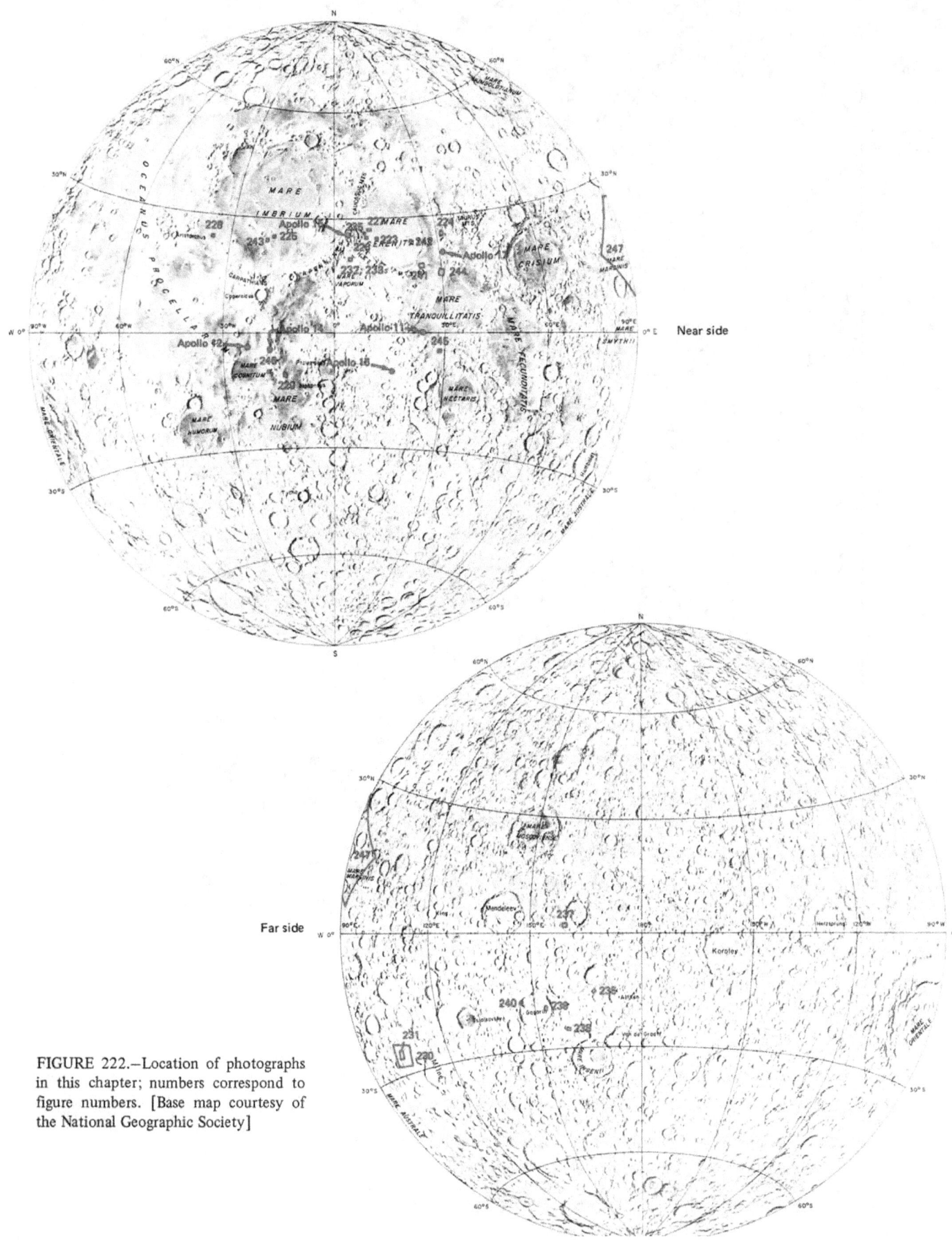

Near side

Far side

FIGURE 222.—Location of photographs in this chapter; numbers correspond to figure numbers. [Base map courtesy of the National Geographic Society]

7

Unusual Features

Included in this chapter is a pot pourri of features that defy easy classification. Many craters with unusual shapes are represented that may be volcanic rather than impact in origin. Other craters have irregular hummocky deposits on their floors; still others may be covered by lava flows. Other features depict collapsed lava tubes or depressions formed by drainage of fragmental surface material into cracks. Together they constitute a fascinating group of features, and they are grouped in this chapter because, at the resolution of the Apollo imagery, more than the usual uncertainty exists in attempting to interpret their origin. Perhaps later Moon landings or a comparison of these pictures with orbiter photography of Earth, Mars, and other planetary bodies will provide additional insight into the geologic processes that formed them.—H.M.

15-9355 (P)

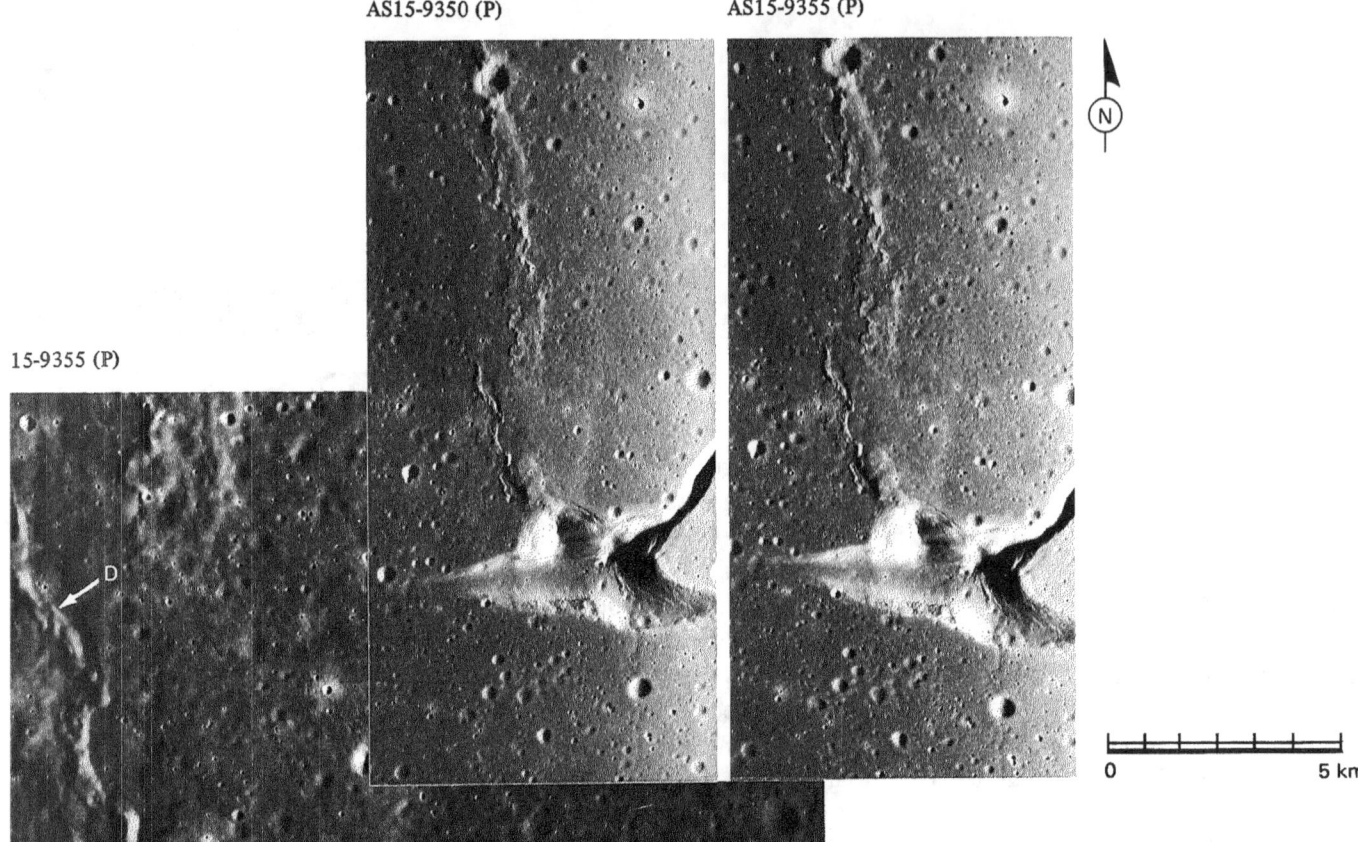

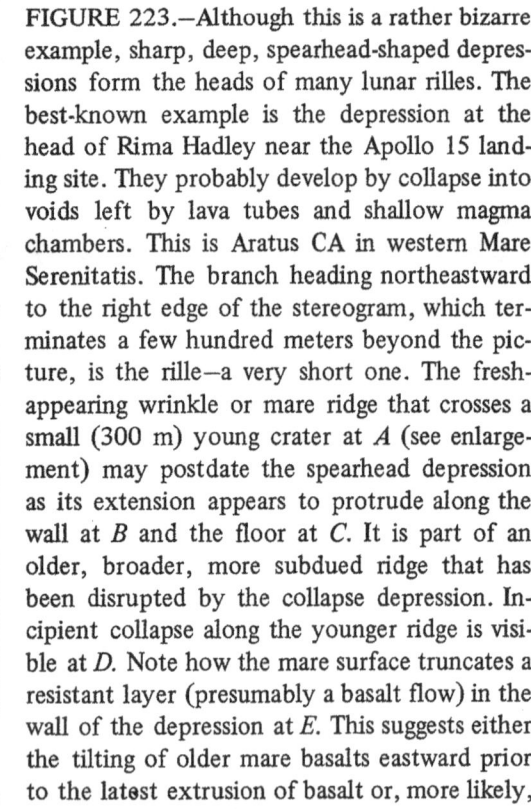

FIGURE 223.—Although this is a rather bizarre example, sharp, deep, spearhead-shaped depressions form the heads of many lunar rilles. The best-known example is the depression at the head of Rima Hadley near the Apollo 15 landing site. They probably develop by collapse into voids left by lava tubes and shallow magma chambers. This is Aratus CA in western Mare Serenitatis. The branch heading northeastward to the right edge of the stereogram, which terminates a few hundred meters beyond the picture, is the rille—a very short one. The fresh-appearing wrinkle or mare ridge that crosses a small (300 m) young crater at *A* (see enlargement) may postdate the spearhead depression as its extension appears to protrude along the wall at *B* and the floor at *C*. It is part of an older, broader, more subdued ridge that has been disrupted by the collapse depression. Incipient collapse along the younger ridge is visible at *D*. Note how the mare surface truncates a resistant layer (presumably a basalt flow) in the wall of the depression at *E*. This suggests either the tilting of older mare basalts eastward prior to the latest extrusion of basalt or, more likely, that the resistant layer is lenticular.—D.H.S.

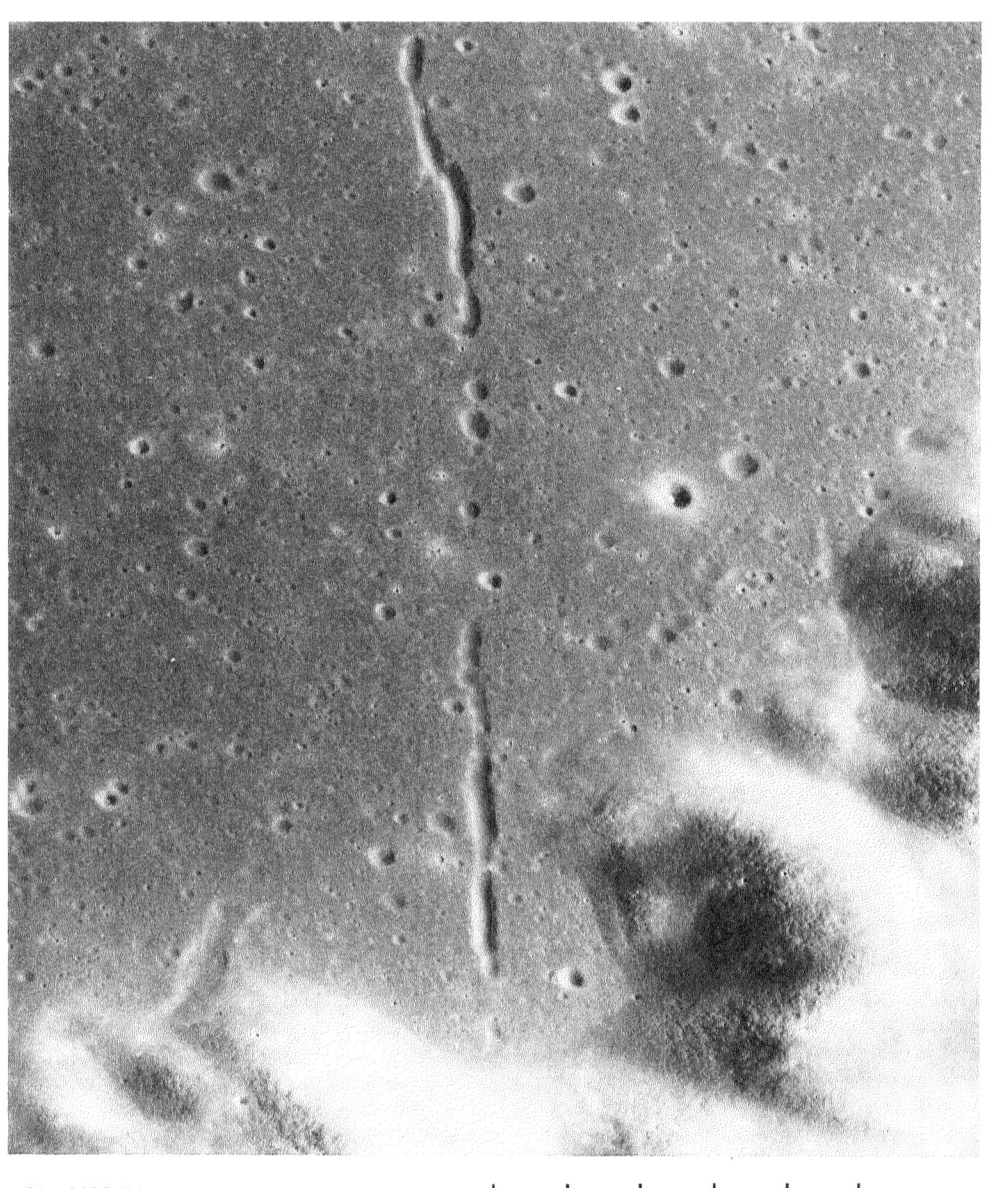

AS15-9299 (P)

Approximate 0   5 km

FIGURE 224.—On January 16, 1973, the Soviet unmanned roving vehicle Lunokhod 2 was landed by Luna 21 in or near this area in the southeastern part of the crater Le Monnier. This crater is a large (61-km) pre-Imbrian crater cut into terra at the eastern edge of Mare Serenitatis before Serenitatis was flooded by mare lavas. Part of Le Monnier's southern wall fills the lower part of the picture. A conspicuous chain of elongate depressions has formed in the lava-filled floor of the crater. The chain trends 22 km northward and its pattern is quite surely controlled by an underlying fracture system. Regionally, the inferred fracture system is concentric to the grossly circular Serenitatis basin, and in this area trends northward. No comparably young structural features having the same trend cut the terrae surrounding Le Monnier. However, older structures having this trend occur in the southern and northern walls and rims of Le Monnier. The alined depressions on the mare are mostly 300 to 400 m wide and 30 to 60 m deep. The three deepest stretches are 1 to 2 km long and about 50 to 65 m deep. These depressions probably were the locus of fissure eruptions of mare basalt. Withdrawal of the last lava back into the fissure may have created subsurface voids into which collapse took place, causing the depressions and accounting for the absence of raised rims on the depressions.—R.E.E.

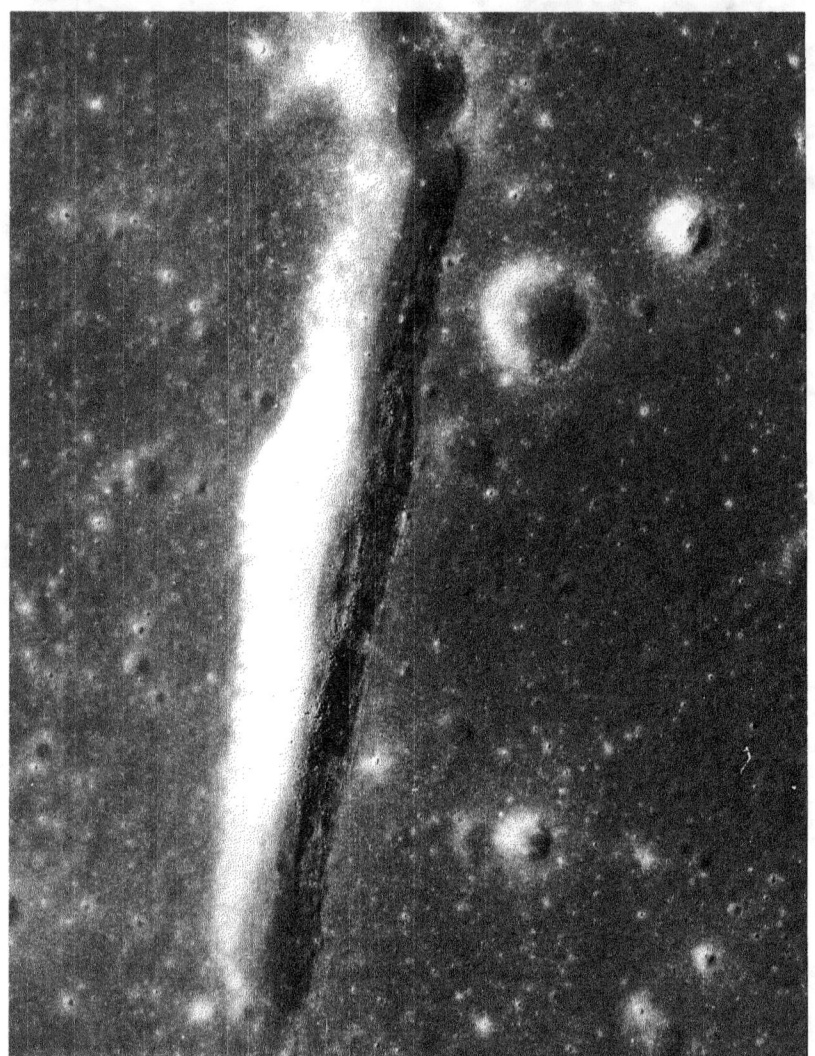

FIGURE 225.—Although considerably wider and longer than the largest of the depressions described in figure 224, this one otherwise closely resembles them. It probably formed in the same way; that is, by collapse into a fissurelike cavity caused by the withdrawal of mare basalt. Closer viewing (and greater enlargement of the original frame) clearly reveals signs of mass wasting. Note the incipient slump fractures along the right rim and large fragments along the right wall and in the bottom of the depression.—G.W.C.

AS15-0244 (P)

0   3 km

FIGURE 226.—Low-rimmed or rimless depressions having irregular outlines are not uncommon on mare surfaces. The one pictured, near the western margin of Mare Serenitatis, is about 10 km long and appears to have been formed by collapse and the coalescence of small craters.—D.H.S.

AS15-9361 (P)

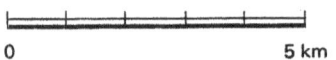

0   5 km

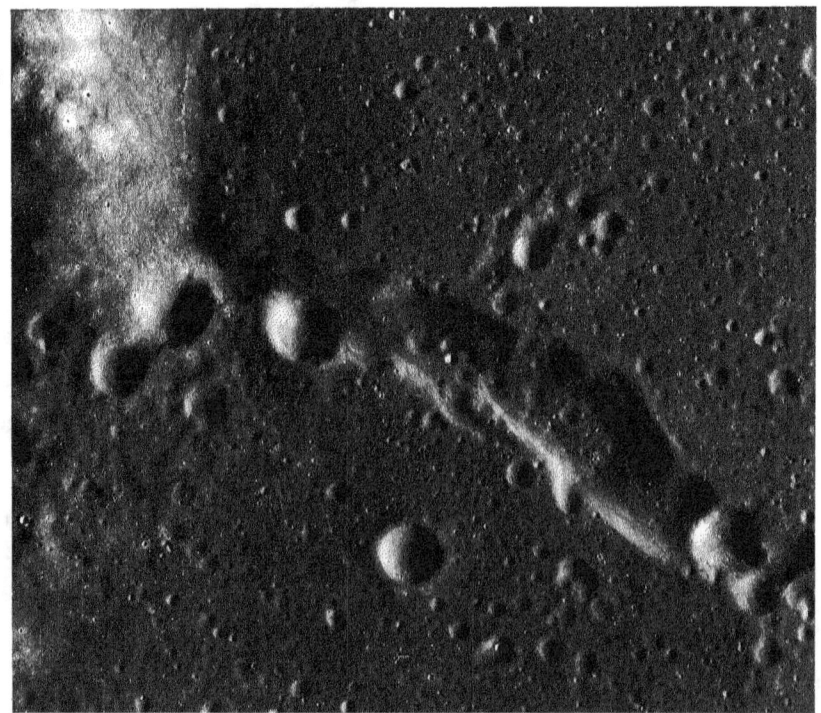

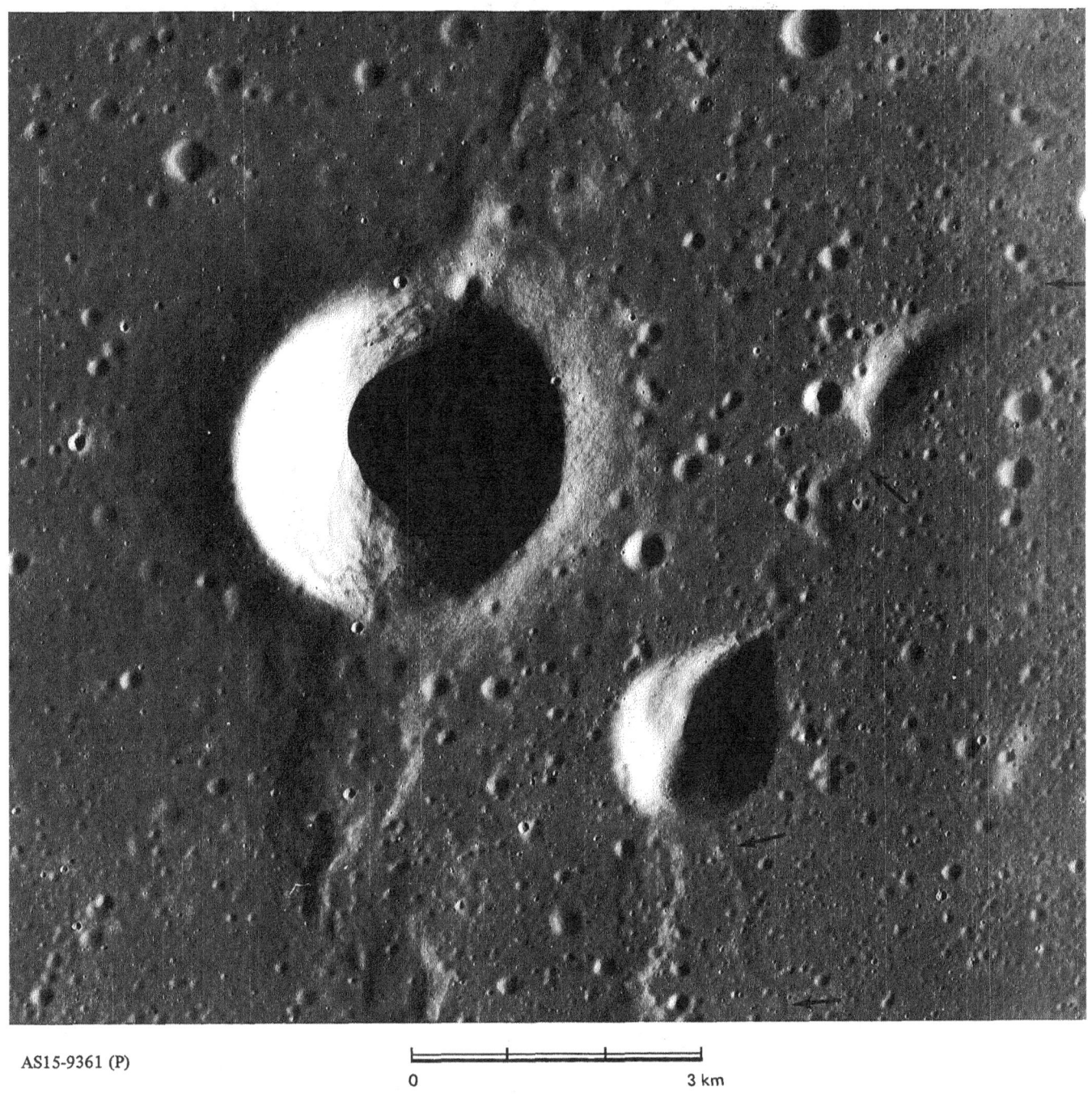

AS15-9361 (P)

0   3 km

FIGURE 227.—The largest crater in this picture is Aratus D in western Mare Serenitatis. Its circular outline, high rim, and deep bowl-shaped interior are characteristic
of many impact craters of comparable size and age. The next largest crater is tear
shaped and nearly rimless. It is, furthermore, associated with a rille (arrows). These
differences strongly suggest that it was formed or at least highly modified by
structural collapse.—D.H.S.

FIGURE 228.—The very young rimless crater near the center of this picture is located near the area where Oceanus Procellarum and Mare Imbrium join. The crater apparently formed in regolith-covered mare basalt. It differs from lunar impact craters of comparable size and age by its lack of a raised rim, surrounding ejecta deposit, or associated secondary impact craters. In addition, its interior walls do not show the steep slopes with craggy outcrops of rock in their upper parts, nor the streams of debris-avalanche deposits and talus that are usually seen in the walls of impact craters of comparable age and size.

Judging from the clear and sharply formed pattern of concentrically curved grooves and scarps that surround the hole, the material near this depression has apparently subsided into a subsurface void. Because of the extreme rarity and inferred short lifetime of steep slopes on the Moon, the latest subsidence must have taken place very recently, after most of the 50- to 300-m diameter craters that densely pepper the nearby mare surface were formed. Movement of the regolithic debris layer during subsidence apparently smoothed out most, if not all, of the craters that must have

existed near the depression. Now the depression is surrounded by low, curved fault scarps and narrow, curved grooves that may be fault troughs (grabens) or may represent drainage of regolithic debris into cracks that opened in the underlying sagging basalt rock. The few craters that have formed on the subsided surface compare in density to the craters formed on the cluster (arrow) of Aristarchus secondary impact craters and associated herring-bone ridges; comparable ages for the Aristarchus secondary features and the depression are thus indicated. The subsidence was triggered either by the ground shock or seismic wave-train generated when Aristarchus was formed 300 km to the west, or by the impacts of the secondary features.

The subdued depression in the upper left may be a similar older feature that was flooded by a later lava flow that now covers the area. The density of craters within the depression and the density on the surrounding lava are comparable. Alternatively, the subsidence there may have been incomplete; however, there is no sign that this subsidence is as young as that in the deeper crater.—R.E.E.

AS17-3125 (P)

0   5 km

FIGURE 229.—This 18-km-long "figure 8" pair of noncircular craters near the crater Guericke probably was not formed by hypervelocity impacts of bodies from space. It could be a secondary impact feature formed by projectiles from the Imbrium basin 700 km to the north. The terrace at the base of the crater walls could be debris from the walls or a "bathtub ring" left by a formerly higher stand of the mare fill. Alternatively, the crater pair and the terrace could have been formed by volcanic eruptions. The superposed bright crater is younger than and unrelated to either the "figure 8" pair or the mare.—D.E.W.

AS16-5410 (P)

0   4 km

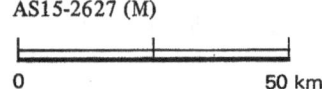

FIGURE 230.—This area of mare material in the far-side uplands centered at 26° S, 103° E shows many features. There are fractures (arrows 1) that may have been vents from which much of the smooth, dark lava was extruded; some of these fractures are bounded by lava levees. Craters containing bulbous material (arrows 2) that may be lavas extruded through the brecciated crater floor are visible. There are breached and partly flooded craters (arrow 3) and barely discernible flooded craters (arrow 4). Mantled remnants of uplands (arrow 5) are still visible. Terraces (arrow 6) mark the level of lava prior to subsidence in the center of the depression.—M.W.

AS15-2627 (M)

0   50 km

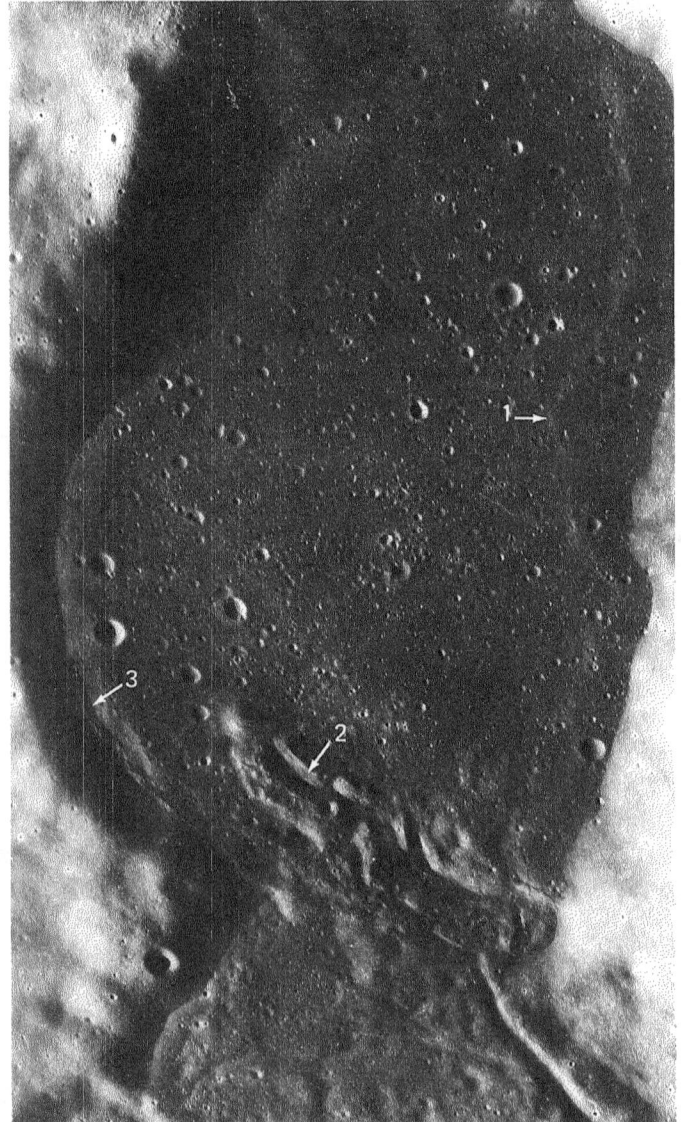

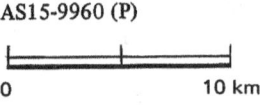

FIGURE 231.—In this part of the area of figure 230, the terrace (arrow 1) is shown again. Other features shown here are a fracture with lava levees (arrow 2), a "strand line" (arrow 3) marking the level of lava before cooling and withdrawal, and a scarp of a fresh-appearing lava flow front (arrow 4).—M.W.

AS15-9960 (P)

0   10 km

FIGURE 232.—The steep-walled but shallow, D-shaped depression near the center of the photograph is apparently a unique feature. It is located in a patch of mare on the foothills of the Montes Haemus, west of Mare Serenitatis. Measured along its straight side, the depression is about 3 km wide. It is situated atop a very gentle circular dome that appears to be somewhat smoother than the surrounding mare surface. As is more clearly shown in the accompanying stereogram (fig. 233), the many bulbous structures on the floor give it a blisterlike appearance. The depression is believed to be volcanic, probably a caldera (El-Baz, 1973b). Figure 234 explains the probable sequence of events leading to the formation of this unusual structure.—F.E.-B.

AS17-1672 (M)

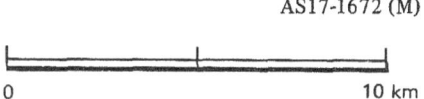

0   10 km

FIGURE 233.—The enlarged view provided by this stereogram shows that there are at least three different types of material within the floor of the D-shaped depression. A brighter annulus parallels the wall, and darker material fills the inner floor. Within both areas numerous bulbous and slightly raised domical structures are easily distinguishable. There are craters on the summits of many of the structures, suggesting that each one is probably an extrusive dome with a summit crater. Many similar features on Earth are caused by the subterranean drainage of lavas following extrusion and release of pressure following the upward movement of a central plug.—F.E.-B.

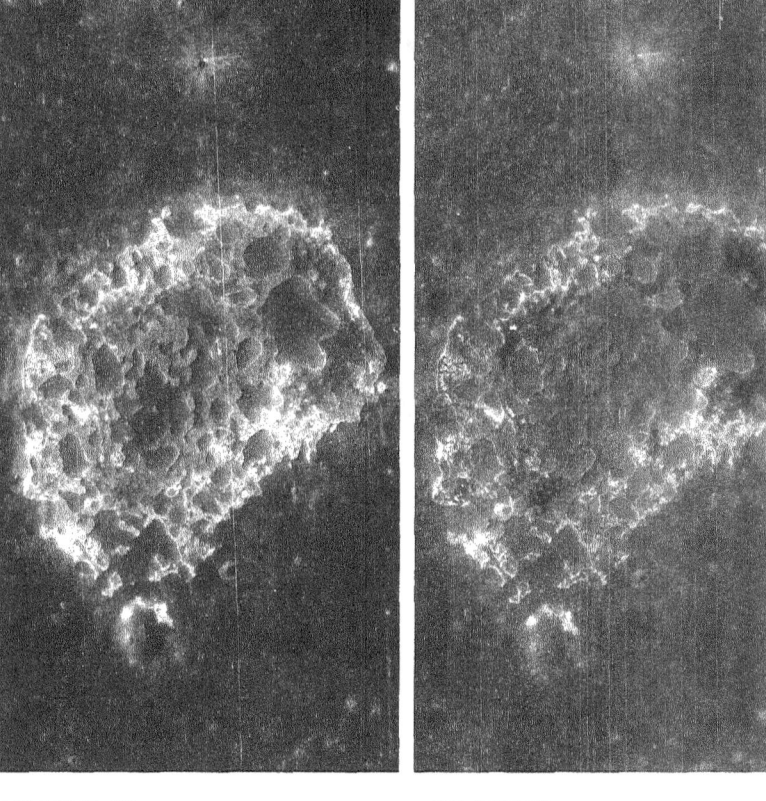

0   2 km   AS15-0181 (P)   AS15-0176 (P)

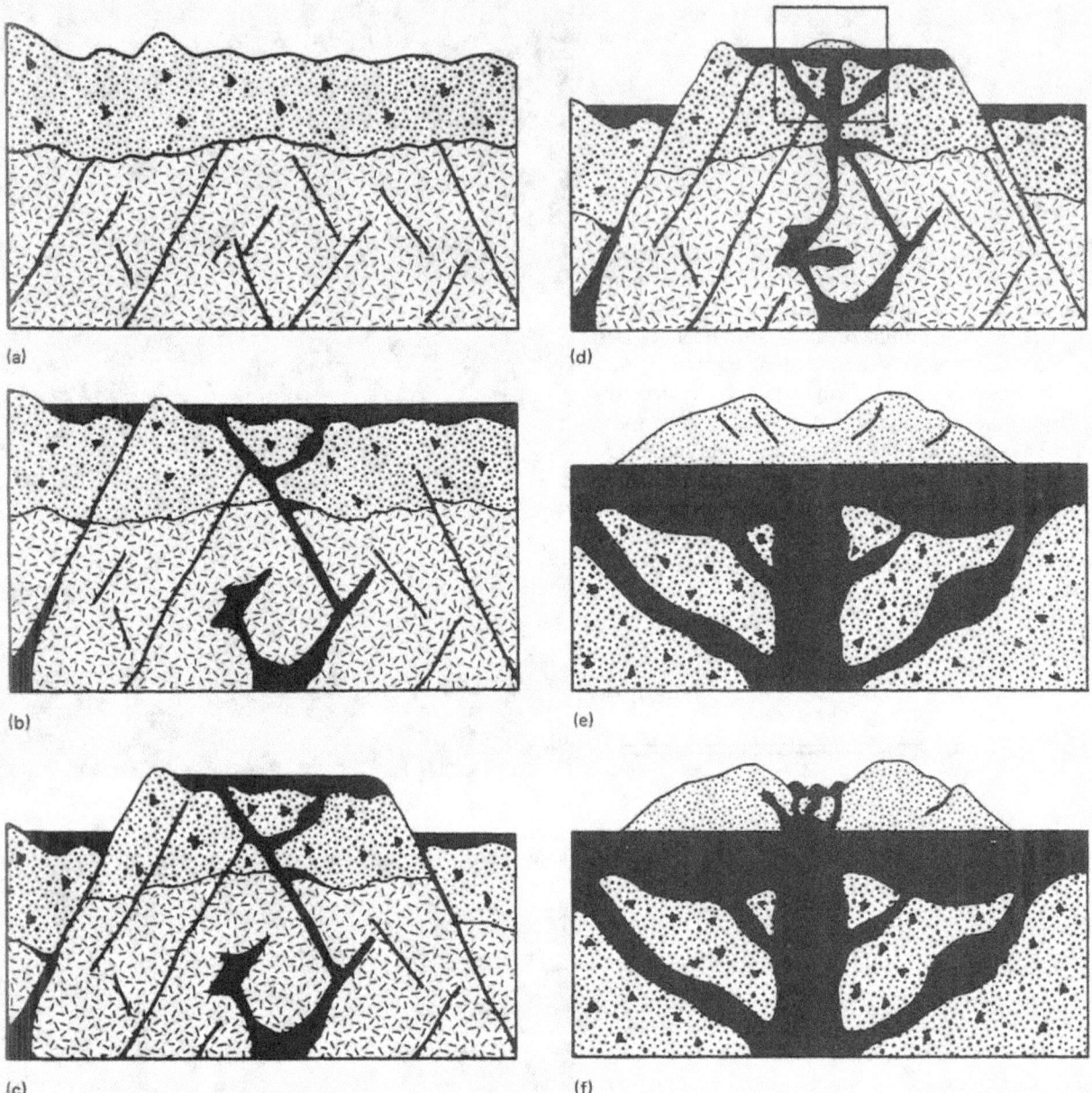

FIGURE 234.—This diagram shows a postulated sequence of events leading to the formation of the D-shaped structure; the events are presented in order of occurrence: (*a*) Numerous faults were generated in the crust and a thick blanket of debris was deposited (upper layer) as a result of the gigantic impact event that formed the Imbrium basin. (*b*) Basaltic lavas migrated to the surface along fractures to form a small, probably thin layer of mare material. Higher areas escaped inundation. (*c*) Vertical displacement then occurred along major faults, and the inundated block was displaced upward relative to the surrounding blocks. (*d*) A broad, gentle volcanic dome formed. It is recognizable as a younger eruptive stage because its surface is less densely cratered than the surrounding mare surface. (*e*) The center of the dome collapsed to form the caldera. This segment is outlined in (*d*). (*f*) Many small extrusions of lava formed the bulbous structures on the caldera floor. The centers of some of these later collapsed to form small summit craters, thus, on a much smaller scale, repeating the earlier collapse caldera sequence.—F.E.-B.

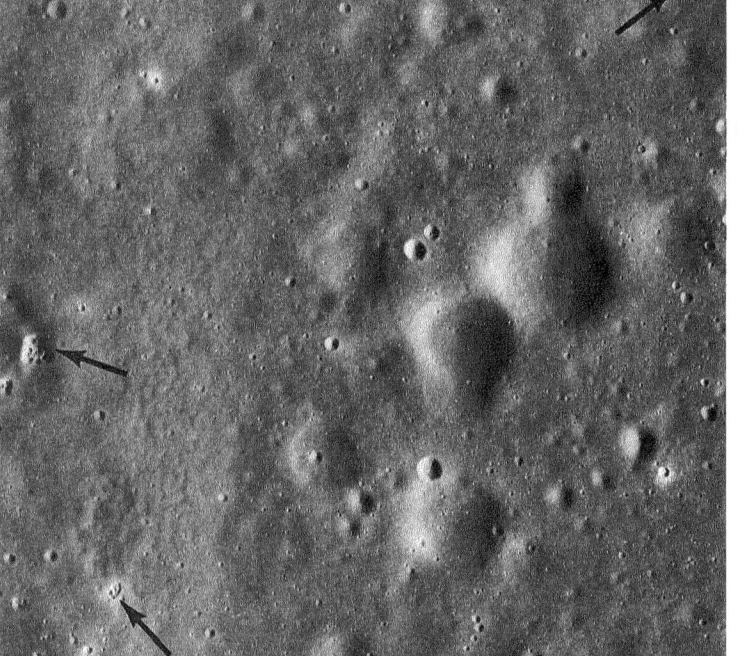

FIGURE 235.—Several extraordinary, tiny, puckered, collapse depressions (arrows) are in this detailed view of a very small area of western Mare Serenitatis. The collapse depressions are distinctly different from the many normal craters in this picture. One normal crater contains two clusters of the depressions. Such depressions are known nowhere else on the Moon; their closest analog is the odd but very much larger feature shown in figures 232 and 233. The collapse depressions here are much fresher and younger than the cratered lavas in which they occur. In fact, they must be among the youngest nonimpact features on the Moon. How the collapses occurred is an intriguing puzzle. Were they formed by the recent escape of gas from beneath the surface?—K.A.H.

AS15-9358 (P)

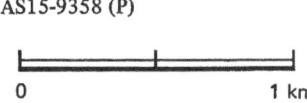

FIGURE 236.—A panoramic camera view of an unusual crater on the lunar far side, west of the large crater Aitken. The crater is polygonal in outline, its rim is not raised, and its walls are relatively smooth. The floor of the crater is occupied by a rather dark material with numerous cracks. This pattern has been informally referred to as a turtleback crater floor (El-Baz and Roosa, 1972). It occurs only in a few craters on the lunar surface and is probably caused by the cooling of a molten or partially molten fill within the crater. However, cracking due to tectonic movement (for example, the upward thrust of a central plug) cannot be excluded.—F.E.-B.

AS17-1931 (P)

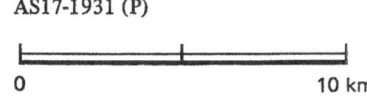

FIGURE 237.—A view looking south into a small 19-km-diameter crater southwest of Mandel'shtam on the lunar far side. The crater has a floor that is heavily lineated and grooved, but this structure is subdued rather than sharp and is contained wholly within the crater. The cracked floor is typical of a variety of craters that occur in the highlands away from the mare basins. They differ from cracked floor craters, such as Humboldt, and rilled craters, such as Goclenius; their origin is unknown.—J.W.H.

AS16-4151 (P)

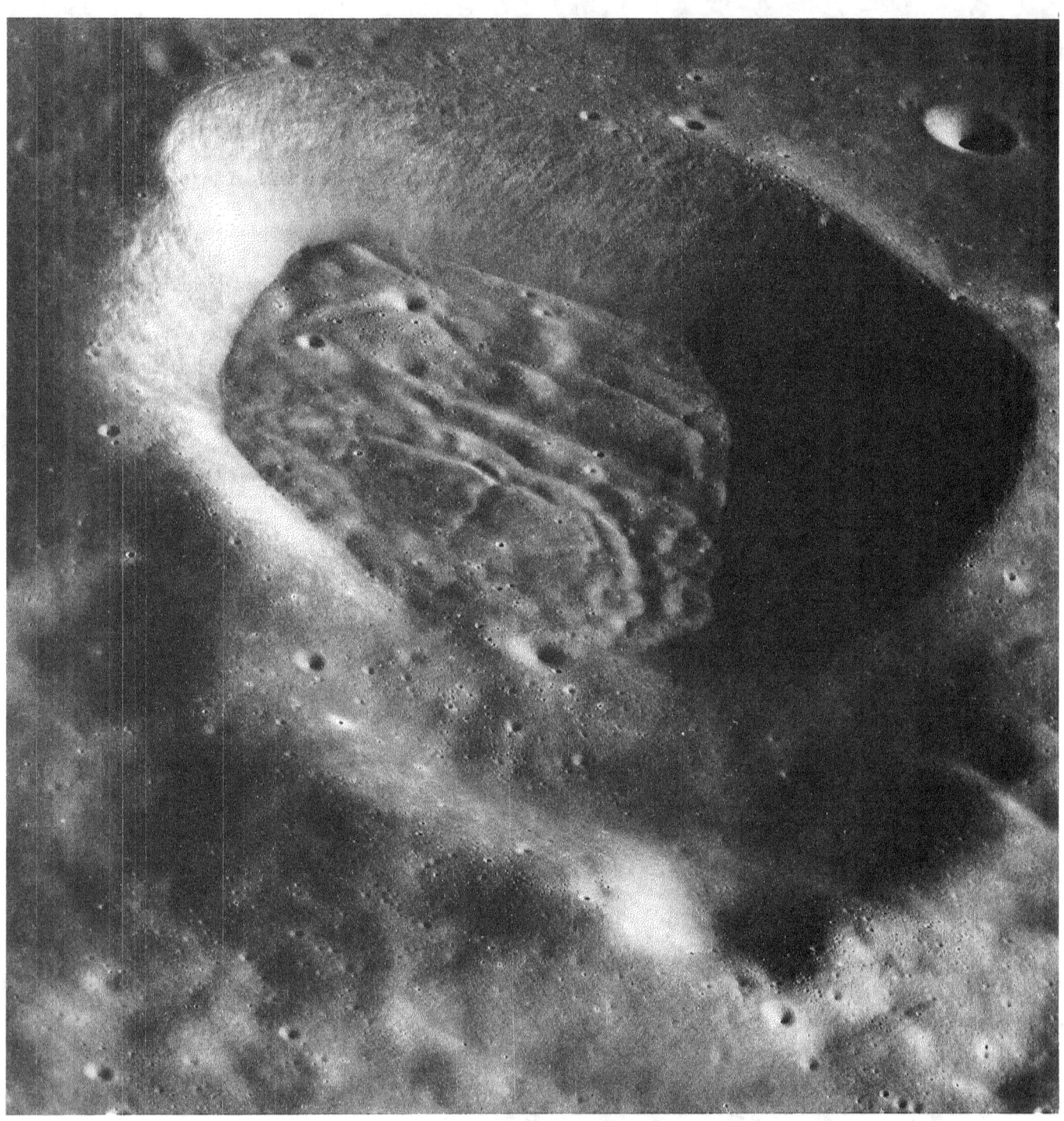

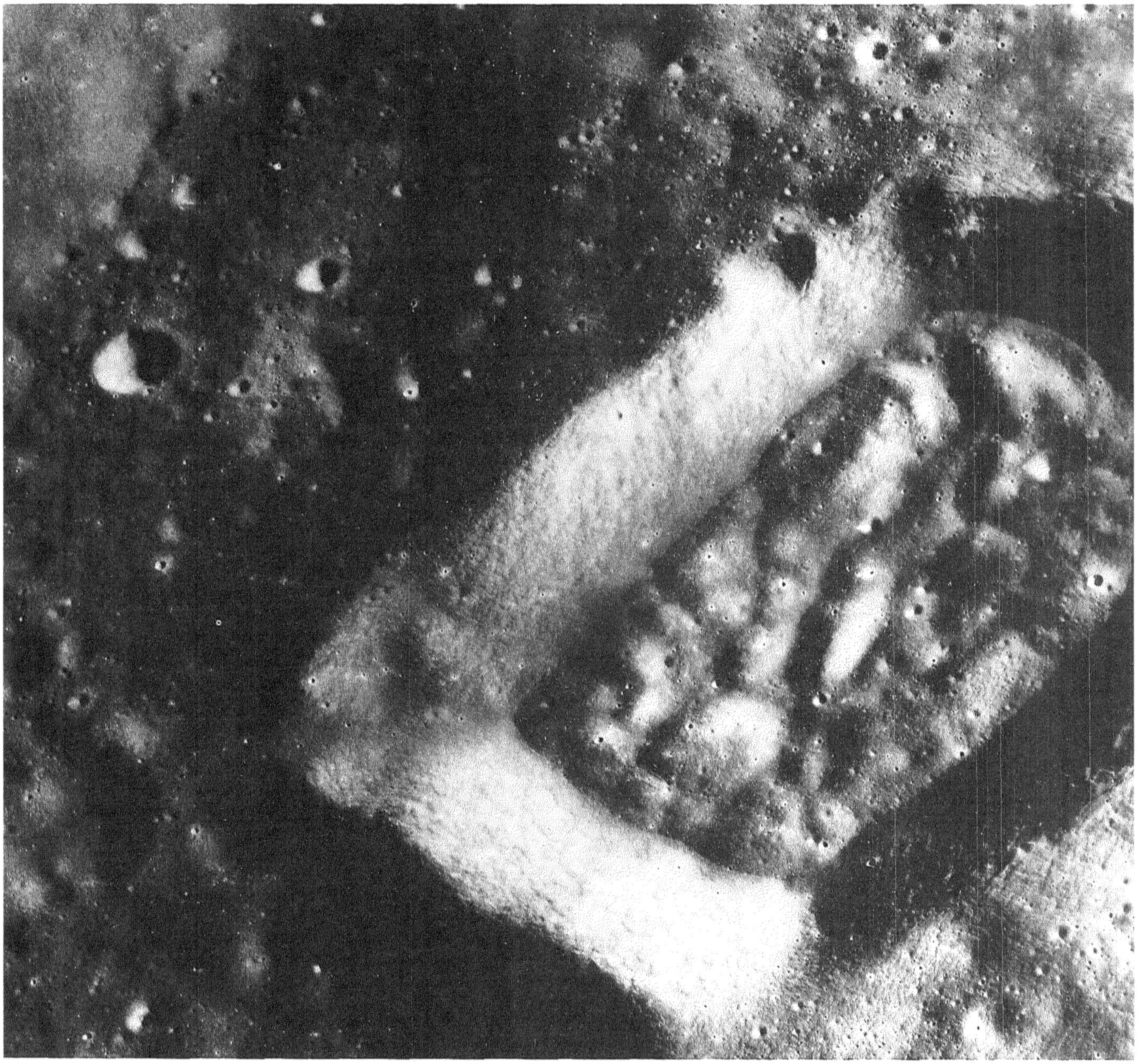

AS15-8893 (P)

FIGURE 238.—The obvious peculiarity of this crater is its shape, the causes for which are open to speculation. The entire area pictured is in the floor of the 60-km-diameter crater Barbier, located on the lunar far side and centered at 23.9° S, 157.7° E. Barbier is a relatively old crater, and the high density of craters on its floor is evident in this high-resolution photograph from the Apollo 15 panoramic camera. The two angular corners of the oddly shaped crater suggest that its present form is partly controlled by faults or joints in the floor of Barbier. The slightly raised rim of the crater—evidenced in the shadow on the rim at the left and the brightening on the rim in the lower right side of the crater—shows that material was ejected from the crater at the time of its formation and that it is not simply a collapsed portion of the floor of Barbier. The long dimension of the pictured crater is 16 km, and the short dimension is 10 km.—M.C.M.

223

FIGURE 239.—The four large depressions in this photograph are part of a cluster of secondary impact craters on the floor of Gagarin, a large (275-km-diameter) crater on the far side of the Moon. The secondaries pictured here lie close to Gagarin's northeastern rim. Mapped as pre-Nectarian in age (Stuart-Alexander, 1976), Gagarin is filled with relatively bright plains-forming deposits. The plains deposits are probably about 1 km thick; Gagarin was originally about 4 to 5 km deep.

Of particular interest in this photograph are the peculiar bumpy floors of the two craters nearest the upper right corner of the photograph. Several other craters in this part of Gagarin have the same morphology. The rounded hills or bumps, ½ to 2 km across, fill the crater floors so that the craters are shallower than is normal for craters of their size and age. For example, the center crater of the four is now the deepest, yet the two eastern ones are as wide or wider and, therefore, should be at least as deep.

A possible explanation for both the surface morphology and the raised floor level of the two eastern craters may be related to the intrusion of magma into the brecciated materials that probably occurred on the floors of these craters. It is known that the bottoms of freshly formed impact craters of this size often contain a deposit of low-density fragmental material. Even though areas of mare material are quite limited on the far side of the Moon, it is reasonable to assume that mare-basalt magma underlies the subsurface in this section of the floor of Gagarin. Mare lava, for example, is exposed in the floor of a larger (90-km-diameter) crater on the west rim of Gagarin. (See fig. 240.) The fracture system associated with the inner walls of a large crater is a likely conduit along which magma may rise and then intrude into the weakest strata. Because the floors of impact craters consist of low-density fragmental material, any crater on the floor of Gagarin, situated near its walls, may have served as a locus for magmatic intrusion. As lunar mare magma rose through the subsurface from a zone of melting at depth and reached the debris in the two crater floors, it may have lifted the debris and fractured it into segments ½ to 2 km across. Surficial fragmental debris from the uplifted segments then drained into the cracks between them to produce the bumpy, rounded topography. The low-density segments would tend to "float" at the same general level on the underlying denser magma, with the largest segments standing highest. The relative amount of uplift of materials in the craters in this photograph correlates both with the original depth of the craters (inferred from their diameters, which are proportional to depth) and with their proximity to the wall of Gagarin, which is located about 1½ km inside the northeastern corner of the picture. Correlation with crater depth suggests that the magma intruded under the deepest crater because the roof was thinnest there and offered the least resistance. Correlation with proximity to Gagarin's wall indicates that the inferred fracture zone near the wall provided a locus for easier subsurface flow and better subsurface plumbing for the upwelling magma. The bumpy material in the northwest depression stands highest of all. Although it was neither the deepest nor the nearest to Gagarin's wall, it may have been centered over the contact between Gagarin's original floor material and the wall and thus been a locus for intrusion. Subsurface flow from the largest (northeastern) depression into the northwestern one may have contributed to uplift of the latter depression; the hill forming the septum between them is the largest uplifted block in the cluster and indicates that intrusion took place under the septum.—R.E.E.

AS15-8910 (P)

0   5 km

FIGURE 240.—This photograph shows part of the floor and walls of the largest (90 km) crater inside the old pre-Imbrian crater Gagarin on the far side of the Moon. The 90-km crater is partly filled with relatively dark, young mare materials and contains the circular depression shown here. The depression is about 4 km in diameter and has a very low raised rim or none at all. Rock layering and narrow terraces along resistant layers are visible. The floor of the depression and the adjacent mare surfaces have an equal density of craters, suggesting that they are the same age.

Morphological evidence supports the hypothesis that the depression was formed by collapse of the mare materials. Two narrow and shallow rilles debouch high on the west and east depression walls; they are interpreted as lava tubes that originated under the depression before it collapsed. This interpretation is further supported by the mare ridge, which extends down the south wall onto the depression floor. A dome adjacent to the ridge on the floor may also record late volcanic activity in the depression.—M.J.G.

AS15-8928 (P)

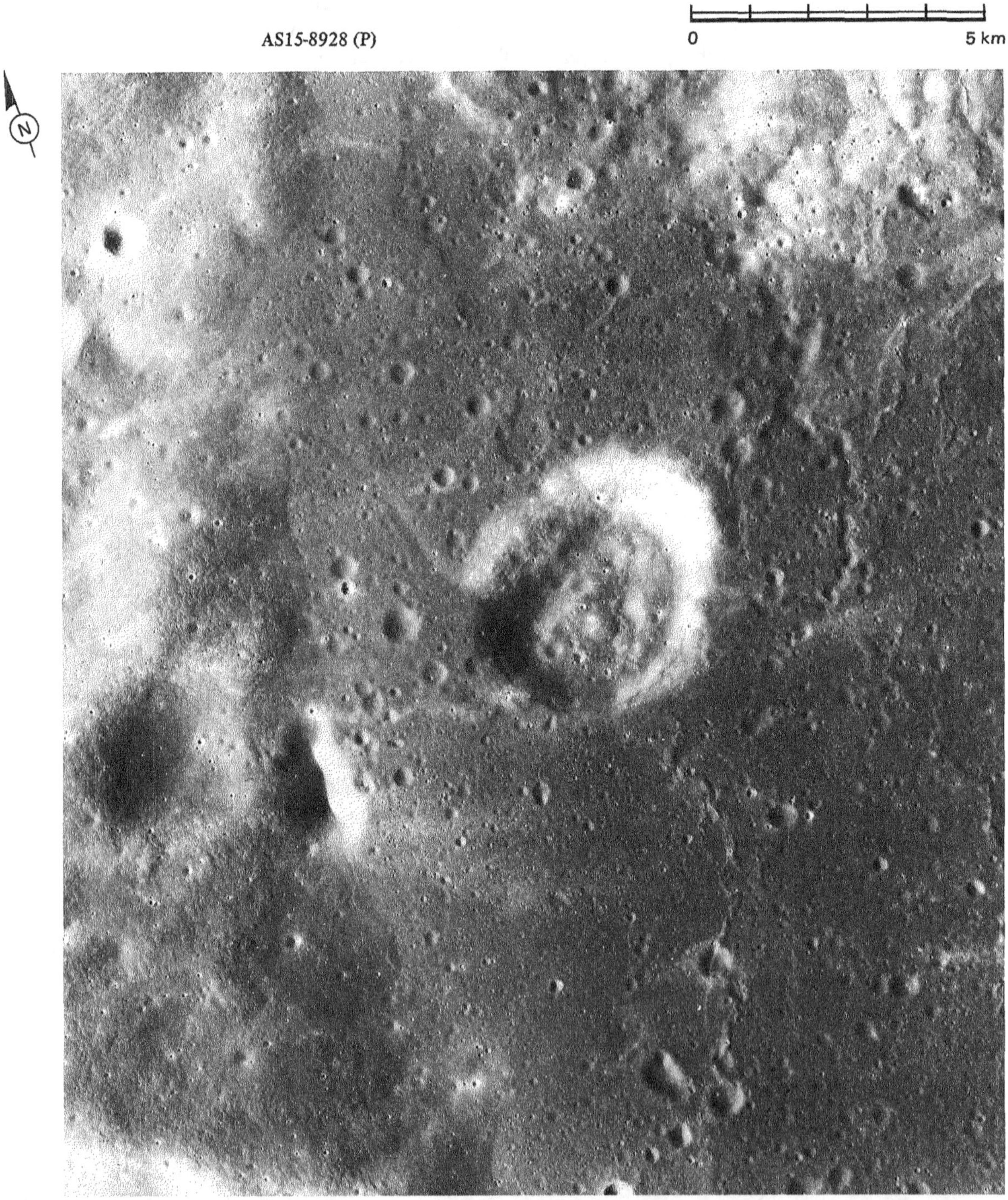

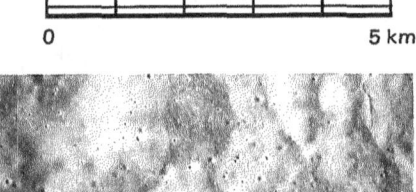

AS17-2334 (P)

0   5 km

N

FIGURE 241.—This crater (approximately 9 km across) is on the juncture between two basalt units at the south border of Mare Serenitatis. The unit south of the dashed line forms a conspicuous dark border around the mare and slopes northward toward the center of the Serenitatis basin. The adjacent mare unit is lighter in color, generally flat, and embays the older dark unit. The straight rille grazing the crater is one of several concentric grabens in the dark border material and probably formed by extension as the lava sagged toward the basin center prior to emplacement of the central mare. The crater is largely filled by the younger mare unit; because the graben transects the crater rim and its trend is influenced by the crater, the graben is probably younger.—C.A.H.

227

AS17-2335 (P)

FIGURE 242.—This picture, showing an area in central Mare Serenitatis 200 km from the nearest outcrops of terra rocks, is an oblique view of a 4-km-diameter crater form about 200 m deep. The crater floor is similar to the surface of the surrounding mare, which is presumably mare basalt covered by 3 to 6 m of regolith. The rim of the crater is raised, although subdued; from this we infer that the present form was developed when mare lava flooded a normal impact crater that had formed on a lower, preexisting surface. The flooding basalt then subsided, more or less in proportion to its thickness, which was greatest inside the crater. The subsidence may have resulted from escape of bubbles from the lava while it was soft, thermal contraction of the lava, and compaction of an underlying relatively loosely packed regolith. Regolith compaction may have occurred when the load and heat of the flooding lava crushed irregular fragments of regolith into more compact shapes and plastically deformed the glassy components of the regolith into more compact shapes or into voids between other particles.

R. J. Pike's (1972) data on the rim heights of lunar craters 90 m to 10 km in diameter indicate this crater probably had a rim about 200 m high when it was buried. The lava plain around the depression presently stands about 25 m below the rim crest, giving a minimum thickness for the flooding lava of 175 m.—R.E.E.

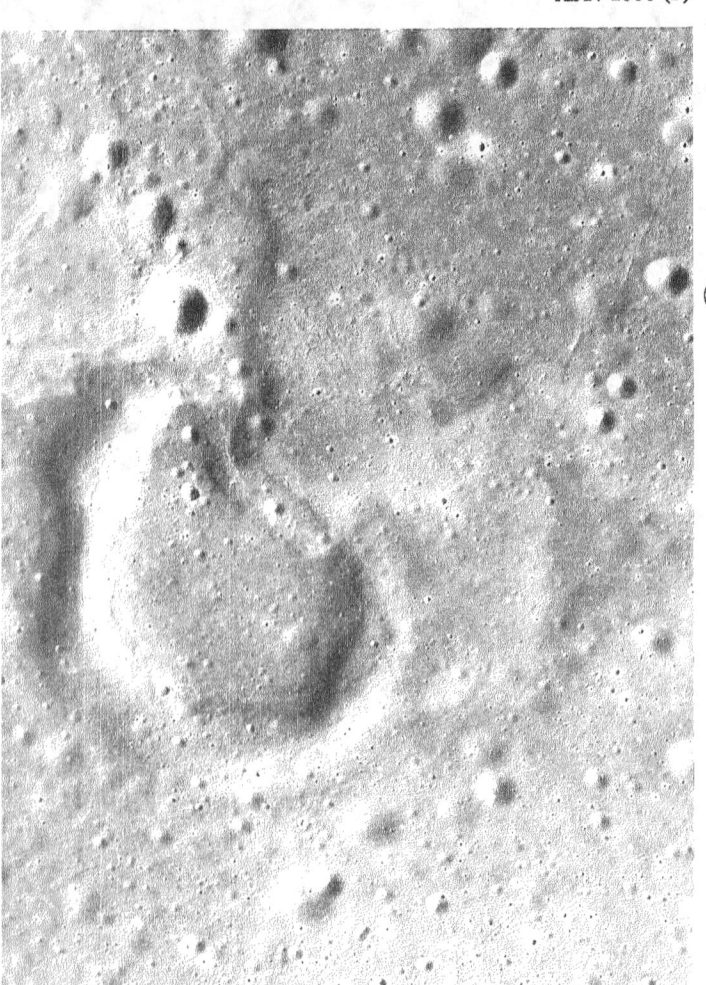

FIGURE 243.—This rimmed depression draped by mare lava is at the end of a mare ridge segment due north of Copernicus in Mare Imbrium. The unusual shape of the crater may indicate a volcanic origin; alternatively, a circular or elliptical impact crater may have been deformed by subsequent development of the mare ridge, a part of which impinges on the crater at the northeast side. In any case, a preexisting rimmed depression seems to have been buried by mare, reappearing as a mantled structure upon solidification and compaction of the lava.—C.A.H.

AS17-3067 (P)

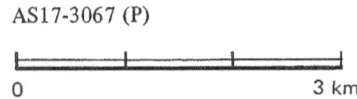

0   3 km

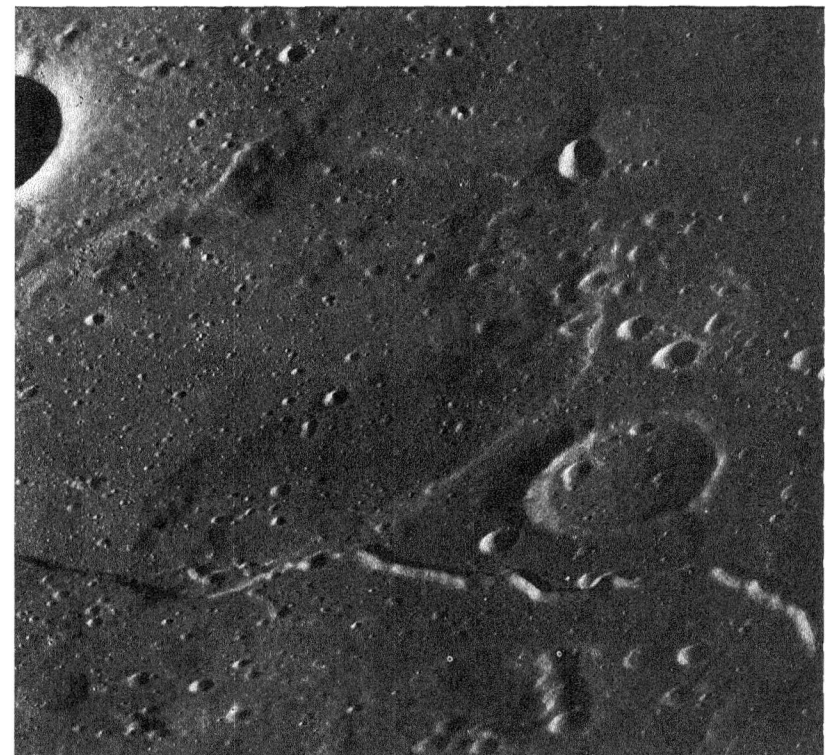

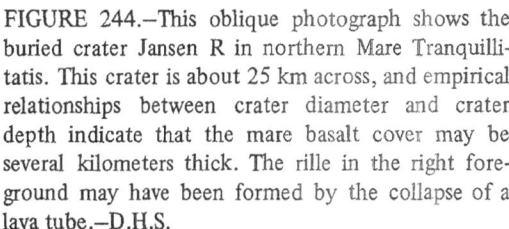

FIGURE 244.—This oblique photograph shows the buried crater Jansen R in northern Mare Tranquillitatis. This crater is about 25 km across, and empirical relationships between crater diameter and crater depth indicate that the mare basalt cover may be several kilometers thick. The rille in the right foreground may have been formed by the collapse of a lava tube.—D.H.S.

AS17-2318 (P)

FIGURE 245.—The elongate crater Torricelli near the north margin of Mare Nectaris was probably formed by two simultaneous impacts as indicated by the partly developed septum at A. The overlapping clusters of small craters at B, C, and elsewhere around the rim were made by secondary impacts of ejecta from the large crater Theophilus more than 150 km to the southwest.—D.H.S.

AS16-4525 (P)

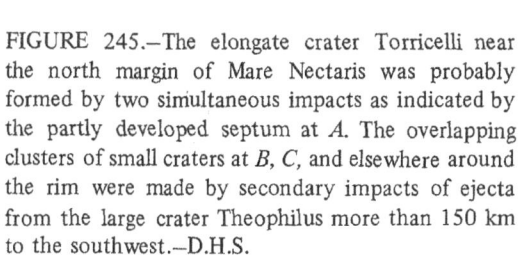

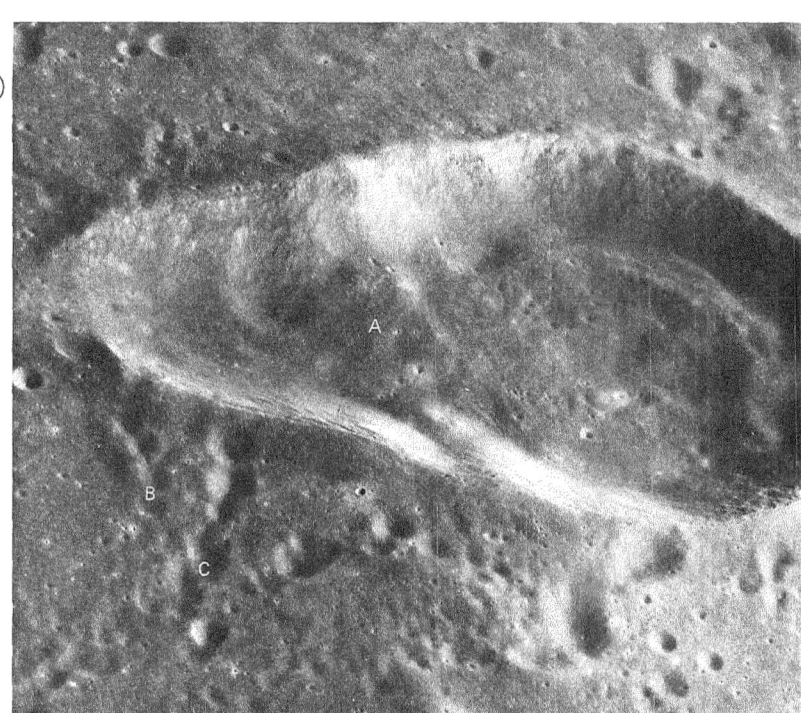

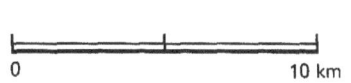

0   10 km

FIGURE 246.—This small unnamed crater in Mare Cognitum is unusual because it appears to be deformed by a fault that also bounds a mare ridge. About 5 km in diameter, the crater is obviously older than the mare materials that have buried the outer part of its ejecta blanket. The visible part of the fault extends between the arrows and clearly transects the western wall of the crater. It also marks the west flank of a small mare ridge north of the crater. Viewed stereoscopically, the fault plane can be seen to dip gently to the west, and the surface west of the fault is lower than that on the east. The fault is, therefore, a low-angle normal fault. The abrupt disappearance of the fault at the south rim of the crater may seem surprising. One of several explanations is that it may lie buried beneath a younger basalt flow that flooded the area immediately south of the crater. Many lunar investigators, including several contributors to this volume, have suggested a relationship between faulting and the development of mare ridges. Although this is a very small and certainly uncommon example, it is a convincing example of a mare ridge that is coincident with a fault and thus lends support to this idea.—G.W.C.

AS16-5429 (P)

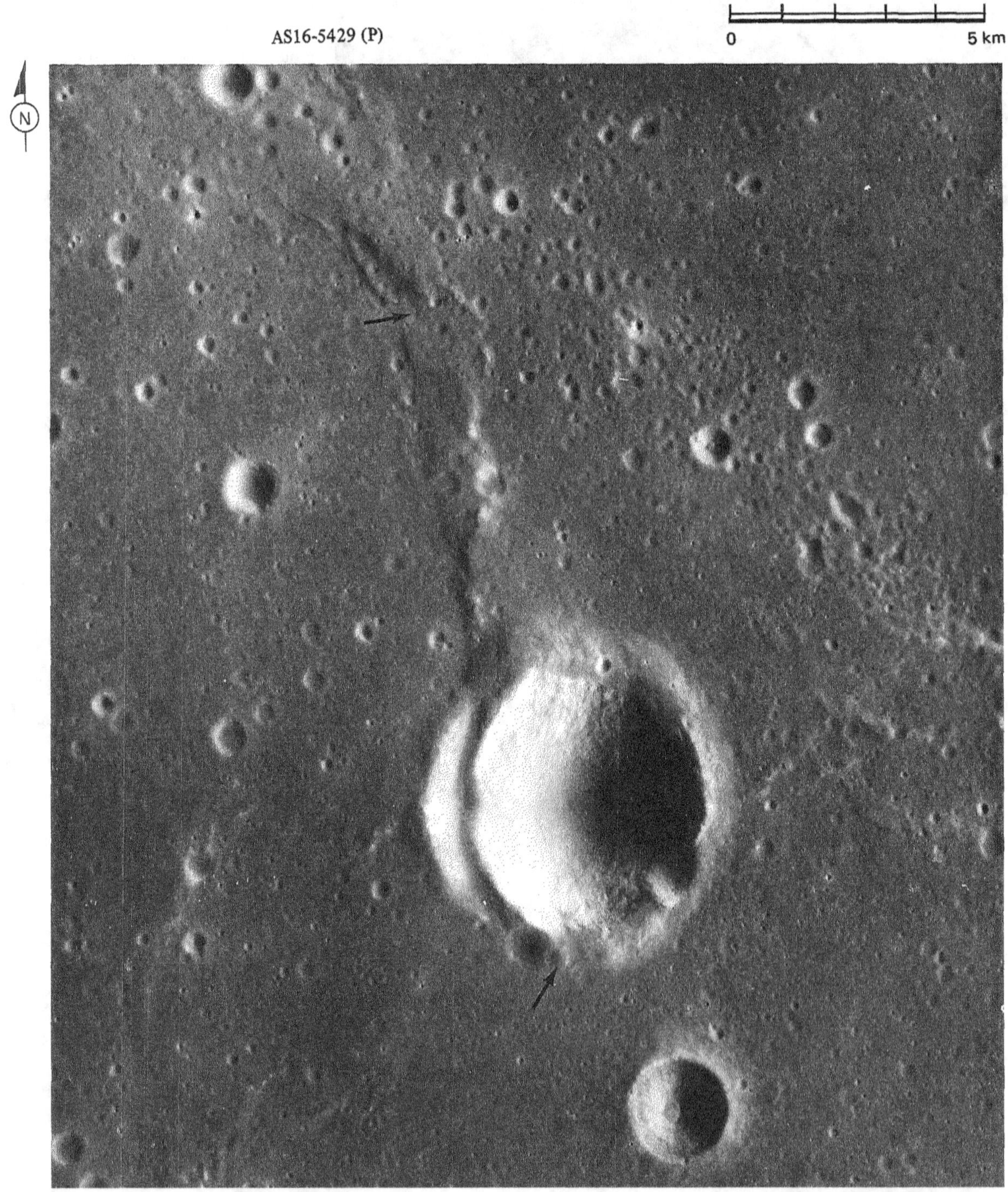

0   5 km

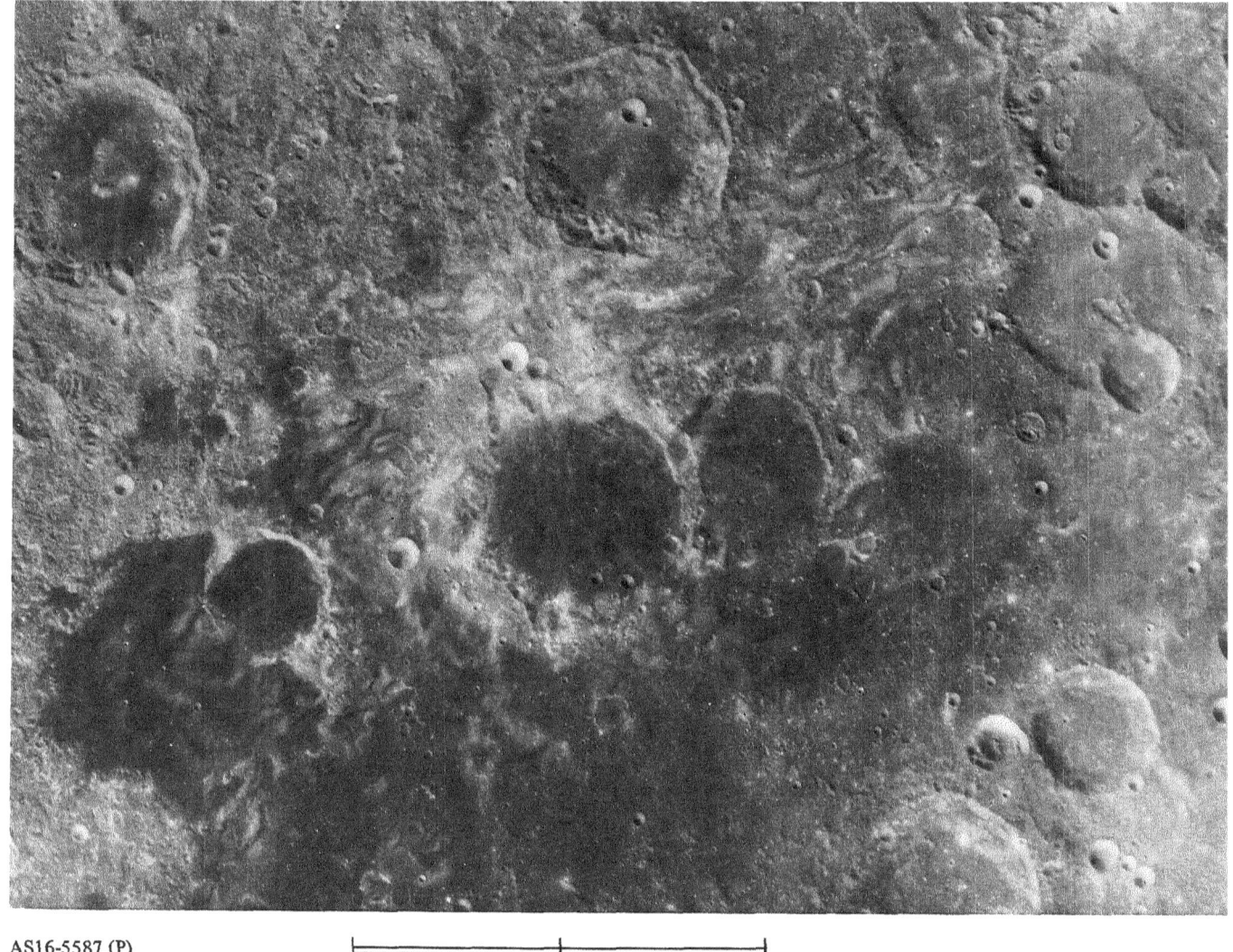

AS16-5587 (P)

0   200 km

FIGURE 247.—Bright swirls in and around Mare Marginis on the eastern limb of the Moon. These peculiar markings that are characterized by very high albedo and sinuous outlines are not fully understood. In this particular field (El-Baz, 1972*a*) the swirls occur in all types of geomorphologic units: They abound in the dark and flat material of Mare Marginis, but they occur also in highland units. A good example of this occurrence is to the southeast of the crater Al-Biruni, upper middle of the photograph. The small fresh crater on the rim of Goddard (center of photograph) cannot be the cause of these swirls that extend far beyond the area shown.

This author previously indicated that the area is antipodal to the location of the Orientale basin on the west limb of the Moon. It is probable that the formation of the Orientale basin is responsible for these bright swirls—either by the meeting of fine ejecta at the antipodal point or by surface disturbance due to seismic waves traveling at or near the surface also meeting on the opposite side of the impact point. The origin of these swirls, however, is still open to discussion.—F.E.-B.

Alternate explanations for the swirls are based on the observations that they have no relief of their own and the crater density within the swirls is similar to surrounding areas. They may represent surface alteration products formed by gases that have escaped from the lunar interior or by some poorly understood process of ejecta deposition. Interaction with the extralunar environment may also have played a part in the development of the swirls.—J.W.H.

231

Photographic Data

Figure number	Frame number	Altitude, km	Sun elevation, deg	Camera		Camera system[a]	Lens focal length, mm	Film type	Frame type[b]
				Tilt, deg	Azimuth, deg				
Introduction									
10	AS17-152-23391	–	–	–	–	M	80	SO-368	FF
Regional views									
17	AS16-3021	–	–	–	–	M	76.2	SO-368	FF
18	AS17-3153	–	–	Vertical	–	P	610	3414	FF
19	AS16-0729	116	15	40	358	M	76.2	3400	FF
20	AS17-150-22959	117	18	57	199	H	80	SO-368	FF
21	AS15-2503	113	15	40	185	M	76.2	3400	FF
22	AS17-2871	113	56	62	292	P	610	3414	FF
23	AS11-37-5437	–	Low	62	–	H	80	SO-168	FF
24	AS16-0566	113	25	25	269	M	76.2	3400	FF
25	AS16-2478	119	28	40	174	M	76.2	3400	FF
26	AS16-2518	124	36	Vertical	–	M	76.2	3400	FF
27	AS17-0940	110	24	40	3	M	76.2	3400	FF
28	AS17-0953	105	6	40	357	M	76.2	3400	FF
29	AS15-1537	106	31	40	5	M	76.2	3400	PF
30	AS15-93-12602	108	19	60	180	H	80	SO-368	FF
31	AS15-2618	105	4	40	180	M	76.2	3400	FF
The Terrae									
33	AS17-155-23702	115	8	61	3	H	80	2485	FF
34	AS17-0274	191	43	Vertical	–	M	76.2	3400	PF–M
	AS17-0278	200	39	Vertical	–	M	76.2	3400	PF–M
	AS17-0281	206	35	Vertical	–	M	76.2	3400	PF–M
	AS17-0286	217	29	Vertical	–	M	76.2	3400	PF–M
	AS17-0289	224	26	Vertical	–	M	76.2	3400	PF–M
	AS17-0293	232	21	Vertical	–	M	76.2	3400	PF–M
	AS17-0294	234	20	Vertical	–	M	76.2	3400	PF–M
35	AS17-1674	104	9	40	177	M	76.2	3400	FF
36	AS17-1817	106	18	Vertical	–	M	76.2	3400	FF–M
	AS17-1819	105	15	Vertical	–	M	76.2	3400	PF–M
37	AS17-2433	114	28	40	183	M	76.2	3400	FF
38	AS17-1825	103	7	Vertical	–	M	76.2	3400	FF
	AS17-1828	103	4	Vertical	–	M	76.2	3400	PF–M
39	AS15-9297	106	39	Aft	–	P	610	3414	PF
40	AS17-136-20694	On surface	16	–	–	H	60	3401	FF
41	AS17-144-22129	On surface	28	–	–	H	500	3401	PF

Note–Footnotes at end of table.

Figure number	Frame number	Altitude, km	Sun elevation, deg	Camera Tilt, deg	Camera Azimuth, deg	Camera system[a]	Lens focal length, mm	Film type	Frame type[b]
42	AS15-9804	107	32	Aft	–	P	610	3414	PF
43	AS15-90-12187	On surface	30	–	–	H	60	3401	PF–S
	AS15-90-12209	On surface	31	–	–	H	60	3401	PF–S
44	AS16-1420	113	2	40	2	M	76.2	3400	FF
45	AS16-0700	114	24	40	180	M	76.2	3400	FF
46	AS16-0974	115	28	Vertical	–	M	76.2	3400	PF–S
	AS16-0976	114	26	Vertical	–	M	76.2	3400	PF–S
47	AS16-124-19829	308	3	50	275	H	250	2485	FF–M
	AS16-124-19830	309	3	45	275	H	250	2485	FF–M
	AS16-124-19831	308	3	50	275	H	250	2485	FF–M
48	AS16-1411	114	15	40	360	M	76.2	3400	FF
49	AS16-0982	113	19	Vertical	–	M	76.2	3400	FF
50	AS17-0304	254	9	Vertical	–	M	76.2	3400	PF
51	AS16-1646	118	52	Vertical	–	M	76.2	3400	PF–S
	AS16-1648	118	49	Vertical	–	M	76.2	3400	PF–S
52	AS17-2302	114	19	Aft	–	P	610	3414	PF
53	AS16-0708	112	13	40	180	M	76.2	3400	FF
54	AS16-0989	112	10	Vertical	–	M	76.2	3400	FF
55	AS16-2965	123	45	Vertical	–	M	76.2	3400	PF
56	AS16-0875	117	29	Vertical	–	M	76.2	3400	FF

The Maria

Figure number	Frame number	Altitude, km	Sun elevation, deg	Camera Tilt, deg	Camera Azimuth, deg	Camera system[a]	Lens focal length, mm	Film type	Frame type[b]
58	AS17-0448	113	12	Vertical	–	M	76.2	3400	PF–M
	AS17-0449	112	10	Vertical	–	M	76.2	3400	PF–M
	AS17-0452	111	7	Vertical	–	M	76.2	3400	PF–M
	AS17-0604	110	5	Vertical	–	M	76.2	3400	PF–M
	AS17-0805	108	8	Vertical	–	M	76.2	3400	PF–M
	AS17-0808	107	4	Vertical	–	M	76.2	3400	PF–M
59	AS17-150-23069	108	20	28	176	H	80	SO-368	FF
60	AS15-1115	111	52	Vertical	–	M	76.2	3400	PF
63	AS17-2102	116	24	Vertical	–	M	76.2	3400	PF–S
	AS17-2103	116	26	Vertical	–	M	76.2	3400	PF–S
64	AS15-1145	103	17	Vertical	–	M	76.2	3400	FF
	AS15-0424	99	2	Vertical	–	M	76.2	3400	FF
65	AS15-1701	100	4	Vertical	–	M	76.2	3400	PF–M
	AS15-2295	115	1	Vertical	–	M	76.2	3400	PF–M
66	AS15-1556	110	4	40	355	M	76.2	3400	FF
67	AS15-1010	99	3	Vertical	–	M	76.2	3400	FF
68	AS16-2836	124	9	Vertical	–	M	76.2	3400	PF–M
	AS16-2839	124	6	Vertical	–	M	76.2	3400	PF–M
69	AS16-120-19244	120	13	Vertical	–	H	250	SO-368	PF
70	AS14-78-10375	–	2.0	–	–	H	80	SO-2485	PF–M
	AS14-78-10376	–	1.5	–	–	H	80	SO-2485	PF–M
	AS14-78-10377	–	.5	–	–	H	80	SO-2485	FF–M
	AS14-78-10378	–	0	Vertical	–	H	80	SO-2485	PF–M
71	AS15-98-13361	99	5	30	180	H	250	2485	FF
72	AS15-2487	106	7	Vertical	–	M	76.2	3400	PF
73	AS15-2399	119	61	Vertical	–	M	76.2	3400	PF–S
	AS15-2401	119	63	Vertical	–	M	76.2	3400	PF–S
74	AS17-0939	111	25	40	4	M	76.2	3400	FF
75	AS17-2313	113	14	Forward	–	P	610	3414	PF
76	AS17-2313	113	14	Forward	–	P	610	3414	PF
77	AS17-2313	113	14	Forward	–	P	610	3414	PF

Figure number	Frame number	Altitude, km	Sun elevation, deg	Camera Tilt, deg	Azimuth, deg	Camera system[a]	Lens focal length, mm	Film type	Frame type[b]
78	AS15-9298	106	37	Forward	–	P	610	3414	PF
79	AS17-2313	113	14	Forward	–	P	610	3414	PF
80	AS17-2309	113	15	Forward	–	P	610	3414	PF–S
	AS17-2314	113	15	Aft	–	P	610	3414	PF–S
81	AS17-2309	113	15	Forward	–	P	610	3414	PF
82	AS15-9361	102	19	Aft	–	P	610	3414	PF
83	AS17-3075	117	23	Forward	–	P	610	3414	PF
84	AS17-1913	118	20	Forward	–	P	610	3414	PF
85	AS17-1676	100	30	Forward	–	P	610	3414	PF
86	AS16-5452	123	17	Forward	–	P	610	3414	PF
87	AS16-4970	107	20	Forward	–	P	610	3414	PF
88	AS16-120-19237	115	11	53	215	H	250	SO-368	FF
	AS16-120-19237	115	11	53	215	H	250	SO-368	FF, PF
89	AS15-0344	108	12	Forward	–	P	610	3414	PF
90	AS17-2317	112	12	Forward	–	P	610	3414	PF
91	AS16-2824	124	24	Vertical	–	M	76.2	3400	FF
92	AS16-5428	124	27	Forward	–	P	610	3414	PF–S
	AS16-5433	124	27	Aft	–	P	610	3414	PF–S
93	AS17-3114	113	71	Aft	–	P	610	3414	PF
94	AS16-5425	123	29	Aft	–	P	610	3414	PF
95	AS17-2224	121	47	Aft	–	P	610	3414	PF

Craters

Figure number	Frame number	Altitude, km	Sun elevation, deg	Camera Tilt, deg	Azimuth, deg	Camera system[a]	Lens focal length, mm	Film type	Frame type[b]
97	AS15-0293	117	17	Vertical	–	M	76.2	3400	FF
98	AS17-150-23102	124	34	46	166	H	250	SO-368	FF
99	AS15-0102	82	34	Vertical	–	M	76.2	3400	FF
100	AS15-8936	81	34	Forward	–	P	610	3414	PF
101	AS15-8937	82	33	Aft	–	P	610	3414	PF
102	AS15-9348	102	21	Forward	–	P	610	3414	PF
103	AS16-4136	119	14	Aft	–	P	610	3414	PF
104	AS16-4136	119	14	Aft	–	P	610	3414	PF
105	AS16-4558	112	17	Forward	–	P	610	3414	PF
107	AS17-1764	95	56	Forward	–	P	610	3414	PF–S
	AS17-1769	95	56	Aft	–	P	610	3414	PF–S
108	AS15-9721	119	55	Forward	–	P	610	3414	PF
109	AS15-9301	106	37	Aft	–	P	610	3414	PF
110	AS16-121-19407	115	47	65	315	H	250	SO-368	FF
111	AS16-4511	115	34	Aft	–	P	610	3414	PF
112	AS15-9337	104	26	Aft	–	P	610	3414	PF
113	AS15-9254	109	50	Forward	–	P	610	3414	PF
114	AS17-2744	112	48	Forward	–	P	610	3414	PF
115	AS15-2405	120	67	Vertical	–	M	76.2	3400	FF
	AS16-4469	118	48	Aft	–	P	610	3414	PF–S
	AS16-4471	118	47	Aft	–	P	610	3414	PF–S
116	AS15-0018	120	94	Forward	–	P	610	3414	PF
117	AS15-9287	107	41	Aft	–	P	610	3414	PF
118	AS16-4559	112	18	Aft	–	P	610	3414	PF
119	AS16-4502	115	35	Forward	–	P	610	3414	PF
120	AS16-5444	144	20	Forward	–	P	610	3414	PF
121	AS16-5444	144	20	Forward	–	P	610	3414	PF
122	AS16-5430	124	25	Forward	–	P	610	3414	PF
123	AS16-4658	116	18	Forward	–	P	610	3414	PF
124	AS17-2120	116	4	Vertical	–	M	76.2	3400	PF

Figure number	Frame number	Altitude, km	Sun elevation, deg	Camera Tilt, deg	Camera Azimuth, deg	Camera system[a]	Lens focal length, mm	Film type	Frame type[b]
125	AS17-2444	118	13	40	177	M	76.2	3400	FF
126	AS17-2291	114	5	Vertical	—	M	76.2	3400	FF
127	AS17-3093	117	18	Vertical	—	P	610	3414	PF
128	AS16-4653	116	22	Aft	—	P	610	3414	PF
129	AS16-4653	116	22	Aft	—	P	610	3414	PF
130	AS16-1973	116	15	Vertical	—	M	76.2	3400	FF
132	AS16-1671	116	22	Vertical	—	M	610	3414	PF
133	AS16-1276	113	15	Vertical	—	M	76.2	3400	FF
134	AS17-149-22838	125	63	57	320	H	250	SO-368	FF
135	AS17-2321	112	11	Forward	—	P	610	3414	PF
136	AS16-4530	114	26	Forward	—	P	610	3414	PF
137	AS17-3103	118	14	Vertical	—	P	610	3414	PF—M
	AS17-3105	118	13	Vertical	—	P	610	3414	PF—M
	AS17-3107	118	13	Vertical	—	P	610	3414	PF—M
138	AS17-2922	118	15	Vertical	—	M	76.2	3400	PF
139	AS15-0274	112	34	Forward	—	P	610	3414	PF
140	AS14-70-9671	—	17	Vertical	—	H	80	3400	FF
141	AS14-72-9975	—	34	50	330	H	500	SO-368	FF
142	AS17-3062	117	28	Aft	—	P	610	3414	PF
143	AS15-9866	117	75	Forward	—	P	610	3414	PF
114	AS15-9874	117	73	Forward	—	P	610	3414	PF
145	AS15-9328	104	27	Forward	—	P	610	3414	PF
146	AS17-3081	117	22	Vertical	—	P	610	3414	PF
147	AS17-2265	117	31	Forward	—	P	610	3414	PF
148	AS17-2006	116	30	Vertical	—	M	76.2	3400	PF—S
	AS17-2007	116	31	Vertical	—	M	76.2	3400	PF—S
	AS17-2773	114	19	Forward	—	P	610	3414	PF
149	AS16-122-19580	—	—	—	270	H	250	SO-368	FF
150	AS16-120-19268	98	14	65	245	H	250	SO-368	FF
151	AS16-4998	109	29	Forward	—	P	610	3414	PF
152	AS16-5000	109	30	Forward	—	P	610	3414	PF
153	AS16-4996	108	25	Forward	—	P	610	3414	PF
154	AS10-30-4349	—	—	—	—	H	250	3400	FF
155	AS16-120-19266	98	13	65	255	H	250	SO-368	FF
156	AS16-5000	109	30	Forward	—	P	610	3414	PF
158	AS16-5000	109	30	Forward	—	P	610	3414	PF
159	AS16-1578	113	35	Vertical	—	M	610	3414	PF
160	AS16-1579	113	36	Vertical	—	M	610	3414	PF
161	AS16-120-19231	114	42	20	260	H	250	SO-368	FF
162	AS16-5006	109	32	Forward	—	P	610	3414	PF
163	AS16-4815	112	33	Forward	—	P	610	3414	PF
164	AS17-151-23260	114	53	64	180	H	250	SO-368	FF
165	AS15-0326	109	18	Forward	—	P	610	3414	PF
166	AS16-0692	115	35	40	176	M	76.2	3400	FF
167	AS16-0154	112	25	Vertical	—	M	76.2	3400	PF
168	AS16-4531	114	27	Aft	—	P	610	3414	PF
169	AS16-0532	120	71	25	267	M	76.2	3400	FF
170	AS15-0757	119	27	25	270	M	76.2	3400	FF
172	AS15-1030	117	21	Vertical	—	M	76.2	3400	FF
173	AS15-9591	115	17	Forward	—	P	610	3414	PF
174	AS15-9596	115	17	Aft	—	P	610	3414	PF
175	AS17-2608	113	18	Vertical	—	M	76.2	3400	FF

Figure number	Frame number	Altitude, km	Sun elevation, deg	Camera		Camera system[a]	Lens focal length, mm	Film type	Frame type[b]
				Tilt, deg	Azimuth, deg				
176	AS15-0892	119	31	Vertical	–	M	76.2	3400	PF–S
	AS15-0893	119	32	Vertical	–	M	76.2	3400	PF–S
178	AS17-0341	121	22	Vertical	–	M	76.2	3400	FF
179	AS15-1541	105	26	40	5	M	76.2	3400	FF
180	AS15-2510	115	27	40	205	M	76.2	3400	FF
181	AS16-120-19295	123	6	55	180	H	250	SO-368	FF
182	AS16-2995	123	4	Vertical	–	M	76.2	3400	FF
183	AS16-0839	112	16	40	359	M	76.2	3400	FF

Rilles

185	AS15-2611	107	15	40	180	M	76.2	3400	FF
186	AS15-0341	108	16	Aft	–	P	610	3414	PF
187	AS15-0342	108	13	Forward	–	P	610	3414	PF
188	AS15-0332	108	16	Forward	–	P	610	3414	PF
190	AS15-9309	106	34	Aft	–	P	610	3414	PF
191	AS15-2606	109	21	40	185	M	76.2	3400	FF
192	AS15-2083	102	9	Vertical	–	M	76.2	3400	PF
193	AS15-0321	109	21	Aft	–	P	610	3414	PF
194	AS15-0320	109	20	Forward	–	P	610	3414	PF
195	AS15-88-11982	114	4	65	280	H	60	SO-168	FF
196	AS15-0349	108	12	Aft	–	P	610	3414	PF
197	AS15-93-12725	109	19	20	0	H	250	SO-368	FF
198	AS17-3128	119	6	Vertical	–	P	610	3414	PF
199	AS15-2076	104	17	Vertical	–	M	76.2	3400	PF
200	AS15-1702	100	3	Vertical	–	M	76.2	3400	FF
201	AS17-23717	114	1	32	352	H	250	2485	PF–M
	AS17-23718	114	1	27	355	H	250	2485	PF–M
202	AS15-1135	106	28	Vertical	–	M	76.2	3400	FF
203	AS15-85-11451	On surface	20	–	–	H	60	3401	FF
204	AS15-11720	12	11	15	260	H	60	SO-168	FF
205	AS15-9432	103	23	Aft	–	P	610	3414	PF
206	AS15-9816	107	42	Aft	–	P	610	3414	PF
207	AS15-91-12366	311	23	35	315	H	250	SO-368	PF
209	AS8-13-2225	–	12	–	South	H	80	3400	FF
210	AS15-93-12641	115	27	60	180	H	80	SO-368	FF
211	AS16-2478	116	83	40	173	M	76.2	3400	FF
216	AS16-4656	110	12	Aft	–	P	610	3414	PF
217	AS10-31-4645	–	Medium	–	–	H	250	3400	FF
218	AS15-9368	101	15	Forward	–	P	610	3414	PF
219	AS15-13345	101	4	55	305	H	250	2485	FF
220	AS17-2313	113	14	Forward	–	P	610	3414	PF
221	AS17-2313	113	14	Forward	–	P	610	3414	PF

Unusual features

223	AS15-9350	102	20	Forward	–	P	610	3414	PF
	AS15-9355	102	20	Aft	–	P	610	3414	PF–S
224	AS15-9299	106	38	Aft	–	P	610	3414	PF
225	AS15-0244	114	44	Forward	–	P	610	3414	PF
226	AS15-9361	102	19	Aft	–	P	610	3414	PF
227	AS15-9361	102	19	Aft	–	P	610	3414	PF
228	AS17-3125	119	7	Vertical	–	P	610	3414	PF
229	AS16-5410	123	32	Forward	–	P	610	3414	PF

Figure number	Frame number	Altitude, km	Sun elevation, deg	Camera		Camera system[a]	Lens focal length, mm	Film type	Frame type[b]
				Tilt, deg	Azimuth, deg				
230	AS15-2627	110	7	Vertical	–	M	76.2	3400	FF
231	AS15-9960	117	16	Forward	–	P	610	3414	PF
232	AS17-1672	105	12	40	178	M	76.2	3400	PF
233	AS15-0176	117	65	Forward	–	P	610	3414	PF–S
	AS15-0181	117	65	Aft	–	P	610	3414	PF–S
235	AS15-9358	102	18	Forward	–	P	610	3414	PF
236	AS17-1931	120	27	Forward	–	P	610	3414	PF
237	AS16-4151	120	21	Forward	–	P	610	3414	PF
238	AS15-8893	91	19	Aft	–	P	610	3414	PF
239	AS15-8910	87	26	Forward	–	P	610	3414	PF
240	AS15-8928	83	32	Forward	–	P	610	3414	PF
241	AS17-2334	111	8	Aft	–	P	610	3414	PF
242	AS17-2335	111	6	Forward	–	P	610	3414	PF
243	AS17-3067	117	25	Forward	–	P	610	3414	PF
244	AS17-2318	113	14	Aft	–	P	610	3414	PF
245	AS16-4525	114	29	Aft	–	P	610	3414	PF
246	AS16-5429	124	28	Aft	–	P	610	3414	PF
247	AS16-5587	–	–	Aft	–	P	610	3414	PF

Epilog

Figure number	Frame number	Altitude, km	Sun elevation, deg	Camera		Camera system[a]	Lens focal length, mm	Film type	Frame type[b]
248	AS11-44-6665	–	–	–	–	H	250	SO-368	FF
249	AS17-148-22725	–	–	–	–	H	80	SO-368	FF

[a]H = Hasselblad; M = metric; P = panoramic.
[b]FF = full frame; PF = partial frame; FF–M = full frame–mosaic; PF–M = partial frame–mosaic; PF–S = partial frame–stereogram.

Lunar Probes, Attempted and Successful

Name	Country	Launch date	Vehicle	Remarks
Thor-Able 1 (Pioneer)	U.S.A.[a]	Aug. 17, 1958	Thor-Able	1st attempt to probe Moon: failed.
Pioneer 1	U.S.A.	Oct. 11, 1958	Thor-Able	Failed to reach Moon; sent 43 hr of data.
Pioneer 2	U.S.A.	Nov. 8, 1958	Thor-Able	Failed to reach Moon.
Pioneer 3	U.S.A.	Dec. 6, 1958	Juno II	Failed to reach Moon; provided radiation data.
Luna 1	U.S.S.R.	Jan. 2, 1959	A-1	Passed within 6000 km of the Moon; went into solar orbit.
Pioneer 4	U.S.A.	Mar. 3, 1959	Juno II	Passed within 60 000 km of the Moon; went into solar orbit.
Luna 2	U.S.S.R.	Sept. 12, 1959	A-1	1st probe to impact on Moon.
Luna 3	U.S.S.R.	Oct. 4, 1959	A-1	1st probe to photograph Moon's far side.
Atlas-Able 4 (Pioneer (P-3))	U.S.A.	Nov. 26, 1959	Atlas-Able	Failed to reach Moon.
Atlas-Able 5A (Pioneer (P-30))	U.S.A.	Sept. 25, 1960	Atlas-Able	Failed to reach Moon.
Atlas-Able 5B (Pioneer (P-3))	U.S.A.	Dec. 15, 1960	Atlas-Able	Failed to reach Moon.
Ranger 3	U.S.A.	Jan. 26, 1962	Atlas-Agena B	Missed Moon by 36 800 km; ended in solar orbit.
Ranger 4	U.S.A.	Apr. 23, 1962	Atlas-Agena B	Impacted on Moon, but cameras and experiments did not operate.
Ranger 5	U.S.A.	Oct. 18, 1962	Atlas-Agena B	Missed Moon by 725 km; ended in solar orbit.
No name	U.S.S.R.	Jan. 4, 1963	A-2-e	Probable lunar probe; failed.
Luna 4	U.S.S.R.	Apr. 2, 1963	A-2-e	Attempted soft landing; missed Moon by 8500 km; went into barycentric orbit.
Ranger 6	U.S.A.	Jan. 30, 1964	Atlas-Agena B	Impacted on Moon; TV system did not function during approach.
Ranger 7	U.S.A.	July 28, 1964	Atlas-Agena B	1st successful Ranger mission; impacted on Moon after taking 4308 pictures of lunar surface during approach.
Ranger 8	U.S.A.	Feb. 17, 1965	Atlas-Agena B	Impacted on Moon after returning 7137 closeup pictures.
Kosmos 60	U.S.S.R.	Mar. 12, 1965	A-2-e	Probable lunar probe: failed.

[a]This project was directed by the Advanced Research Projects Agency and launched by the U.S. Air Force. All subsequent U.S. lunar projects were directed by the National Aeronautics and Space Administration.

Name	Country	Launch date	Vehicle	Remarks
Ranger 9	U.S.A.	Mar. 21, 1965	Atlas-Agena B	Impacted in crater Alphonsus after returning 5814 pictures.
Luna 5	U.S.S.R.	May 9, 1965	A-2-e	Attempted soft landing by U.S.S.R.: impacted on Moon.
Luna 6	U.S.S.R.	June 8, 1965	A-2-e	Attempted soft landing: missed Moon by 161 000 km and went into solar orbit.
Zond 3	U.S.S.R.	July 18, 1965	A-2-e	Transmitted 25 pictures of lunar far side taken during lunar flyby; went into solar orbit.
Luna 7	U.S.S.R.	Oct. 4, 1965	A-2-e	Attempted soft landing: impacted on Moon.
Luna 8	U.S.S.R.	Dec. 3, 1965	A-2-e	Attempted soft landing: impacted on Moon.
Luna 9	U.S.S.R.	Jan. 31, 1966	A-2-e	1st successful soft landing by U.S.S.R.; returned 27 pictures of lunar surface in 3 days.
Kosmos III	U.S.S.R.	Mar. 1, 1966	A-2-e	Suspected lunar probe: failed.
Luna 10	U.S.S.R.	Mar. 31, 1966	A-2-e	Achieved lunar orbit; transmitted physical measurements for nearly 2 months.
Surveyor 1	U.S.A.	May 30, 1966	Atlas-Centaur	Successful soft landing; transmitted 11 237 pictures of lunar surface.
Explorer 33	U.S.A.	July 1, 1966	Thrust-augmented Delta (TAD)	Attempted lunar orbit not achieved; in barycentric orbit.
Lunar Orbiter 1	U.S.A.	Aug. 10, 1966	Atlas-Agena D	Successful orbital photographic mission; photographed Moon until Aug. 29, 1966; impacted on Moon Oct. 29, 1966.
Luna II	U.S.S.R.	Aug. 24, 1966	A-2-e	Achieved lunar orbit; transmitted data until Oct. 1, 1966.
Surveyor 2	U.S.A.	Sept. 20, 1966	Atlas-Centaur	Attempted soft landing; impacted southeast of Crater Copernicus.
Luna 12	U.S.S.R.	Oct. 22, 1966	A-2-e	Achieved lunar orbit; transmitted lunar pictures and data.
Lunar Orbiter 2	U.S.A.	Nov. 6, 1966	Atlas-Agena D	After transmitting 422 pictures from orbit, impacted on the Moon.
Luna 13	U.S.S.R.	Dec. 21, 1966	A-2-e	Successful soft landing; transmitted pictures and soil density data.
Lunar Orbiter 3	U.S.A.	Feb. 4, 1967	Atlas-Agena D	After transmitting 307 pictures from orbit, impacted on the Moon.
Surveyor 3	U.S.A.	Apr. 17, 1967	Atlas-Centaur	Successful soft landing; transmitted 6315 pictures and soils data.
Lunar Orbiter 4	U.S.A.	May 4, 1967	Atlas-Agena D	Photographed all of near side; after returning 326 pictures from orbit, impacted on the Moon.
Surveyor 4	U.S.A.	July 14, 1967	Atlas-Centaur	Impacted on the Moon; experiments did not function.

Name	Country	Launch date	Vehicle	Remarks
Explorer 35	U.S.A.	July 19, 1967	TAD	In lunar orbit; measures Earth's magnetic tail every 29.5 days.
Lunar Orbiter 5	U.S.A.	Aug. 1, 1967	Atlas-Agena D	Last of the highly successful Orbiter missions; impacted on the Moon after completing photographic coverage of far side.
Surveyor 5	U.S.A.	Sept. 8, 1967	Atlas-Centaur	Successful soft landing; transmitted 18 006 pictures and chemical analysis of soil.
Surveyor 6	U.S.A.	Nov. 7, 1967	Atlas-Centaur	Successful soft landing; transmitted 30 065 pictures and chemical and mechanical soil studies; performed first rocket takeoff on Moon.
Surveyor 7	U.S.A.	Jan. 7, 1968	Atlas-Centaur	Successful soft landing; transmitted 21 274 pictures and chemical analysis of soil.
Zond 4	U.S.S.R.	Mar. 2, 1968	D-1-e	Launched into undisclosed trajectory from initial parking orbit, probable precursor to manned lunar mission.
Apollo 6	U.S.A.	Apr. 4, 1968	Saturn V	Failed to reach Moon.
Luna 14	U.S.S.R.	Apr. 7, 1968	A-2-e	In lunar orbit; measuring lunar gravity field and Earth-Moon mass relationship.
Zond 5	U.S.S.R.	Sept. 15, 1968	D-1-e	1st circumlunar mission after which spacecraft returned to Earth; recovered from Indian Ocean Sept. 21, 1968.
Zond 6	U.S.S.R.	Nov. 10, 1968	D-1-e	2d unmanned circumlunar mission; spacecraft recovered in U.S.S.R.
Apollo 8	U.S.A.	Dec. 21, 1968	Saturn V	1st manned circumlunar flight; 10 orbits completed; first mission in which true photographs taken from lunar orbit were returned to Earth (Astronauts Borman, Lovell, and Anders).
Apollo 10	U.S.A.	May 18, 1969	Saturn V	LM detached from the CM and piloted to within 14.9 km of the Moon and returned to the CM; first color TV from space (Astronauts Stafford, Young, and Cernan).
Luna 15	U.S.S.R.	July 13, 1969	D-1-e	After completing 52 orbits, impacted on Moon.
Apollo 11	U.S.A.	July 16, 1969	Saturn V	1st manned lunar landing July 20, 1969, western Mare Tranquillitatus; 21.6 hr spent on the Moon; first samples of lunar rocks returned to Earth (Astronauts Armstrong, Aldrin, and Collins).

Name	Country	Launch date	Vehicle	Remarks
Zond 7	U.S.S.R.	Aug. 8, 1969	D-1-e	3d successful unmanned circumlunar flight; spacecraft recovered in U.S.S.R.
Apollo 12	U.S.A.	Nov. 14, 1969	Saturn V	2d manned lunar landing Nov. 19, 1969, eastern Oceanus Procellarum; 31.6 hr on the Moon (Astronauts Conrad, Bean, and Gordon).
Apollo 13	U.S.A.	Apr. 11, 1970	Saturn V	Attempted manned landing mission aborted 56 hr after launch; returned after crossing the far side of the Moon (Astronauts Lovell, Swigert, and Haise).
Luna 16	U.S.S.R.	Sept. 12, 1970	D-1-e	Unmanned lander arrived in Mare Fecunditatis Sept. 20, 1970; 1st unmanned mission to return lunar samples to Earth.
Zond 8	U.S.S.R.	Sept. 20, 1970	D-1-e	Circled Moon; recovered Oct. 27, 1970.
Luna 17	U.S.S.R.	Nov. 10, 1970	D-1-e	2d unmanned lander arrived in western Mare Imbrium Nov. 17, 1970; automated roving vehicle Lunokhod 1 traveled 10.54 km, took more than 20 000 pictures, and conducted experiments during 11 lunar days.
Apollo 14	U.S.A.	Jan. 31, 1971	Saturn V	3d successful manned landing Feb. 5, 1971, Fra Mauro site in eastern Oceanus Procellarum (Astronauts Shepard, Mitchell, and Roosa).
Apollo 15	U.S.A.	July 26, 1971	Saturn V	4th manned landing July 30, 1971, Apennine-Hadley site in eastern Mare Imbrium; first manned lunar roving vehicle; SIM permitted larger number of scientific experiments and more sophisticated photography to be conducted from orbit.
Luna 18	U.S.S.R.	Sept. 2, 1971	D-1-e	Impacted on Moon after 54 orbits.
Luna 19	U.S.S.R.	Sept. 28, 1971	D-1-e	Lunar orbital photographic mission; completed more than 4000 orbits; filmed selected areas; still in orbit.
Luna 20	U.S.S.R.	Feb. 14, 1972	D-1-e	Achieved soft landing Feb. 21, 1972, in northeastern Mare Fecunditatis; returned to Earth Feb. 25, 1972, with lunar samples.
Apollo 16	U.S.A.	Apr. 16, 1972	Saturn V	5th manned landing, Apr. 21, 1972, Descartes region of Central Highlands (Astronauts Young, Duke, and Mattingly).

Name	Country	Launch date	Vehicle	Remarks
Apollo 17	U.S.A.	Dec. 7, 1972	Saturn V	6th and last manned lunar landing of the Apollo Program, Taurus-Littrow region; 79 hr on the lunar surface; 113 kg of rock samples returned to Earth (Astronauts Cernan, Schmitt, and Evans).
Luna 21	U.S.S.R.	Jan. 8, 1973	D-1-e	Achieved successful soft landing in crater Le Monnier; automated roving vehicle Lunokhod 2 traveled 37 km and conducted many experiments during 5 lunar days.
Explorer 49	U.S.A.	June 10, 1973	Thorad Delta	Radio astronomy from far side of Moon.
Luna 22	U.S.S.R.	May 29, 1974	D-1-e	Lunar orbital mission; obtained photographs and other data; ran out of attitude control gas Sept. 2, 1975; last communication Nov. 6, 1975.
Luna 23	U.S.S.R.	Oct. 28, 1974	D-1-e	Achieved successful soft landing on Mare Crisium; sampling prevented by damage to rock drill; conducted restricted research until Nov. 9, 1975.
Luna 24	U.S.S.R.	Aug. 9, 1976	D-1-e	Achieved successful soft landing in Mare Crisium; complete core sample from depth of 2 m returned to Earth on Aug. 22, 1976.

Basic Data Analysis
Scheme for Preparation
of Lunar Maps*

(1) Identification of star images on the stellar photographs, measurement of their image coordinates, and computation of the orientation of the camera in the right ascension and declination Earth-centered stellar coordinate system

(2) Computation of the orientation of the mapping camera in the stellar coordinate system by using the precalibrated relationship between the stellar and mapping cameras

(3) Transfer of the orientation of the mapping camera from the Earth-centered stellar coordinate system to the Moon-centered geographic system—this transformation involves the ephemeris and physical librations of the Moon

(4) Selection of lunar-surface points (approximately 30 per frame) on the mapping camera photographs and identification of them on all overlapping frames on which they appear

(5) Measurement of the coordinates of the selected points and correction of the points for film deformation, lens distortion, and displacement of the focal-plane reseau

(6) Triangulation of groups of photographs—this computation includes the measured image coordinates, the attitudes obtained with the stellar camera, the laser altimeter data, a state vector obtained from the tracking data, and a gravity model[a]

(7) Assembly of the triangulated groups into a single adjusted network—the output of this computer program will be a geometrically homogeneous set of coordinate values for all selected points on the lunar surface, plus the position and orientation of each photograph in the same Moon-centered coordinate system[a]

(8) Preparation of small-scale maps from the mapping camera photographs

(9) Transformation of the panoramic photographs into equivalent vertical photographs for interpretation and mapping

(10) Preparation of large-scale maps from the panoramic photographs

---

*From Doyle (1972).

[a]The procedure in steps 6 and 7 was not actually followed because errors in the spacecraft tracking data proved to be much larger (up to 1.5 km) than those in the photogrammetric measurements (about 30 m).

In the control solution performed by the Defense Mapping Agency Aerospace Center, each orbital pass was triangulated separately using spacecraft tracking data as constraints. Pass 44 on Apollo 15 gave the best fit between photogrammetry and tracking data. It was adopted as the fundamental control, and all other passes were subsequently adjusted to fit.

In the control solution performed by the U.S. Geological Survey, the entire photogrammetric network was computed in a single simultaneous solution without tracking data constraints. This network was subsequently adjusted to the tracking data from pass 44 on Apollo 15.

Glossary

Many of the geological terms listed in this glossary, which were originally defined for terrestrial use, have been modified or shortened for this volume in keeping with their commonly accepted usage by lunar geologists.

Albedo—a measure of the reflectivity of a surface; using the Moon as an example, the ratio of sunlight reflected from the Moon to that reaching it.

Allochthonous—as used here, a part of the lunar crust that has moved from its original position by displacement along a fault.

Anorthosite—a light-colored igneous rock composed almost entirely of the mineral group plagioclase feldspar.

Asteroid—a subplanetary body within the solar system, synonymous with "planetoid."

Autointrusion (or autoinjection)—the movement of magma from the lower, still liquid, part of a flow into cracks in the hardened crust of the flow.

Avalanche—a mass of rock material that has slid or fallen rapidly under the influence of gravity; one form of mass wasting.

Basalt—a dark-colored igneous rock that most commonly solidifies on the surface in the form of lava flows. It is the dominant rock type in the lunar maria.

Base surge—a cloud of gas and suspended debris that moves radially outward across the surface at high velocity; may result from a violent volcanic explosion or from the explosion caused by a body traveling at high velocity when it impacts on the surface of a planetary body.

Basin—a large circular area on the Moon, typically 300 or more km in diameter, surrounded by one or more mountainous rings; may be occupied to varying extent by mare material, and may or may not be lower in elevation than the surrounding terrain. Basins are considered by most workers to be impact scars.

Bedrock—in situ solid rock.

Bistatic radar—a method of studying the electrical properties of the surface by reflected radio waves. In the lunar experiment the waves were emitted from the CSM and received on Earth both directly and after reflection by the lunar surface.

Breccia—a rock composed of fragments of preexisting rocks.

Cartography—the science or art of making maps.

Central peak (or central uplift)—a mountainous mass in the center of many impact craters more than 40 km in diameter; formed by the inward and upward movement of material originally below the level of the crater floor.

Cinder cone—a conical hill composed of volcanic cinders, ash, and larger fragments of ejecta.

Colluvium—a general term to include loose rock and soil material that accumulates at the base of a slope as the result of mass wasting processes. See *talus*.

Comet—seen as a light-giving body having a bright head and a luminous tail moving through space under the gravitational influence of the Sun. Mass-to-size ratio is low. Apparently composed of frozen gases, dust, and other cosmic debris.

Cosmic debris—material that originates anywhere in the universe beyond Earth's atmosphere; includes material believed to represent primordial condensation or sublimation products and debris resulting from collisions of meteorites, asteroids, and comets with each other, and with the planets and the Moon.

Crater—a hole or depression. Most are roughly circular or oval in outline, and, typically, depth is much less than diameter. On Earth most natural craters are of volcanic origin, whereas on the Moon most are of impact origin. Secondary craters are produced by the impact of material ejected from the parent or primary crater.

Creep—the slow, more or less continuous, permanent deformation and displacement of material under the influence of gravity; one form of mass wasting.

Crystalline rock—igneous rock consisting mainly or entirely of crystals.

Deceleration dunes—dunelike lobes of ejecta from impact craters formed as the velocity of the base surge cloud decreases; most commonly formed on slopes facing toward the source of base-surge flow.

Degradation—the wearing down and general lowering of an area or a feature by any process of weathering and erosion.

Differentiation, **magmatic**—a general term for the various physicochemical processes that lead to the formation of two or more rock types from a common igneous melt.

Dike—a tabular body of intrusive rock that cuts across the planar structure of the surrounding (and older) rocks. See *intrusion.*

Diurnal (adj.)—recurring daily; in the case of the Moon, recurring every 28 Earth days.

Doppler tracking—a system for measuring the trajectory of spacecraft from Earth, using continuous radiowaves and the Doppler effect. Because of this effect, the frequency of the radiowaves received on Earth is changed slightly by the velocity of the spacecraft.

Drag fold—a subsidiary fold developed in response to movement along or within a larger structural feature.

Dune—a low mound or ridge of loose rock material. Most dunes on Earth are formed by wind action, whereas most of those on the Moon apparently are formed during the ejection of material from an impact crater.

Earthshine—sunlight reflected from Earth that illuminates the lunar surface.

Ejecta—rock material ejected during the process of impact (as from a meteorite impact crater) or by explosive volcanic action.

Erosion—a general term to include all processes whereby rock materials are disintegrated or dissolved and transported from one place to another, whether the agency be water, ice, wind, gravity, or impact cratering. In the case of the Moon, impact cratering is the dominant erosional process.

Extrusion—the process of emitting volcanic material (as liquid lava, particulate matter suspended in bases, or as solid fragments) onto the surface of a planetary body; also, the rock so formed.

Fault—a fracture along which rock masses have been displaced.

Fault scarp—a steep slope or cliff caused by displacement along a fault and, if unmodified by erosion, representing the exposed surface of the fault.

Flux—the rate of transfer of some quantity across a unit area. As used here it applies to the rate at which bodies impact the lunar surface.

Gamma ray—highly penetrating rays emitted by radioactive substances. Gamma radiation from the lunar surface was measured by gamma-ray spectrometers aboard the Apollo 15 and 16 spacecraft.

Gardening—mechanical mixing of the unconsolidated surface debris that occupies most of the Moon's surface, the regolith, or "lunar soil," by various processes, including meteorite impact and mass wasting.

Geodesy—the science of determining the exact size and shape of bodies in the solar system, and of the distribution of mass within the bodies.

Geophysics—the study of the physical properties of Earth, the Moon, and planetary bodies. Basic divisions are solid-Earth, atmospheric, hydrospheric, and magnetospheric. Apollo lunar geophysical experiments included studies of gravity, magnetism, heat flow, radioactivity, seismology, space physics, geodesy, and meteorology.

Glass—a form of igneous rock lacking crystal structure, produced by the rapid cooling of a magma.

Graben—an elongate depression formed by the downward displacement of a block of crust along faults bordering its long sides.

Igneous (adj.)—pertaining to or describing a rock that has solidified from molten material (magma), or the processes and conditions related to the formation of such rocks.

Imbrian—a unit of geologic time that describes the interval of time between the formation of the Imbrium basin and the end of the accumulation deposition of the lavas that occupy most of the maria on the Moon's near side.

Imbrian sculpture—a system of scarps, ridges, and troughs radial to the center of the Imbrium basin and transecting much of the lunar surface. The features are a response to the event that formed the basin, and, because of their wide extent, are useful in determining the relative age of rock units far from the basin.

Impact—a forceful collision. For example, the impact of a meteoroid traveling at high velocity with the surface of Earth or the Moon.

Intrusion—the process of emplacing magma into preexisting rock; also, the rock so formed (for example, a dike).

Isostatic equilibrium—the adjustment of the crust to maintain equilibrium among blocks of different density; examples: blocks of dense material will "sink" more than less dense blocks; excess mass or density in the upper part of a block is compensated by a deficit of mass in the lower part.

Laccolith—an igneous intrusion: top, domical; bottom or floor, essentially flat; and outline (when viewed from above), roughly circular.

Laser altimeter—an instrument used to measure distance between two points by means of the traveltime of a pulse of light. In the lunar laser altimeter, light is transmitted from the CSM and reflected from the lunar surface back to the detector in the CSM. Knowing the position and elevation of the spacecraft from orbital data, differences in elevation of the lunar surface were measured along the ground track.

Lava—molten rock material (magma) that has reached the surface; also, the solidified rock.

Lava channel—a channel on the upper surface of a partly or completely solidified body of lava through which liquid lava has flowed. Its rims may be higher than surrounding terrain, like the natural levees along some rivers on Earth.

Lava tube—a tube within a body of partly or completely solidified lava through which liquid lava has flowed. If near the surface, rocks above the tube may collapse, resulting in a channellike depression on the upper surface of the lava body.

Levee—on Earth, a raised embankment bordering a river channel; on the Moon, a raised embankment along a presumed lava channel.

Limb—as used here, the east or west edge of the Moon when viewed from the direction of Earth. This term generally applies to the outer edge of the apparent disk of any celestial body.

Lineament—a broad term used to include any visible linear trend. It is commonly, but not always, of regional extent. It may consist of a single, more or less continuous feature; an alined series of a particular type of feature; or an alined series of unlike features. It is commonly interpreted as marking points of major dislocations of the crust.

Lithology—the physical character of a rock.

Magma—molten rock material generated within the Earth or Moon that cools to form igneous rocks.

Mare—a dark, level, relatively smooth part of the lunar surface (so distinct from the lunar highlands or terrae that most large mare areas on the near side are visible from Earth with the unaided eye). Most geologists now agree that they are underlain by solidified (basaltic) lava flows. (plural = maria)

Mare ridge—a ridge on a mare surface. The morphology varies considerably, but typically length is much greater than width, and width is much greater than height. (Also called "wrinkle ridge.")

Maria—plural of mare.

Mascon—literally, mass concentration; an area of the lunar crust characterized by an excess of mass. Those detected to date coincide with the circular maria, indicating the presence of relatively dense materials (basaltic lava) at shallow depth.

Massif—as used here, a discrete mountain mass; typically is bright and composes part of the uplifted mountainous rings around circular basins.

Mass spectrometer—an instrument for determining chemical species in terms of isotopic mass and relative abundances of isotopes within a compound. On the Apollo 15 and 16 missions a mass spectrometer was used to measure composition and density of the lunar atmosphere from the CSM in orbit.

Mass wasting—a general term for the downslope movement of rock material solely under the influence of gravity; includes slow displacement such as creep and rapid displacements such as earth flows, rock slides, and avalanches.

Metamorphism—the mineralogic, textural, and structural adjustment of rocks to physical and chemical conditions different from those under which the rocks originally formed. Metamorphism by impact-generated shock is the dominant type of metamorphism in lunar rocks.

Meteorite—a meteoroid that has arrived on the surface of a moon or planet from outer space. Composition ranges from silicate rock to nickel-iron metal; size ranges from that of a submicroscopic particle to that of a body approaching the size of an asteroid or planetesimal.

Meteoroid—one of the countless small solid bodies in the solar system.

Morphology—as used here, the external shape and arrangement of landforms.

Mosaic—a composite picture formed by assembling overlapping aerial or orbital spacecraft photographs taken from different camera positions, or from the same camera position but at different angles.

Orthophotograph—a photographic copy, normally of an aerial or orbital photograph, that has been processed to remove the effects of camera tilt and the distortion caused by perspective viewing so that all distances measured on the orthophotograph are proportional by the same factor to horizontal distances measured on the ground.

Outcrop—the exposed part of a unit of bedrock; rock not covered by surface debris or vegetation.

Plains—a general term to describe the relatively level areas of the lunar surface. They range from light to dark and may be smooth or rough. The maria are commonly included as one variety of plains.

Primordial (adj.)—as used here, the oldest lunar rocks—those created during the Moon's formative stages.

Projectile—specifically, in this volume, a body that strikes the lunar surface. A projectile may be a meteoroid or other object from outer space, rarely a spacecraft or spacecraft component, or, most commonly, a discrete rock fragment explosively ejected from a crater.

Ray—narrow light or dark streaks that extend radially outward from some lunar craters. They are a natural result of the impact process and form when ejected material covers or disturbs the preexisting surface.

Regolith—unconsolidated fragmental rock debris, regardless of origin, that almost everywhere forms the surface of the Moon; also called the "lunar soil."

Rille—a trenchlike valley on the Moon. Rilles vary widely in size, but width and depth are small compared to length. Viewed from above they may be sinuous, straight, or angular.

S-Band transponder—a device aboard the CSM that uses the traveltime of radiowaves transmitted from Earth and returned to it to aid in tracking the spacecraft. As an experiment on board Apollos 13 to 17, it measured small variations in the Moon's gravity under the ground track of the spacecraft.

Scarp—a relatively straight clifflike face or slope that separates terrain lying at different levels.

Scree—loose fragmental rock debris derived from and mantling a slope. See *talus*.

Seismic (adj.)—related to mechanical vibrations within Earth or the Moon. A common probable cause of seismic vibrations on the Moon is the impact of meteorites.

Shock *metamorphism*—the permanent changes (physical and chemical) produced in rocks by the passage of a transient high-pressure shock wave. The only known natural cause is by hypervelocity impact, thus the expression is essentially synonymous with impact metamorphisms.

Slickensides—the polished striations on a rock surface caused by friction generated by faulting.

Specific gravity—the ratio of the density of a substance to the density of another—commonly water. The average specific gravity of lunar basaltic rock samples is about 3.4, which means that a unit volume weighs about 3.4 times as much as the same volume of water.

Squeeze-up—a small extrusion of viscous lava on the solidified surface of a lava flow.

Stellar—of or pertaining to the stars.

Stereoscope—an optical device to facilitate obtaining a stereoscopic image. (See next definition.)

Stereoscopic view, image—the impression of the third dimension, normally depth, which can be obtained by viewing two photographs of the same area taken from slightly different points.

Structure—the general disposition (attitude, arrangement, or position) of the rock masses of a region or area. The term "structure" also is applied to individual structural features, such as that of a graben, fault, or basin.

Summit crater—a crater occupying the crest of a volcanic cone or dome.

Superposed—that condition wherein one stratified rock unit overlies, and hence is younger than, another such unit; also, a physical feature such as a crater located on, and younger than, another feature.

Talus—loose fragmental rock material derived from a cliff or slope and lying at its base.

Tectonic movement—the displacement of large masses of the crust, whether by uplift, subsidence, or large-scale folding and faulting. On the Moon it is considered to include the displacement caused by large-scale impact events.

Terminator—the line separating the illuminated and darkened areas of a nonluminous planetary body such as Earth or the Moon. In the absence of an atmosphere, as on the Moon, this line is very sharply defined.

Terra—an older, lighter, more densely cratered area of the Moon; encompasses all the lunar surface except the maria (plural = terrae).

Thrust fault—a relatively low-angle fracture along which one rock mass has moved upward and over another.

Topographic (adj.)—pertaining to the three-dimensional configuration of the solid surface of a planetary body and to its graphical description, usually on maps or charts.

Trajectory—the path of a moving body through space or the atmosphere.

Transient (adj.)—passing quickly into and out of existence; that is, of short duration.

Transverse fault—a fault that strikes obliquely or perpendicularly to the general structural trend.

Vesicle—a cavity in a lava formed by the entrapment of a gas bubble during solidification of the lava.

Viscosity—the property of a fluid that resists internal flow; its internal friction.

Volcanism—includes all the processes whereby magma and its associated fluids rise in the crust and are extruded onto the surface and ejected into the atmosphere.

Wrinkle ridge—synonymous with mare ridge.

X-ray fluorescence experiment—an experiment carried onboard the Apollo 15 and 16 spacecraft for determining the chemical composition of the lunar surface. It records the fluorescent X-rays that are emitted from the Moon's surface as a result of its bombardment by X-rays from the Sun.

References and Bibliography

Adler, I., Podwysocki, C. A., Trombka, R., Schmadebeck, L. Y., and Yinn, L. 1974, "The Role of Horizontal Transport—As Evaluated From the Apollo 15 and 16 Orbital Experiments." Proc. Lunar Sci. Conf., 5th, *Geochim. Cosmochim. Acta* 2, suppl. 3, pp. 975-979.

Baldwin, R. B. 1968, "Rille Pattern in the Lunar Crater Humboldt." *J. Geophys. Res.* 73(10), 3227-3229.

Carr, M. H. 1966, Geologic Map of the Mare Serenitatis Region of the Moon. U.S. Geological Survey Miscellaneous Geologic Investigations Map I-489 (LAC-42).

Carr, M. H., Howard, K. A., and El-Baz, F. 1971, Geologic Maps of the Apennine-Hadley Region of the Moon: Apollo 15 Pre-Mission Maps. U.S. Geological Survey Miscellaneous Geologic Investigations Map I-723.

Coleman, P. J., Jr., Lichtenstein, B. R., Russell, C. T., Schubert, G., and Sharp, L. R. 1972a, "The Particles and Fields Subsatellite Magnetometer Experiment." *Apollo 16 Preliminary Science Report*, pp. 23-1 to 23-13. NASA SP-315.

Coleman, P. J., Jr., Schubert, G., Russell, C. T., and Sharp, L. R. 1972b, "The Particles and Field Subsatellite Magnetometer Experiment." *Apollo 15 Preliminary Science Report*, pp. 22-1 to 22-9. NASA SP-289.

Coleman, P. J., Jr., Schubert, G., Russell, C. T., and Sharp, L. R. 1972c, "Satellite Measurements of the Moon's Magnetic Field: A Preliminary Report." *The Moon* 5(3/4), 419-429.

Doyle, F. J. 1972, "Photogrammetric Analysis of Apollo 15 Records." *Apollo 15 Preliminary Science Report*, pp. 25-27 to 25-36. NASA SP-289.

Eggleton, R. E., and Marshall, C. H. 1962, "Notes on the Apenninian Series and Pre-Imbrium Stratigraphy in the Vicinity of Mare Humorum and Mare Nubium." Astrogeologic Studies Semi-Annual Progress Report, Feb. 26, 1961, to Aug. 24, 1961, U.S. Geological Survey Open-File Report, pp. 132-137.

El-Baz, F. 1972a, "The Alhazen to Abul Wafa Swirl Belt: An Extensive Field of Light-Colored, Sinuous Markings. *Apollo 16 Preliminary Science Report*, pp. 29-93 to 29-97. NASA SP-315.

El-Baz, F. 1972b, "King Crater and Its Environs." *Apollo 16 Preliminary Science Report*, pp. 29-62 to 29-70. NASA SP-315.

El-Baz, F. 1973a, "Aitken Crater and Its Environs." *Apollo 17 Preliminary Science Report*, pp. 32-8 to 32-12. NASA SP-330.

El-Baz, F. 1973b, "Al-Khwarizmi: A New Found Basin on the Lunar Far-Side." *Science* 180(409), 1173-1176.

El-Baz, F. 1973c, "'D-Caldera': New Photographs of a Unique Feature." *Apollo 17 Preliminary Science Report*, pp. 30-13 to 30-17. NASA SP-330.

El-Baz, F., and Roosa, S. A. 1972, "Significant Results From Apollo 14 Lunar Orbital Photography." *Geochim. Cosmochim. Acta* 1 suppl. 3, pp. 63-83.

Fielder, G., and Fielder, J. 1971, "Lava Flows and the Origin of Small Craters in Mare Imbrium." *Geology and Physics of the Moon* (ed., G. Fielder). Elsevier Pub. Co., New York.

Greeley, R. 1971, "Lunar Hadley Rille: Considerations of Its Origin." *Science* 172(3984), 722-725.

Hackman, R. J. 1966, Geologic Map of the Montes Apenninus Region of the Moon. U.S. Geological Survey Miscellaneous Geologic Investigations Map I-463 (LAC-41).

Howard, K. A. 1972, "Ejecta Blankets of Large Craters Exemplified by King Crater." *Apollo 16 Preliminary Science Report*, pp. 29-70 to 29-77. NASA SP-315.

Howard, K. A., 1973, "Avalanche Mode of Motion: Implications From Lunar Examples." *Science* 180, 1052-1055.

Howard, K. A., Carr, M. H., Muehlberger, W. R. 1973, "Basalt Stratigraphy of Southern Mare Serenitatis." *Apollo 17 Preliminary Science Report*, pp. 29-1 to 29-12. NASA SP-330.

Hulme, G. 1973, "Turbulent Lava Flow and the Formation of Lunar Sinuous Rilles." *Mod. Geol.* 4(2), 107-119.

Kaula, W. M., Schubert, G., Lingenfelter, R. E., Sjogren, W. L., and Wollenhaupt, W. R. 1974, "Apollo Laser Altimetry Inferences as to Lunar Structure." Proc. Lunar Sci. Conf., 5th, *Geochim. Cosmochim. Acta* 3, suppl. 5, pp. 3049-3058.

Kosofsky, L. J. 1973, "Moon Revisited in Stereo." *PSA J.* 39, 1-11.

Kosofsky, L. J., and El-Baz, F. 1970, *The Moon as Viewed by Lunar Orbiter*. NASA SP-200.

Kozyrev, N. A. 1971, "Relationships of Tectonic Processes

of the Earth and Moon." *Geological Problems in Lunar and Planetary Research* (ed., Jack Green), AAAS Science and Technology Series 25, American Astronautical Society, pp. 213-227.

Mattingly, T. K., El-Baz, F., and Laidley, R. A. 1972, "Observations and Impressions From Lunar Orbit." *Apollo 16 Preliminary Science Report,* pp. 28-1 to 28-16. NASA SP-315.

Metzger, A. E., Trombka, J. I., Peterson, L. E., Reedy, R. C., and Arnold, J. R. 1973, "Lunar Surface Radioactivity: Preliminary Results of the Apollo 15 and 16 Gamma-Ray Spectrometer Experiments." *Science* 179(4975), 800-803.

Metzger, A. G., Trombka, J. I., Reedy, R. D., and Arnold, J. K. 1974, Element Concentrations From Lunar Orbital Gamma Ray Measurements." Proc. Lunar Sci. Conf., 5th, *Geochim. Cosmochim. Acta* 2, suppl. 3, pp. 1067-1078.

Milton, D. J. 1968, Geologic Map of the Theophilus Quadrangle of the Moon. U.S. Geological Survey I-546 (LAC-78).

Mitchell, J. K., Carrier, W. D., III, Costes, N. C., Houston, W. N., Scott, R. R., and Hovland, H. J. 1973, "Soil Mechanics." *Apollo 17 Preliminary Science Report,* pp. 8-1 to 8-21. NASA SP-330.

Moore, H. J. 1971, "Craters Produced by Missile Impacts." *J. Geophys. Res.* **76**(23), 57-50 to 57-55.

Moore, H. J. 1972, "Ranger and Other Impact Craters Photographed by Apollo 16." *Apollo 16 Preliminary Science Report,* pp. 29-45 to 29-50. NASA SP-315.

Moore, H. J. 1976, "Missile Impact Craters (White Sands Missile Range, New Mexico) and Applications to Lunar Research." U.S. Geological Survey Professional Paper 812-813.

NASA. 1969, *Apollo 11 Preliminary Science Report.* NASA SP-214.

NASA Langley Research Center. 1971, *Lunar Orbiter Photographic Atlas of the Moon* (eds., David E. Bowker and J. Kendrick Hughes). NASA SP-206.

NASA Lyndon B. Johnson Space Center. 1973, *Apollo 17 Preliminary Science Report.* NASA SP-330.

NASA Manned Spacecraft Center. 1969, *Analysis of Apollo 8 Photography and Visual Observations.* NASA SP-201.

NASA Manned Spacecraft Center. 1970, *Apollo 12 Preliminary Science Report.* NASA SP-235.

NASA Manned Spacecraft Center. 1971a, *Analysis of Apollo 10 Photography and Visual Observations.* NASA SP-232.

NASA Manned Spacecraft Center. 1971b, *Apollo 14 Preliminary Science Report.* NASA SP-272.

NASA Manned Spacecraft Center. 1972, *Apollo 16 Preliminary Science Report.* NASA SP-315.

National Geographic Society. 1969, The Earth's Moon [map]. Washington, D.C.

Nicks, O. W. 1970, *This Island Earth.* NASA SP-250.

Pike, R. J. 1972. "Geometric Similitude of Lunar and Terrestrial Craters." Int. Geol. Congr. 24th, Sec. 15, pp. 41-47.

Roddy, D. J. 1968, "The Flynn Creek Crater, Tennessee." *Shock Metamorphism of Natural Materials* (eds., B. French and N. Short), pp. 291-322. Mono Book Co., Baltimore, Md.

Schaber, G. G. 1973, "Lava Flows in Mare Imbrium: Geologic Evaluation From Apollo Orbital Photography." Proc. Lunar Sci. Conf., 4th, *Geochim. Cosmochim. Acta* 1, suppl. 4, pp. 73-92.

Scott, D. H., Lucchitta, B. K., and Carr, M. H. 1972, Geologic Maps of the Taurus-Littrow Region of the Moon: Apollo 17 Pre-Mission Maps. U.S. Geological Survey Miscellaneous Geologic Investigations Map I-800.

Shoemaker, E. M. 1962, "Interpretation of Lunar Craters." *Physics and Astronomy of the Moon* (ed., Zdeněk Kopal), pp. 283-359. Academic Press, Inc., London.

Simmons, G. 1971, "On the Moon With Apollo 15: A Guidebook to Hadley Rille and the Apennine Mountains." NASA TM-X-68638.

Simmons, G. 1972a, "On the Moon With Apollo 16: A Guidebook to the Descartes Region." NASA EP-95.

Simmons, G. 1972b, "On the Moon With Apollo 17: A Guidebook to Taurus-Littrow." NASA EP-101.

Sjogren, W. L., Wimberly, R. N., and Wollenhaupt, W. R. 1974, "Apollo 17 Gravity Results" (abstract). Proc. Lunar Sci. Conf., 5th, pt. 2, pp. 712-713.

Soderblom, L. A., and Lebofsky, L. A. 1972, "Technique for Rapid Determination of Relative Ages of Lunar Areas From Orbital Photography." *J. Geophys. Res.* 77(2), 279-296.

Strom, R. G. 1964, "Analysis of Lunar Lineaments, 1. Tectonic Maps of the Moon." *Univ. Ariz. Lunar Planet. Lab. Commun.* 2(39), 205-221.

Stuart-Alexander, D. E. 1976, Geologic Map of the Central Far Side of the Moon. U.S. Geological Survey Interagency Report—Astrogeology 79.

Stuart-Alexander, D. E., and Howard, K. A. 1970, "Lunar Maria and Circular Basins—A Review." *Icarus* 12(3), 440-456.

Whitaker, E. A. 1972, "Artificial Lunar Impact Craters: Four New Identifications." *Apollo 16 Preliminary Science Report,* pp. 29-39 to 29-45. NASA SP-315.

Wu, S. S. C. 1972, "Photogrammetry of Apollo 15 Photography." *Apollo 15 Preliminary Science Report,* p. 25-43, fig. 25-38. NASA SP-289.

Wu, S. S. C., Schafer, F. J., Jordan, R., and Nakata, G. M. 1972, "Photogrammetry Using Apollo 16 Orbital Photography." *Apollo 16 Preliminary Science Report,* pp. 30-5 to 30-10. NASA SP-315.

Young, R. A., Brennan, W. J., and Wolfe, W. R. 1972, "Selected Volcanic and Surficial Features." *Apollo 16 Preliminary Science Report,* pp. 29-78 to 29-79. NASA SP-315.

Epilog

AS11-44-6665 (H)

FIGURE 248.—A view of the full lunar disk taken by the Apollo 11 astronauts upon completion of their exploration mission, on the way back to Earth. In this view we see the dark lunar maria that dot the brighter highlands which form most of the lunar surface. From the samples, photographs, and other data gathered by the Apollo lunar exploration missions, we have learned a great deal about the Moon.

In a nutshell, the Moon appears to have formed about 4.6 billion years ago, along with the rest of the solar system. Accretional energy generated by the collisional impacts may have melted the upper 100 to 300 km, allowing differentiation that resulted in a lighter (anorthositic-gabbroic) crust and a denser (pyroxene-rich) interior. A few remaining large bodies collided with the solid crust, the last of them about 4.0 billion years ago, forming the circular basins. After that, between about 3.9 and 3.2 billion years ago, a period of volcanism occurred. For 700 million years, basaltic lavas poured from the lunar interior to cover nearly 15 percent of its surface. In the last 3 billion years or so, the only tangible modifications of the lunar surface were due to meteoroid impacts; there may have been some volcanic eruptions, but they were not globally significant. The only satellite of our dynamic Earth (facing page) is a silent Moon.—F.E.-B.

AS17-148-22725 (H)

Quietly, like a night bird, floating, soaring, wingless
We glide from shore to shore, curving and falling
  but not quite touching;
Earth: a distant memory seen in an instant of repose,
  crescent-shaped, ethereal, beautiful,
I wonder which part is home, but I know it doesn't matter . . .
  the bond is there in my mind and memory;
Earth: a small, bubbly balloon hanging delicately
  in the nothingness of space

—Alfred M. Worden
Apollo 15 Astronaut

FIGURE 249.—This magnificent view of the Earth was taken during the last manned journey to the Moon. For the first time, on Apollo 17, the Antarctic icecap was visible to the astronauts. The view also encompasses much of the South Atlantic Ocean; virtually all of the Indian Ocean, Africa, the Arabian Peninsula; part of Iran and India; and, on the horizon, Indonesia and the western edge of Australia.

The history of the Moon summarized on the preceding page may provide significant information to increase our understanding of the Earth, because the Earth and the Moon were most probably formed at the same time (4.6 billion years ago). Because of the dynamism of the Earth and the many changes that have affected its surface since its creation, the oldest rocks that we find on it are about 3.5 billion years old (in Greenland and South Africa). This means that the first 1 billion years of Earth's history have been obliterated. Therefore, the Moon may be used as a window to view the early history of our own Earth.—F.E.-B.

255